Closing the Power Gap Between ASIC & Custom

Tools and Techniques for Low Power Design

David Chinnery · Kurt Keutzer

Closing the Power Gap Between ASIC & Custom

Tools and Techniques for Low Power Design

 Springer

David Chinnery
AMD
Sunnyvale, CA
USA

Kurt Keutzer
University of California, Berkeley
Berkeley, CA
USA

Library of Congress Control Number: 2007929117

ISBN 978-0-387-25763-1   e-ISBN 978-0-387-68953-1

Printed on acid-free paper.

9 8 7 6 5 4 3 2 1

springer.com

PREFACE

We were very pleased with the warm reception that industry analysts, the trade press, and most importantly, hard working circuit designers, gave to our book entitled *Closing the Gap Between ASIC and Custom* that was first published in 2002. In that book, we focused on identifying the factors that cause a significant speed differential between circuits designed in an ASIC methodology and those designed with a "no holds barred" custom approach. We also sought to identify and describe design tools and techniques that could close the gap between the speeds of ASIC and custom circuits. That book wasn't even in press before designers and fellow researchers came forward to challenge us to investigate a gap of growing importance: the gap in power dissipation and energy efficiency between circuits designed in ASIC and custom methodologies.

We learned a lot from our first book. In the content of our work we learned that circuit design and layout tricks were unlikely to be the source of sustained advantages of custom design. Instead clocking methodologies and microarchitecture were more likely to be areas where custom circuits sustained their advantage over ASICs. In our presentation of our research we found that technical conferences such as the Design Automation Conference were good venues for trying out our material and getting valuable feedback. Finally, in the production of the book itself we learned that putting high-level surveys and detailed descriptions of current research together with illustrative design examples was a good formula for creating a book of broad interest.

Like its predecessor we envision three main audiences for this book. The first audience is ASIC and ASSP designers who are restricted to a high productivity ASIC design methodology but still need to produce low-power circuits with high-energy efficiency. The second audience is custom designers who are seeking to design low power circuits with a more productive design flow. While the perspective of these two groups is different, the solutions they are seeking are very similar. In this book we account for the relative power impact of different elements of a custom-design methodology. We believe that this analysis should help custom design groups to determine where their limited design resources are best spent and help ASIC-oriented design groups understand where they most need improvement. Secondly, we identify specific tools and methodologies targeted to reduce the power of ASICs that are consistent with an ASIC design methodology, but which can also be usefully employed in custom circuit design.

The third audience for this book is researchers in electronic design automation who are looking for a broader survey of contemporary low-power tools, methodologies, and design techniques. We hope that this book offers a more complete presentation of the battery of techniques that can be brought to bear to save power than is typically offered in conference publications or even survey articles. We also hope that the design examples used in this book will help researchers to contextualize their own research.

Occasionally at technical conferences you will hear someone say: "Another power paper? Isn't that a solved problem?" Low power design has indeed been a focal research area for fifteen years. However, a look at the power challenges of today's industrial designs indicates that the topic of this book has never been timelier.

David Chinnery
Kurt Keutzer

ACKNOWLEDGMENTS

Many people have given us advice, feedback and support over the years. We will endeavor to acknowledge the majority of those people here, but there are also numerous others with whom we have discussed research and who have made helpful suggestions.

The Semiconductor Research Corporation supported our research on low power. Our thanks to STMicroelectronics for access to their 0.13um process technology and to the contacts at STMicroelectronics, Bhusan Gupta and Ernesto Perea. For the algorithmic portion of our research, we collaborated extensively with David Blaauw, Sarvesh Kulkarni, Ashish Srivastava, and Dennis Sylvester. Sarvesh Kulkarni and Ashish Srivastava provided characterized asynchronous level converters and Synopsys PowerArc characterized libraries for STMicroelectronics 0.13um process. We would like to thank the Intel industrial liaisons, in particular Vijay Pitchumani and Desmond Kirkpatrick, for their advice.

We would like to thank researchers at the Berkeley Wireless Research Center: Stephanie Augsburger, Rhett Davis, Sohrab Emami-Neyestanak, Borivoje Nikolić, Fujio Ishihara, Dejan Markovic, Brian Richards, Farhana Sheikh, and Radu Zlatanovici. Laurent El Ghaoui also helped with convex optimization research.

We would like to acknowledge the contributors to sessions on Closing the Gap between ASIC and Custom and the two books on the topic. Fruitful discussions with them have helped clarify our assumptions and delve into the details: Ameya Agnihotri, Debashis Bhattacharya, Subhrajit Bhattacharya, Vamsi Boppana, Andrew Chang, Pinhong Chen, John Cohn, Michel Cote, Michel Courtoy, Wayne Dai, William Dally, David Flynn, Jerry Frenkil, Eliot Gerstner, Ricardo Gonzalez, Razak Hossain, Lun Bin Huang, Bill Huffman, Philippe Hurat, Anand Iyer, Srikanth Jadcherla, Michael Keating, Earl Killian, George Kuo, Yuji Kukimoto, Julian Lewis, Pong-Fei Lu, Patrick Madden, Murari Mani, Borivoje Nikolić, Greg Northrop, Satoshi Ono, Michael Orshansky, Barry Pangrle, Matthew Parker, Ruchir Puri, Stephen Rich, Nick Richardson, Jagesh Sanghavi, Kaushik Sheth, Jim Schwartz, Naresh Soni, David Staepelaere, Leon Stok, Xiaoping Tang, Chin-Chi Teng, Srini Venkatraman, Radu Zlatanovici, and Tommy Zounes.

We would also like to thank others within our department who have helped with low power research and editing: Abhijit Davare, Masayuki Ito, Trevor Meyerowitz, Matthew Moskewicz, David Nguyen, Kaushik Ravindran, Nadathur Satish, and Brandon Thompson.

David thanks his wife, Eleyda Negron, for her help and support. Kurt thanks Barbara Creech for her patience and support.

The cover was designed by Steven Chan. It shows the Soft-Output Viterbi Algorithm (SOVA) chip morphed with a custom 64-bit datapath. The SOVA chip picture is courtesy of Stephanie Ausberger, Rhett Davis, Borivoje Nikolić, Tina Smilkstein, and Engling Yeo. The SOVA chip was fabricated with STMicroelectronics. The 64-bit datapath is courtesy of Andrew Chang and William Dally. GSRC and MARCO logos were added.

CONTENTS

DESIGN TECHNIQUES

DESIGN EXAMPLES

Chapter 1

INTRODUCTION

David Chinnery, Kurt Keutzer
Department of Electrical Engineering and Computer Sciences
University of California at Berkeley
Berkeley, CA 94720, USA

This book examines the power consumption of ASIC and custom inte-rated-circuits. In particular, we examine the relationship between custom circuits designed without any significant restriction in design methodology and ASIC circuits designed in a high-productivity EDA tool methodology. From analysis of similar ASIC and custom designs, we estimate that the power consumption of typical ASICs may be 3 to 7× that of custom ICs fabricated in process technology of the same generation. We consider ways to augment and enhance an ASIC methodology to bridge this power gap between ASIC and custom.

Reducing circuit power consumption has been a hot topic for some time; however, there has not been detailed analysis of the power gap between an automated design methodology and custom design. This work gives a quan-iative analysis of the factors contributing to the power gap. By identifying the largest contributing factors, and which of these can be automated, we aim to help close the power gap.

We examine design approaches and tools to reduce the power con-sumption of designs produced in an automated design flow. In particular, we focus on microarchitectural techniques, improvements in algorithms for gate sizing and place and route, voltage scaling and use of multiple supply voltages to reduce dynamic power, power gating to reduce leakage power, and statistical power minimization. Design examples illustrate the use of these techniques and show that energy efficiency can be improve by a factor of 2 to 3×.

1.1   DEFINITIONS: ASIC AND CUSTOM

The term *application-specific integrated-circuit* (ASIC), has a wide variety of associations. Strictly speaking, it simply refers to an *integrated circuit* (IC) that has been designed for a particular *application.* This defines a portion of the semiconductor market. Other market segments include memories, microprocessors, and field programmable gate arrays (FPGAs).

Two industries grew to support the development of ASICs: vendors fabricating chips, and companies offering electronic design automation (EDA) software. The ASIC semiconductor-vendor industry, established by companies such as LSI Logic, provides the service of fabricating ASICs designed by other independent design groups. EDA companies such as Cadence and Synopsys provide commercial tools for designing these ASICs. Another key element of the ASIC design process is *ASIC libraries.* ASIC libraries are carefully characterized descriptions of the primitive logic-level building blocks provided by the ASIC vendors. Initially these libraries targeted gate-array implementations, but in time the higher-performance standard-cell targets became more popular.

ASIC vendors then offered complete design flows for their fabrication process. These consisted of ASIC tools, ASIC libraries for the process, and a particular design methodology. These embodied an *ASIC methodology* and were known as *ASIC design kits.* Smith's book on ASICs [6] is a great one-stop reference for ASICs.

Generally, ASICs are designed at the register-transfer level (RTL) in Verilog or VHDL, specifying the flow of data between registers and the state to store in registers. Commercial EDA tools are used to map the higher level RTL description to standard cells in an ASIC library, and then place the cells and route wires. It is much easier to migrate ASIC designs to a new process technology, compared to custom designs which have been optimized for a specific process at the gate or transistor-level. ASIC designers generally focus on high level designs choices, at the microarchitectural level for example.

With this broader context, let us pause to note that the use of the term ASIC can be misleading: it most often refers to an IC produced through a standard cell ASIC methodology and fabricated by an ASIC vendor. That IC may belong to the application-specific standard product (ASSP) portion of the semiconductor market. ASSPs are sold to many different system vendors [6], and often may be purchased as standard parts from a catalog, unlike ASICs.

The term *custom integrated-circuit,* or *custom IC,* also has a variety of associations, but it principally means a circuit produced through a custom-design methodology. More generally, custom IC is used synonymously with the semiconductor market segments of high-performance microprocessors and digital signal processors.

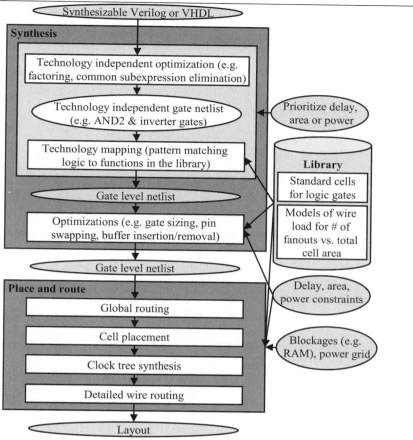

Figure 1.1 A typical EDA flow from a high level hardware design language (HDL) description through to layout.

Custom ICs are typically optimized for a specific process technology and take significantly more time to design than ASICs, but can achieve higher performance and lower power by higher quality design and use of techniques that are not generally available to ASICs. For example, custom designers may design logic gates at the transistor-level to provide implementations that are optimal for that specific design; whereas an ASIC designer is limited by what is available in the standard cell library.

1.2 WHAT IS A STANDARD CELL ASIC METHODOLOGY?

A standard cell ASIC methodology incorporates a standard cell library and automated design tools to utilize this library, in order to achieve higher designer productivity. The designer specifies the circuit behavior in a hardware description language (HDL) such as Verilog or VHDL [5]. This high level description is then mapped to a library of standard cells that implement various logic functions, as shown in Figure 1.1. Various optimizations are performed to try and meet delay, power, or area constraints specified by the designer. The final layout of the chip is not known at the synthesis stage, so the wire capacitances are estimated using a wire load model. Then the standard cells are placed, wires are routed between them, and a clock tree network is inserted to distribute the clock signal.

The EDA flow may be iterated through many times as a design is changed to meet performance constraints. Small changes may be made at the layout level, but significant changes like resizing gates on a delay-critical path may require redoing place and route. After place and route, wire load models for later iterations may be updated based on the resulting layout.

There are also verification steps to try to ensure that the final circuit that is fabricated performs correctly. These include verifying that the gate level logic corresponds to the HDL description; gate level simulation to check correct functional behavior; verifying the layout meets design rules; checking that supply and ground voltage (IR) drops are within tolerances for the standard cell library or design; cross-talk analysis to check signal interference between wires on the chip; and electromagnetic interference analysis to check signal interference with the surrounding environment.

Custom designers sometimes use an ASIC methodology, in particular for portions of the chip that are not timing critical, such as control logic. For performance-critical datapath logic, it is highly advantageous in terms of speed, power, and area to manually lay out the semi-regular logic. If their position is known, cells can have less guard banding, or input and output ports in a particular place to reduce wire lengths, and so forth. Custom design of individual cells and manual placement is laborious, increasing the time-to-market and requiring much larger design teams. Such design-specific optimizations are seldom useful on other designs except for commonly used structures such as memory, and also may not be usable if the technology for the design changes. There have been several attempts by EDA companies to sell datapath synthesis tools, but they have not been successful. It is very difficult for tools to identify the appropriate layout, as a datapath does not usually a regular structure that can be identified by a general purpose tool, though some design companies do have in-house datapath generation tools.

Using a vendor-provided standard cell library for a given fabrication process technology improves designer productivity. Lower transistor-level circuit design issues are abstracted to gate-level power and delay characteristics, and standard cells are designed robustly with guard-banding to ensure correct behavior. A library typically has several drive strengths of cells that implement a given logic function. These drive strengths correspond to the capacitive load that a cell can drive without excessive delay and with acceptable signal characteristics. Cell placement is simplified by using a fixed height for all the cells. Rows of cells are placed on the chip, with contiguous supply voltage (V_{dd}) rails and ground voltage (GND) rails at the top and bottom of the rows. This makes it possible for automated placement of standard cells in the manner shown in Figure 1.2. A new standard cell library can be used by iterating an RTL design through the design flow again, which makes it much easier to migrate between process technologies.

We will discuss later optimizing the drive strength of logic gates in a circuit and similar issues. Thus it is useful to briefly examine transistor-level layouts. The left of Figure 1.3 shows a detailed circuit schematic for an inverter. There is some ambiguity when we refer to a "gate", whether it is a logic gate such as an inverter, or the transistor gate shown labeled G – this will be clarified where appropriate in the text.

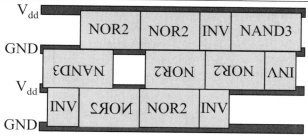

Figure 1.2 Placement of standard cells on standard cell rows are shown, with cells on alternate rows rotated 180° to share power rails. Standard cell height is fixed, but width and placement along a row may vary, and cells may also be mirrored horizontally.

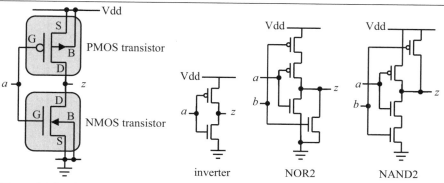

Figure 1.3 On the left is shown a detailed circuit schematic for an inverter. Transistor gate G, source S, drain D and bulk B (also referred to as substrate) nodes are noted. Connections to the substrate are generally omitted, in which case it is assumed that the NMOS p-well connects to ground (0V), and the PMOS n-well connects to the supply voltage (Vdd). On the right are shown the circuit schematics for three logic gates.

In Figure 1.3, note that the NOR2 gate has two PMOS transistors in series in the pull-up portion, whereas the NAND2 gate has two NMOS transistors in series in the pull-down portion. The more transistors in series, the slower the logic gate is due to increased series resistance. PMOS transistors are slower than NMOS transistors, so the NOR2 is slower than a NAND2, assuming the same transistor sizes. Wider transistors may be used to reduce the delay. A typical inverter PMOS to NMOS ratio to have equal pull-up and pull-drive strengths is 2:1. To reduce the additional delay of transistors in series, for a NOR2 gate this becomes 4:1, and for a NAND2 gate this is 2:2. Skewed P to N ratios, substrate biasing, and other circuit issues will be discussed briefly in later chapters. However, increasing the transistor widths increases the power used in charging and discharging the logic gate.

This is an example of low-level power-performance trade-offs that would be considered by a custom circuit designer. To reduce the time to design a circuit, an ASIC circuit designer typically avoids such issues by using a fixed library of standard cells. It is assumed that the library has appropriate PMOS to NMOS ratios and a range of sizes for logic gates to optimally drive the circuit in different load capacitance conditions. However, this assumption is not necessarily true. Such factors may contribute to suboptimal ASIC designs. This and other power-performance trade-offs are examined in this book.

1.3   WHO SHOULD CARE ABOUT THIS BOOK?

1.3.1   ASIC and ASSP designers seeking high performance

Power consumption has become a major design constraint for high performance circuits and limits performance for high end microprocessor chips in today's technologies. Our book titled *Closing the Gap between ASIC & Custom* [1] detailed how to achieve high performance for ASICs in an EDA design flow, but we did not focus on the limitations imposed by power consumption. Some of the techniques used to achieve lower power in high performance custom designs can be automated for use in an ASIC design methodology.

While many ASIC designers may be power-budget limited when seeking higher performance, we quickly acknowledge that not all ASIC designers are seeking higher performance. Many ASIC designs need only to be cheaper than FPGAs (field programmable gate arrays) or faster than general-purpose processor solutions to be viable. For these designs, the desire for higher performance is dominated by final part cost, low non-recurring engineering cost, and time-to-market concerns. Non-recurring engineering costs for ASICs have grown substantially with increased transistor density and deep-submicron design issues. Mask-set costs are now exceeding one million dollars. Both the number and cost of tools required to do ASIC design are rising. In order to recoup their massive investment in fabrication facilities, semiconductor vendors for ASICs are "raising the bar" for incoming ASIC designs. Specifically, ASIC semiconductor vendors are raising the minimum volumes and expected revenues required to enter into contract for fabricating ASICs. These different factors are causing more ASIC design groups to rethink their approach. Some groups are migrating to using FPGA solutions. Some groups are migrating to application-specific standard parts (ASSPs) that can be configured or programmed for their target application.

Those groups that retain their resolve to design ASICs have a few common characteristics. First, these groups aim to amortize increasing non-recurring engineering costs for ASIC designs by reusing a design across multiple applications. Thus they are no longer designing "point solution" ASICs, but are tending toward more sustainable IC *platforms* with software that can be updated as application requirements change [2]. Secondly, as transistor density increases with Moore's law [4], more and more devices are integrated onto a single chip to reduce the net production cost. Multiple processor cores are integrated onto a chip to increase performance or allow for more programmability to achieve retargetable IC platforms. However, the power consumption also increases with more devices on a chip. Finally, given the effort and attention required to design a highly complex ASIC, design groups are demanding more out of their investment. In short, this book targets ASIC and ASSP designers seeking high-performance and low power within an automated design methodology, and we contend that this number is *increasing* over time.

1.3.2 ASIC and ASSP designers seeking lower power

Power consumption is of primary importance in chips designed for embedded and battery powered applications. To reduce part costs, cheap plastic packaging is preferred, which limits the maximum heat dissipation. For many applications such as mobile phones, a long battery lifetime is desirable, so low power is important. ASIC implementations are often chosen for low power, as they can be an order of magnitude or more lower power than applications implemented on an FPGA [3] or in software running on a general purpose processor.

The main approaches to reducing power consumption are scaling down supply voltage and using smaller gate sizes to reduce dynamic power, and increasing threshold voltage to reduce static leakage power; however, these techniques to reduce power also substantially slow down a circuit. Thus we focus on reducing the power gap between ASIC and custom designs subject to some performance constraint. In ASIC designs with tight performance constraints and a tight power budget, ASIC designers must use high performance techniques to create some timing slack for power minimization.

1.3.3 Custom designers seeking higher productivity

An equally important audience for this book is custom designers seeking low power ICs in a design methodology that uses less human resources, such as an ASIC design methodology. Without methodological improvements, custom design teams can grow as fast as Moore's Law to design the most complex custom ICs. Even the design teams of the most commercially successful microprocessors cannot afford to grow at that rate.

We hope to serve this audience in two ways. First, we account for the relative power impact of different elements of a custom-design methodology. Projects have limited design resources and must be used judiciously. Therefore, design effort should be applied where it offers the greatest benefit. We believe that our analysis should help to determine where limited design resources are best spent.

Secondly, specific tools targeted to reduce the power of ASICs can be applied to custom design. The custom designer has always lacked adequate tool support. Electronic Design Automation (EDA) companies have never successfully found a way to tie their revenues to the revenues of the devices they help design. Instead, EDA tool vendors get their revenues from licensing design tools for each designer, known as a "design seat". It doesn't matter if the chip designed with an EDA tool sells in volumes of ten million parts or one, the revenue to the EDA company is the same. It has been estimated that there are more than ten times as many ASIC designers (50,000 – 100,000 worldwide) as custom designers (3,000 – 5,000 worldwide). As a result EDA tool vendors naturally "follow the seats" and therefore have focused on tools to support ASIC designers rather than custom designers. Companies using custom design augment tools from EDA vendors with their own in-house tools. These in-house tools can be improved by identifying where gaps exist in the standard approaches that have been used for circuit design, or replaced in cases where EDA tools perform sufficiently well.

1.4   ORGANIZATION OF THE REST OF THE BOOK

This book examines the power gap between ASIC and custom design methodologies, techniques to reduce the power gap, and design examples illustrating these techniques. The remaining chapters in this book are organized into these three groups.

The first set of chapters discusses the contributing factors to power consumption in ASICs being larger than in custom designs, with power and performance models. The power gap is estimated in Chapter 2, then we provide a detailed overview of the contributing factors, and discuss the design difficulties associated with exploiting these methods in an automated design flow and the extent to which they may be automated. A high-level pipeline power-performance model is combined with a low-level model of gate sizing and voltage scaling power-delay trade-offs in Chapter 3. This enables estimates of the benefit of using microarchitectural techniques to provide timing slack for power minimization at later design stages, and quantitative analysis of the influence of different design factors. Chapter 4 compares analytical and empirical models of circuit power and delay with voltage scaling, discussing the dynamic power, leakage power and delay trade-offs with gate sizing and optimization of supply and threshold voltages.

The second group of chapters details a variety of design techniques and tools to help minimize power consumption. Chapter 5 gives examples of microarchitectural optimizations that can increase energy efficiency by more than $10\times$ for specific applications. Chapter 6 shows that the typical greedy heuristics for gate sizing are suboptimal, and that a linear programming formulation with a global circuit can provide greater power reductions. The linear programming approach is applied to gate-level supply and threshold voltage assignment in Chapter 7 to analyze how much power may be saved by using these approaches. Alternative algorithms for supply voltage assignment are examined in Chapter 8. Chapter 9 details improved tools for automated placement and discusses the placement issues when using multiple supply voltages. Results for reducing leakage power with an automated tool for power gating are presented in Chapter 10. Design verification issues and verification tool support needed for use of multiple voltage and sleep domains are examined in Chapter 11. Then Chapter 12 details power minimization with statistical timing and power analysis.

The last set of chapters presents two design examples utilizing low power techniques. Chapter 13 reports the power savings achieved with standard cell library improvements, arithmetic optimizations, bit slicing, and voltage scaling on DSP (digital signal processor) blocks for a satellite communications chip. Chapter 14 presents a low power design flow that was used to minimize power consumption of an ARM 1136JF-S processor, utilizing multiple supply and threshold voltages. These design examples show that using the low power techniques discussed in this book can provide increased energy efficiency by a factor of 2 to $3\times$.

1.5 WHAT'S NOT IN THIS BOOK

This book focuses on power consumption of integrated circuits and the tools and techniques by which lower power can be achieved. ASIC and custom performance and approaches to increase circuit speed were discussed extensively in our book on the topic [1]. Other than in the context of place and route tools in Chapter 11, area minimization is not a direct focus, as that is a less critical design constraint compared to speed and power in today's technologies that allow billions of transistors on a chip. Where the power minimization techniques that we suggest here negatively impact on circuit delay or area we have made every effort to point that out.

1.6   REFERENCES

[1] Chinnery, D., and Keutzer, K., *Closing the Gap Between ASIC & Custom: Tools and Techniques for High-Performance ASIC Design*, Kluwer Academic Publishers, 2002, 432 pp.

[2] Keutzer, K. et al., "System-level Design: Orthogonalization of Concerns and Platform-Based Design," *IEEE Transactions on Computer-Aided Design*, vol. 19, no. 12, December 2000, pp. 1523-1543.

[3] Kuon, I., and Rose, J., "Measuring the Gap Between FPGAs and ASICs," *International Symposium on Field Programmable Gate Arrays*, 2006, pp. 21-30.

[4] Moore, G., "Cramming more components onto integrated circuits," *Electronics*, vol. 38, no. 8, 1965, pp. 114-117.

[5] Smith, D., *HDL Chip Design: A Practical Guide for Designing, Synthesizing and Simulating ASICs and FPGAs Using VHDL or Verilog*, Doone Publications, 1998, 464 pp.

[6] Smith, M., *Application-specific Integrated Circuits*, Addison-Wesley, Berkeley, CA, 1997.

Chapter 2

OVERVIEW OF THE FACTORS AFFECTING THE POWER CONSUMPTION

David Chinnery, Kurt Keutzer
Department of Electrical Engineering and Computer Sciences
University of California at Berkeley
Berkeley, CA 94720, USA

We investigate differences in power between application-specific integrated circuits (ASICs) and custom integrated circuits, with examples from 0.6um to 0.13um CMOS. A variety of factors cause synthesizable designs to consume 3 to 7× more power. We discuss the shortcomings of typical synthesis flows, and changes to tools and standard cell libraries needed to reduce power. Using these methods, we believe that the power gap between ASICs and custom circuits can be closed to within 2.6× at a tight performance constraint for a typical ASIC design.

2.1 INTRODUCTION

In the same technology generation, custom designs can achieve 3 to 8× higher clock frequency than ASICs [18]. Custom techniques that are used to achieve high speed can also be used to achieve low power [62]. Custom designers can optimize the individual logic cells, the layout and wiring between the cells, and other aspects of the design. In contrast, ASIC designers generally focus on optimization at the RTL level, relying on EDA tools to map RTL to cells in a standard cell library and then automatically place and route the design. Automation reduces the design time, but the resulting circuitry may not be optimal.

Low power consumption is essential for embedded applications. Power affects battery life and the heat dissipated by hand-held applications must be limited. Passive cooling is often required, as using a heat sink and/or fan is larger and more expensive.

Power is also becoming a design constraint for high-end applications due to reliability, and costs for electricity usage and cooling. As technology scales, power density has increased with transistor density, and leakage power is becoming a significant issue even for high end processors. Power consumption is now a major problem even for high end microprocessors. Intel canceled the next generation Tejas Pentium 4 chips due to power consumption issues [100].

In this chapter, we will discuss the impact of manual and automated design on the power consumption, and also the impact of process technology and process variation. Our aim is to quantify the influence of individual design factors on the power gap. Thus, we begin by discussing a process technology independent delay metric in Section 2.2. Section 2.3 discusses the contribution to a chip's power consumption from memory, control and datapath logic, and clocking, and also provides an overview of dynamic and leakage power.

In Section 2.4, we compare full custom and synthesizable ARM processors and a digital signal processor (DSP) functional unit. We show that ASICs range from 3 to 7× higher power than custom designs for a similar performance target. To date the contribution of various factors to this gap has been unclear. While automated design flows are often blamed for poor performance and poor energy efficiency, process technology is also significant. Section 2.5 outlines factors contributing to the power gap. We then examine each factor, describing the differences between custom and ASIC design methodologies, and account for its impact on the power gap. Finally, we detail approaches that can reduce this power gap. We summarize our analysis in Section 2.6.

2.2   PROCESS TECHNOLOGY INDEPENDENT FO4 DELAY METRIC

At times we will discuss delay in terms of FO4 delays. It is a useful metric for normalizing out process technology dependent scaling of the delay of circuit elements.

The fanout-of-4 inverter delay is the delay of an inverter driving a load capacitance that has four times the inverter's input capacitance [38]. This is shown in Figure 2.1. The FO4 metric is not substantially changed by process technology or operating conditions. In terms of FO4 delays, other fanout-of-4 gates have at most 30% range in delay over a wide variety of process and operating conditions, for both static logic and domino logic [38].

If it has not been simulated in SPICE or tested silicon, the FO4 delay in a given process technology can be estimated from the channel length. Based on the effective gate length L_{eff}, the rule of thumb for FO4 delay is [39].

$360 \times L_{eff}$ ps for typical operating and typical process conditions   (2.1)

$500 \times L_{eff}$ ps for worst case operating and typical process conditions   (2.2) where the effective gate length L_{eff} has units of micrometers. Typical process conditions give high yield, but are not overly pessimistic. Worst case operating conditions are lower supply voltage and higher temperature than typical operating conditions. Typical operating conditions for ASICs may assume a temperature of 25°C, which is optimistic for most applications. Equation (2.2) can be used to estimate the FO4 delay in silicon for realistic operating conditions [39].

L_{eff} is often assumed to be about 0.7 of the drawn gate length for a process technology – for example, 0.13um for a 0.18um process technology. However, many foundries are aggressively scaling the channel length to increase the speed. Thus, the FO4 delay should be calculated from the effective gate length, if it is known, rather than from the process technology generation.

From previous analysis [18], typical process conditions are between 17% and 28% faster than worst case process conditions. Derating worst case process conditions by a factor of 1.2× gives

$$600 \times L_{eff} \text{ ps for worst case operating and worst case process conditions} \quad (2.3)$$

Equation (2.3) was used for estimating the FO4 delays of synthesized ASICs, which have been characterized for worst case operating and worst case process conditions. This allows analysis of the delay per pipeline stage, independent of the process technology, and independent of the process and operating conditions.

Note: these rules of thumb give approximate values for the FO4 delay in a technology. They may be inaccurate by as much as 50% compared to simulated or measured FO4 delays in silicon. These equations do not accurately account for operating conditions. Speed-binning and process improvements that do not affect the effective channel length are not accounted for. Accurate analysis with FO4 delays requires proper calibration of the metric: simulating or measuring the actual FO4 delays for the given process and operating conditions.

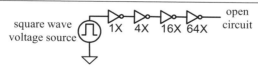

Figure 2.1 This illustrates a circuit to measure FO4 delays. The delay of the 4X drive strength inverter gives the FO4 delay. The other inverters are required to appropriately shape the input waveform to the 4X inverter and reduce the switching time of the 16X inverter, which affect the delay of the 4X inverter [38].

2.3   COMPONENTS OF POWER CONSUMPTION

Designers typically focus on reducing both the total power when a circuit is active and its standby power. There is usually a minimum performance target, for example 30 frames/s for MPEG. When performance is less important, the energy per operation to perform a given task can be minimized.

Active power includes both dynamic power consumption, when the logic evaluates or the clock transitions, and current leakage when logic is not switching. There is no computation in logic in standby, the clock must be gated to prevent it switching, and leakage is the dominant source of power consumption in standby.

The major sources of power consumption in circuitry are the clock tree and registers, control and datapath logic, and memory. The breakdown of power consumption between these is very application and design dependent. The power consumption of the clock tree and registers ranged from 18% to 36% of the total power for some typical embedded processors and microprocessors (see Section 3.2.4). In custom cores for discrete cosine transform (DCT) and its inverse (IDCT), contributions to the total power were 5% to 10% from control logic, about 40% from the clock tree and clock buffers, and about 40% from datapath logic [101][102]. Memory can also account for a substantial portion of the power consumption. For example, in the StrongARM caches consume 43% of the power [62].

2.3.1   Dynamic power

Dynamic power is due to switching capacitances and short circuit power when there is a current path from supply to ground.

The switching power is proportional to $\alpha f C V_{dd}^2$, where α is the switching activity per clock cycle, f is the clock frequency, C is the capacitance that is (dis)charged, and V_{dd} is the voltage swing. The switching activity is increased by glitches, which typically cause 15% to 20% of the activity in complementary static CMOS logic [77].

Short circuit power typically contributes less than 10% of the total dynamic power [14], and increases with increasing Vdd, and with decreasing Vth. Short circuit power can be reduced by matching input and output rise and fall times [96].

As the dynamic power depends quadratically on Vdd, methods for reducing active power often focus on reducing Vdd. Reducing the capacitance by downsizing gates and reducing wire lengths is also important.

2.3.2   Leakage power

In today's processes, leakage can account for 10% to 30% of the total power when a chip is active. Leakage can contribute a large portion of the average power consumption for low performance applications, particularly when a chip has long idle modes without being fully off.

Optimally choosing Vdd and Vth to minimize the total power consumption for a range of delay constraints in 0.13um technology, the leakage varied from 8% to 21% of the total power consumption in combinational logic, as discussed later in Section 4.6.1. However, the possible Vdd and Vth values depend on the particular process technology and standard cell libraries available. For example for a delay constraint of 1.2× the minimum delay, the best library choice had Vdd of 0.8V and Vth of 0.08V (see Table 7.7 with 0.8V input drivers), and leakage contributed on average 40% of total power.

Leakage power in complementary static CMOS logic in bulk CMOS is primarily due to subthreshold leakage and gate leakage. Subthreshold leakage increases exponentially with decrease in Vth and increase in temperature. It can also be strongly dependent on transistor channel length in short channel devices. Gate leakage has increased exponentially with reduction in gate oxide thickness. There is also substrate leakage. Leakage has become increasingly significant in deep submicron process technologies.

2.4   ASIC AND CUSTOM POWER COMPARISON

To illustrate the power gap, we examine custom and ASIC implementtations of ARM processors and dedicated hardware to implement discrete cosine transform (DCT) and its inverse (IDCT). ARM processors are general purpose processors for embedded applications. ASICs often have dedicated functional blocks to achieve low power and high performance on specific applications – for example, media processing. JPEG and MPEG compression and decompression of pictures and video use DCT and IDCT. There is a similar power gap between ASIC and custom for the ARM processors and for DCT and IDCT blocks.

2.4.1   ARM processors from 0.6 to 0.13um

We compare chips with full custom ARM processors, soft, and hard ARM cores. Soft macros of RTL code may be sold as individual IP (intellectual property) blocks and are portable between fabrication processes. In a hard macro, the standard cell logic used, layout and wiring have been specified and optimized then fixed for a particular fabrication process. A hard macro may be custom, or it may be "hardened" from a soft core. A complete chip includes additional memory, I/O logic, and so forth.

Table 2.1 Full custom and hard macro ARMs [11][31][32][43][70]. The highlighted full custom chips have 2 to 3× MIPS/mW.

Processor	Technology (um)	Voltage (V)	Frequency (MHz)	MIPS	Power (mW)	MIPS/mW
ARM710	0.60	5.0	40	36	424	0.08
Burd	0.60	1.2	5	6	3	1.85
Burd	0.60	3.8	80	85	476	0.18
ARM810	0.50	3.3	72	86	500	0.17
ARM910T	0.35	3.3	120	133	600	0.22
StrongARM	0.35	1.5	175	210	334	0.63
StrongARM	0.35	2.0	233	360	950	0.38
ARM920T	0.25	2.5	200	220	560	0.39
ARM1020E	0.18	1.5	400	500	400	1.25
XScale	0.18	1.0	400	510	150	3.40
XScale	0.18	1.8	1000	1250	1600	0.78
ARM1020E	0.13	1.1	400	500	240	2.08

Table 2.2 The highlighted ARM7TDMI hard macros have 1.3 to 1.4× MIPS/mW versus the synthesizable ARM7TDMI-S cores [5].

ARM Core	Technology (um)	Frequency (MHz)	Power (mW)	MIPS/mW
ARM7TDMI	0.25	66	51	1.17
ARM7TDMI-S	0.25	60	66	0.83
ARM7TDMI	0.18	100	30	3.00
ARM7TDMI-S	0.18	90	35	2.28
ARM7TDMI	0.13	130	10	11.06
ARM7TDMI-S	0.13	120	13	8.33

   To quantify the power gap between ASIC and custom, we first examined hard macro and full custom ARMs, listed in Table 2.1. Compared to the other designs, the three full custom chips in bold achieved 2 to 3× millions of instructions per second per milliwatt (MIPS/mW) at similar MIPS, as shown in Figure 2.2. The inverse of this metric, mW/MIPS, is the energy per operation. The Dhrystone 2.1 MIPS benchmark is the performance metric [98]. It fits in the cache of these designs, so there are no performance hits for cache misses or additional power to read off-chip memory.
   Lower power was achieved in several ways. The DEC StrongARM used clock-gating and cache sub-banking to substantially reduce the dynamic power [62]. The Intel XScale and DEC StrongARM used high speed logic styles to reduce critical path delay, at the price of higher power consumption on these paths. To reduce pipeline register delay, the StrongARM used pulse-triggered flip-flops [62] and the XScale used clock pulsed latches [22]. Shorter critical paths allow the same performance to be achieved with a lower supply voltage (Vdd), which can lower the total power consumption. Longer channel lengths were used in the StrongARM caches to reduce the

leakage power, as the two 16kB caches occupy 90% of the chip area [62]. The XScale used substrate biasing to reduce the leakage [24].

For the same technology and similar performance (MIPS), the Vdd of the full custom chips is lower than that of the hard macros – reducing Vdd gives a quadratic reduction in dynamic power. The StrongARM can operate at up to 233MHz at 2.0V and the XScale can operate at up to 1GHz at 1.65V [43]. If operating at higher performance was not required, it is likely that even higher MIPS/mW could have been achieved.

Energy efficiency can be improved substantially if performance is sacrificed. Burd's 0.6um ARM8 had software controlled dynamic voltage scaling based on the processor load. It scaled from 0.18MIPS/mW at 80MHz and 3.8V, to 2.14MIPS/mW at 5MHz and 1.2V [11]. Voltage scaling increased the energy efficiency by 1.1× for MPEG decompression which required an average clock frequency of 50MHz, and increased the energy efficiency by 4.5× for audio processing which required a clock frequency of only 17MHz [12].

There is an additional factor of 1.3 to 1.4× between hard macro and synthesizable ARM7 soft cores, as shown in Table 2.2. These MIPS/mW are higher than those in Table 2.1, as they exclude caches and other essential units. The ARM7TDMI cores are also lower performance, and thus can achieve higher energy efficiency.

Overall, there is a factor of 3 to 4× between synthesizable ARMs and the best full custom ARM implementations.

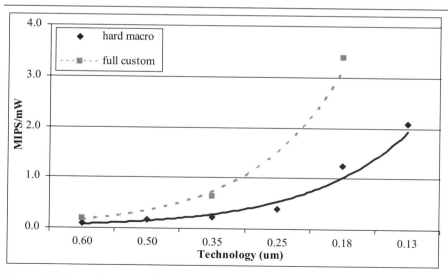

Figure 2.2 This graph compares MIPS/mW of custom and hard macro ARMs in Table 2.1.

2.4.1.1 Other full custom ARM implementations

There are two other noteworthy higher performance full custom ARMs, though they are less energy efficient than the 0.18um XScale.

Samsung's Halla is a full custom 0.13um implementation of the ARM1020E with power consumption from 0.26W at 400MHz and Vdd of 0.7V to 1.8W at 1200MHz and Vdd of 1.1V [50]. Achieving 1480MIPS at 1200MHz clock frequency, the energy efficiency ranged from 0.82MIPS/mW at 1200MHz to 1.90MIPS/mW at 400MHz. Differential cascode voltage switch logic (DCVSL) was used for high performance, but DCVSL has substantial power consumption compared to complementary static CMOS logic that is used in ASICs. Sense amplifiers were used with the low voltage swing dual rail bus to detect voltage swings of less than 200mV, achieving high bus speeds at lower power consumption [60]. The die area of the Halla was 74% more than ARM's 0.13um ARM1020E.

Intel's 90nm implementation of the XScale, codenamed Monahans, has 770mW dynamic power consumption at 1500MHz and Vdd of 1.5V with performance of 1200MIPS at this point [72]. The energy efficiency of Monahans is 1.56MIPS/mW at 1500MHz – data for improved energy efficiencies at lower Vdd has not been published. Clock pulsed latches were also used in this implementation of the XScale. The hold time for the clock gating enable signal was the duration of the clock pulse, and thus did not require latching. Domino logic was used for high performance in the shifter and cache tag NOR comparators. 75% of instruction cache tag accesses were avoided by checking if the instruction cache request line was the same as the previous one. Selective accesses and avoiding repeated accesses reduced power by 42% in the dynamic memory management unit [21].

2.4.2   Comparison of DCT/IDCT cores

Application-specific circuits can reduce power by an order of magnitude compared to using general purpose hardware [77]. Two 0.18um ARM9 cores were required to decode 30 frames/s for MPEG2, consuming 15× the power of a synthesizable DCT/IDCT design [28]. However, the synthesizable DCT/IDCT significantly lags its custom counterparts in energy efficiency.

Table 2.3 Comparison of ASIC and custom DCT/IDCT core power consumption at 30 frames/s for MPEG2 [28][101][102].

Design	Technology (um)	Voltage (V)	DCT (mW)	IDCT (mW)
ASIC	0.18	1.60	8.70	7.20
custom DCT	0.6 (L_{eff} 0.6)	1.56	4.38	
custom IDCT	0.7 (L_{eff} 0.5)	1.32		4.65

Fanucci and Saponara designed a low power synthesizable DCT/IDCT core, using similar techniques to prior custom designs. Despite being three technology generations ahead, the synthesizable core was 1.5 to 2.0× higher power [28]. Accounting for the technology difference by conservatively assuming power scales linearly with device dimensions [71], the gap is a factor of 4.3 to 6.6×. The data is shown in Table 2.3.

2.5   FACTORS CONTRIBUTING TO ASICS BEING HIGHER POWER

Various parts of the circuit design and fabrication process contribute to the gap between ASIC and custom power. Our analysis of the most significant design factors and their impact on the total power when a chip is active is outlined in Table 2.4. The "typical" column shows the maximum contribution of individual factors comparing a typical ASIC to a custom design. In total these factors can make power an order of magnitude worse. In practice, even the best custom designs can't fully exploit all these factors simultaneously. Low power design techniques that can be incorporated within an EDA flow can reduce the impact of these factors in a carefully designed ASIC as per the "excellent" column in Table 2.4.

Most low power EDA tools focus on reducing the dynamic power in control logic, datapath logic, and the clock tree. The design cost for custom memory is low, because of the high regularity. Several companies provide custom memory for ASIC processes. Optimization of memory hierarchy, memory size, caching policies, and so forth is application dependent and beyond the scope of this book, though they have a substantial impact on the system-level performance and power consumption. We will focus on the power consumption in a processor core.

Table 2.4 Factors contributing to ASICs being higher power than custom. The excellent column is what ASICs may achieve using low power and high performance techniques. This table focuses on the total power when a circuit is active, so power gating and other standby leakage reduction techniques are omitted. The combined impact of these factors is not multiplicative – see discussion in Section 2.5.1.

Contributing Factor	Typical ASIC	Excellent ASIC
microarchitecture	5.1×	1.9×
clock gating	1.6×	1.0×
logic style	2.0×	2.0×
logic design	1.2×	1.0×
technology mapping	1.4×	1.0×
cell and wire sizing	1.6×	1.1×
voltage scaling	4.0×	1.0×
floorplanning and placement	1.5×	1.1×
process technology	1.6×	1.0×
process variation	2.0×	1.3×

Microarchitectural techniques such as pipelining and parallelism increase throughput, allowing timing slack for gate downsizing and voltage scaling. The microarchitecture also affects the average instructions per cycle (IPC), and hence energy efficiency. The power and delay overheads for microarchitectural techniques must be considered. With sufficient timing slack, reducing the supply voltage can greatly increase the energy efficiency. For example in Table 2.1, scaling the XScale from Vdd of 1.8V to 1.0V increases the efficiency from 0.78MIPS/mW to 3.40MIPS/mW, a factor of 4.4×, but the performance decreases from 1250MIPS to 510MIPS.

Process technology can reduce leakage by more than an order of magnitude. It also has a large impact on dynamic power. Process variation results in a wide range of the leakage power for chips and some variation in the maximum operating clock frequency for a given supply voltage. For high yield, a higher supply voltage may be needed to ensure parts meet the desired performance target, resulting in a significant spread in power consumption. Limiting process variation and guard-banding for it without being overly conservative help reduce the power consumption.

Using a high speed logic style on critical paths can increase the speed by 1.5× [18]. Circuitry using only slower complementary static CMOS logic at a tight performance constraint may be 2.0× higher power than circuitry using a high speed logic style to provide timing slack for power reduction by voltage scaling and gate downsizing.

Other factors in Table 2.4 have smaller contributions to the power gap. We will discuss the combined impact of the factors and then look at the individual factors and low power techniques to reduce their impact.

2.5.1   Combined impact of the contributing factors

The combined impact of the factors is complicated. The estimate of the contribution from voltage scaling assumes that timing slack is provided by pipelining, so this portion is double counted. The timing slack depends on the tightness of the performance constraint, which has a large impact on the power gap. We assumed a tight performance constraint for both the typical ASIC and excellent ASIC for the contributions from microarchitecture, logic style, and voltage scaling in Table 2.4. If the performance constraint is relaxed, then the power gap is less. For example, from our model of pipelining to provide timing slack for voltage scaling and gate sizing, the power gap between a typical ASIC and custom decreases from 5.1× at a tight performance constraint for the typical ASIC to 4.0× if the constraint is relaxed by 7%.

Chapter 3 details our power and delay model that incorporates pipelining, logic delay, voltage scaling and gate sizing. The logic delay is determined by factors such as the logic style, wire lengths, process technology, and process variation which affects the worse case delay.

From analysis with this model, an excellent ASIC using the low power techniques that we recommend below may close the power gap to a factor of 2.6 at a tight performance constraint for a typical ASIC [16].

2.5.2 Microarchitecture

Algorithmic and architectural choices can reduce the power by an order of magnitude [77]. We assume that ASIC and custom designers make similar algorithmic and architectural choices to find a low power implementation that is appropriate for the required performance and target application. Pipelining and parallelism are the two major microarchitectural techniques that can be used to maintain throughput (see Figure 2.3), when other power reduction techniques increase critical path delay. With similar microarchitectures, how do ASIC and custom pipelining and parallelism compare?

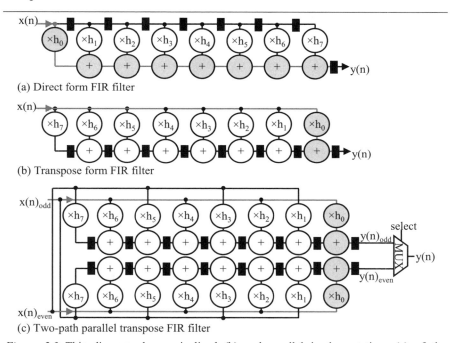

(a) Direct form FIR filter

(b) Transpose form FIR filter

(c) Two-path parallel transpose FIR filter

Figure 2.3 This diagram shows pipelined (b) and parallel implementations (c) of the unpipelined direct form finite input response (FIR) filter in (a) [19][79]. The FIR filter calculates $y_n = h_0 x_n + h_1 x_{n-7} + ... + h_7 x_{n-7}$. The critical paths are shown in grey. The minimum clock period decreases as the registers break the critical path up into separate pipeline stages. Computation in each pipeline stage proceeds concurrently. The parallel implementation doubles the throughput, but the area is more than doubled. The multiplexer to select the odd or even result from the two parallel datapaths at each clock cycle is denoted by MUX.

On their own, pipelining and parallelism do not reduce power. Pipelining reduces the critical path delay by inserting registers between combinational logic. Glitches may not propagate through pipeline registers, but the switching activity of the combinational logic is otherwise unchanged. Additional pipeline registers add to the leakage power and especially to the dynamic power, because the clock signal going to the registers has high activity. Pipelining may reduce the instructions per cycle (IPC) due to branch misprediction and other hazards; in turn this reduces the energy efficiency. Parallelism trades off area for increased throughput, with overheads for multiplexing and additional wiring [6]. Both techniques enable the same performance to be met at lower supply voltage with smaller gate sizes, which can provide a net reduction in power.

Bhavnagarwala et al. [6] predict a 2 to 4× reduction in power with voltage scaling by using 2 to 4 parallel datapaths. Generally, ASICs can make as full use of parallelism as custom designs, but careful layout is required to minimize additional wiring overheads.

Delay overheads for pipelining include: register delay; register setup time; clock skew; clock jitter; and any imbalance in pipeline stage delays that cannot be compensated for by slack passing or useful clock skew. For a given performance constraint, the pipelining delay overheads reduce the slack available to perform downsizing and voltage scaling.

In the IDCT, the cost of pipelining was about a 20% increase in total power, but pipelining reduced the critical path length by a factor of 4. For the same performance without pipelining, Vdd would have to be increased from 1.32V to 2.2V. Thus pipelining helped reduce power by 50% [102].

2.5.2.1   What's the problem?

The timing overhead per pipeline stage for a custom design is about 3 FO4 delays, but it may be 20 FO4 delays for an ASIC, substantially reducing the timing slack available for power reduction. For a typical ASIC, the budget for the register delay, register setup time, clock skew and clock jitter is about 10 FO4 delays. Unbalanced critical path delays in different pipeline stages can contribute an additional 10 FO4 delays in ASICs. If the delay constraint is tight, a little extra timing slack can provide substantial power savings from downsizing gates – for example, a 3% increase in delay gave a 20% reduction in energy for a 64-bit adder [104].

For pipeline registers, most ASICs use slow edge-triggered D-type flip-flops that present a hard timing boundary between pipeline stages, preventing slack passing. The clock skew between clock signal arrivals at different points on the chip must be accounted for. Faster pulse-triggered flip-flops were used in the custom StrongARM [62]. Some pulse-triggered flip-flops have greater clock skew tolerance [80]. Custom designs may use level-sensitive latches to allow slack passing, and latches are also less sensitive to clock skew [19].

The custom XScale used clock-pulsed transparent latches [22]. A D-type flip-flop is composed of a master-slave latch pair. Thus a clock-pulsed latch has about half the delay of a D-type flip-flop and has a smaller clock load, which reduced the clock power by 33%. Clock-pulsed latches have increased hold time and thus more problems with races. The pulse width had to be carefully controlled and buffers were inserted to prevent races. The clock duty cycle also needs to be carefully balanced.

To estimate the impact of worse ASIC pipelining delay overhead, we developed a pipeline performance and power model, with power reduction from gate downsizing and voltage scaling versus timing slack (see Chapter 3). At a tight performance constraint for the ASIC design, we estimate that ASIC power consumption can be $5.1\times$ that of custom, despite using a similar number of pipeline stages. While there is no timing slack available to the ASIC design, the lower custom pipeline delay overhead allows significant power reduction by gate downsizing and voltage scaling.

2.5.2.2 What can we do about it?

Latches are well-supported by synthesis tools [83], but are rarely used other than in custom designs. Scripts can be used to convert timing critical portions of an ASIC to use latches instead of flip-flops [17]. High-speed flip-flops are now available in some standard cell libraries and can be used in an automated design methodology to replace slower D-type flip-flops on critical paths [33]. Useful clock skew tailors the arrival time of the clock signal to different registers by adjusting buffers in the clock tree and can be used in ASIC designs for pipeline balancing [26]. With these methods, the pipeline delay overhead in ASICs can be reduced to as low as 5 FO4 delays [18]. This enables more slack to be used for downsizing, voltage scaling, or increasing the clock frequency. From our pipeline model, ASICs can close the gap for the microarchitecture and timing overhead factor to within $1.9\times$ of custom.

2.5.3   Clock gating

In typical operation, pipeline stages and functional units are not always in use. For example, during a sequence of integer operations, the floating point unit may be idle. Providing the logical inputs to the idle unit are held constant, there are only two sources of power dissipation in the idle unit: static leakage; and switching activity at registers and any other clocked elements due to the clock signal – for example, precharge of domino logic.

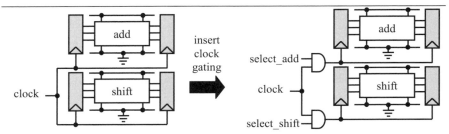

Figure 2.4 This is a simple illustration of clock gating. The clock signal to the registers is gated with a control signal that selects which functional unit is in use. A transparent low latch is usually inserted to de-glitch the enable signal [51].

Architectural or gate-level signals can turn off the clock to portions of the clock tree that go to idle units. This can be done with a clock gating control signal and clock signal at an AND gate, as illustrated in Figure 2.4. As the clock tree and registers can contribute 20% to 40% of the total power, this gives substantial dynamic power savings if units are often idle. The power overheads for logic to generate clock gating signals and the clock gating logic need to be compared versus the potential power savings. Usually clock gating signals can be generated within only one clock cycle, and there is only a small delay increase in the arrival of the gated clock signal at the register.

The StrongARM's total power when active would be about 1.5× worse without clock gating [62]. The StrongARM uses a 12 bit by 32 bit multiply-accumulate (MAC) unit. For some applications, one multiply operand will be 24-bit or less, or 12-bit or less, thus the number of cycles the 12×32 MAC is required is less than the three cycles for a full 32×32 multiply. This saves power by avoiding unnecessary computation. Typical code traces had shift operations of zero, so power could be saved by disabling the shifter in this case [62].

The custom DCT core uses clock gating techniques extensively. In typical operation, consecutive images are highly correlated. Calculations using the significant bits of pixels in common between consecutive images can be avoided. This reduced the number of additions required by 40%, and gave on average 22% power savings for typical images [101]. After the discrete cosine transform on a typical image, there are many coefficients of value zero. This was exploited in the custom IDCT to separately clock gate pipeline stages processing coefficients of zero [102].

We estimate that clock gating techniques can increase energy efficiency when the chip is active by up to 1.6×. Note that the power savings from clock gating vary substantially with the application.

2.5.3.1 What's the problem?

Clock gating requires knowledge of typical circuit operation over a variety of benchmarks. If a unit is seldom idle, clock gating would increase power consumption. Until recently, clock gating was not fully supported by commercial tools. Retiming to reposition the registers [75] can be essential to better balance the pipeline stages, but EDA tools did not support retiming of registers with clock gating.

Care must be taken with gated clock signals to ensure timing correct operation of the registers. Glitches in the enable signal must not propagate to the clock gate while the clock gate is high. This results in a long hold time for the enable signal, which may be avoided by inserting a transparent low latch to de-glitch the enable signal [51]. The transparent low latch prevents the signal that goes to the clock gate from changing while the clock is high. The setup time for the enable signal is longer to account for the clock gate and de-glitching latch. The clock signal arrives later to the register due to the delay of the clock gate, which increases the hold time for that register. The clock tree delay of the clock signal to the clock gate can be reduced to compensate for this, but that may require manual clock tree design.

2.5.3.2 What can we do about it?

An ASIC designer can make full use of clock gating techniques by carefully coding the RTL for the desired applications, or using automated clock-gating. The techniques used in custom DCT and IDCT designs were used in the synthesizable DCT/IDCT [28]. In the synthesizable DCT/IDCT, clock gating and data driven switching activity reduction increased the energy efficiency by 1.4× for DCT and 1.6× for IDCT [28].

In the last few years, commercial synthesis tools have become available to automate gate-level clock-gating, generating clock gating signals and inserting logic to gate the clock. There is now support for retiming of flip-flops with gated clock signals. Power Compiler was able to reduce the power of the synthesizable ARM9S core by 41% at 5% worse clock frequency [30] – primarily by gate downsizing, pin reordering, and clock gating. Useful clock skew tools can compensate for the additional delay on the gated clock signal [26].

There are tools for analyzing the clock-tree power consumption and activity of functional units during benchmark tests. These tools help designers identify signals to cut off the clock signal to logic when it is not in use, and to group logic that is clock gated to move the clock gating closer to the root of the clock tree to save more power.

As ASICs can make effective use of clock gating, there should be no power gap due to clock gating in comparison with custom.

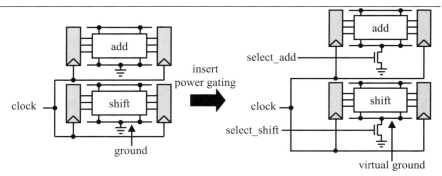

Figure 2.5 This is a simple illustration of power gating. The sleep transistors are turned on by a control signal that selects which functional unit is in use, reducing leakage from supply to ground. Registers may not be disconnected in this manner without losing state information.

2.5.4   Power gating and other techniques to reduce leakage in standby

After clock gating idle units, only static leakage remains. The leakage can be substantially reduced by several methods: reducing the supply voltage (see Section 2.5.9); disconnecting the power rails with sleep transistors [64], known as *power gating*; increasing Vth via substrate biasing to reduce subthreshold leakage; and assigning logic gate inputs to a lower leakage state [56]. All these methods take a significant amount of power and thus are only worthwhile when a unit will be idle for tens to thousands of clock cycles or more [27] – for example, when most of a mobile phone's circuitry is idling while awaiting an incoming call. This requires architectural or software level signals to transition between normal operation and sleep mode.

Reducing the supply voltage reduces the subthreshold leakage current as there is less drain induced barrier lowering (DIBL), and also reduces the gate-oxide tunneling leakage [57]. For example, leakage decreases by 3× when Vdd is reduced from 1.2V to 0.6V with our 0.13um libraries (see Section 4.5.1). Dynamic voltage scaling is discussed further in Section 2.5.9.

Subthreshold leakage and gate leakage vary substantially depending on which transistors in a gate are off, which is determined by the inputs. Leakage in combinational logic can be reduced by a factor of 2 to 4× by assigning primary inputs to a lower leakage state [56][58]. Additional circuitry is required in the registers to store the correct state, while outputting the low leakage state; or state information may be copied from the registers and restored on resumption from standby. There is also dynamic power consumption in the combinational logic in the cycle that inputs are assigned to a low leakage state. Thus, units must be idle for on the order of ten cycles to justify going to a low leakage state.

Normally, the p-well of the NMOS transistors is connected to ground and the n-well of the PMOS transistors is connected to the supply. The subthreshold leakage can be reduced by increasing the threshold voltages by reverse biasing the substrate, connecting the n-well to more than 0V and connecting the p-well to less than Vdd. This requires charge pump circuitry to change the voltage, additional power rails, and a twin well or triple well process [24]. The advantage of reverse body bias is that the state is retained. Reverse body bias is less effective for reducing leakage in shorter channel transistors, for example providing a 4× reduction in leakage in 0.18um technology and 2.5× reduction in 0.13um [49], making it a less useful technique in deeper submicron technologies.

An alternate method of reverse body bias is to connect both the NMOS transistor source and well to a virtual ground V_{ss} (see Figure 2.5) which is raised to reduce leakage in standby. This avoids the need for charge pump circuitry and twin well or triple well process [24]. To avoid losing state information, the reduction in $V_{dd} - V_{ss}$ must be limited by circuitry to regulate the voltage [23]. Reducing $V_{dd} - V_{ss}$ also helps reduce the leakage. This reverse body bias and voltage collapse approach gave a 28× reduction in leakage in the 0.18um XScale with minimal area penalty [24]. Returning from "drowsy" mode took 20us, corresponding to 18,000 cycles at 800MHz, as the phase-locked loop (PLL) was also turned off to limit power consumption. In comparison, using sleep transistors in the XScale would have reduced leakage by only about 5×, if power gating was not applied to latches and other memory elements that need to retain state, as they comprise about a sixth of the total transistor width [24].

Power gating with sleep transistors to disconnect the supply and/or ground rail (Figure 2.5) can provide more than an order of magnitude leakage reduction in circuitry that uses leaky low Vth transistors on critical paths and high Vth sleep transistors [64]. This is often referred to as MTCMOS, multi-threshold voltage CMOS. The "virtual" supply and "virtual" ground rails, which are connected to the actual power rails via sleep transistors, may be shared between logic gates to reduce the area overhead for the sleep transistors. Disconnecting the power rails results in loss of state, unless registers contain a latch connected to the actual supply and ground rails [64]. Registers also have connections to the virtual supply and virtual ground rails to limit leakage.

Leakage was reduced by 37× in a 0.13um 32-bit arithmetic logic unit (ALU) using PMOS sleep transistors at the expense of a 6% area overhead and 2.3% speed decrease [89]. The leakage was reduced 64× by using reverse body bias in conjunction with PMOS sleep transistors in sleep mode, and forward body bias in active mode reduced the speed penalty to 1.8%. The total area overhead for the sleep transistors and the body bias circuitry was 8%. Using sleep transistors saved power if the ALU was idle for at least a hundred clock cycles. Two clock cycles were required to turn the transistors back on from sleep mode, and four cycles were required to change from reverse body bias. With only forward body bias, Vdd could be reduced from 1.32V to 1.28V with no speed penalty, and leakage was reduced by 1.9× at zero bias in standby [89].

2.5.4.1   What's the problem?

Reducing leakage via state assignment, substrate biasing, reducing supply voltage, and sleep transistors requires architectural or software level signals to specify when units will be idle for many cycles. These techniques cannot be automated at the gate level and require architectural level support for signals to enter and exit standby over multiple cycles.

For state assignment, registers that retain data instead output a 0 or 1 in sleep mode. For registers that don't retain data in standby, extra circuitry is required if the reset output differs from the low leakage state output. These registers are larger and consume more power than a standard register.

Substrate biasing and reducing the supply voltage require a variable supply voltage from a voltage regulator. The cell libraries need to have delay, dynamic power, leakage power and noise immunity characterized at the different supply and substrate voltages. If functional units enter standby at different times, additional power rails may be required and wells biased at different potentials must be spatially isolated. These techniques are often used in low power custom designs, but are complicated to implement in ASICs.

There is a voltage drop across sleep transistors when they are on, degrading the voltage swing for logic. Wider sleep transistors degrade the voltage swing less, but have higher capacitance. Power up of sleep transistors takes substantial energy due to their large capacitance [64]. Standard cells must be characterized for the degraded supply voltage. Layout tools must cluster the gates that connect to the same virtual power rail that is disconnected by a given sleep signal, as having individual sleep transistors in each gate is too area expensive. As the registers that retain state connect to the virtual power rails and directly to the power rails, the standard cell rows on which registers are placed must be taller to accommodate the extra rails. The virtual and actual supply and ground voltages differ in standby. Thus, the substrates of the transistors connected to virtual power rails and those connected directly to the power rails are at different voltages and must be isolated spatially, increasing the area overhead. The floating output of a power-gated cell can cause large currents if it connects directly to a cell which is not power gated, so additional circuitry is required to drive the output of the power-gated cell [92].

2.5.4.2 What can we do about it?

ASICs seldom use standby power reduction techniques other than full power down, but there is now better tool support for power gating. An EDA flow with power gating can provide two orders of magnitude reduction in leakage if state is not retained, at the cost of 10% to 20% area overhead and 6% higher delay (see Chapter 10). The ARM1176JZ-S synthesizable core supports dynamic voltage scaling, allowing the supply voltage to be scaled from 1.21V to 0.69V in the 0.13um process, but this requires additional hardware support [35].

To date state assignment and reverse substrate biasing have not been implemented in an EDA methodology. As state assignment cannot be effectively used with combinational logic that is power gated and provides far less leakage reduction than using sleep transistors, it is unlikely to be useful except for circuits that have only short idle periods, on the order of tens of clock cycles. Substrate biasing nicely complements power gating with forward body bias reducing the delay penalty for voltage drop across the sleep transistors, and with reverse body bias reducing the leakage in registers that are on to retain state information. As reverse substrate bias is less effective at shorter channel lengths, ASICs may have from 4× higher standby leakage than custom designs that use reverse body bias in 0.18um to 2× worse than custom in deeper submicron technologies.

2.5.5 Logic style

ASICs almost exclusively use complementary static CMOS logic for combinational logic, because it is more robust to noise and Vdd variation than other logic styles. Pass transistor logic (PTL), dynamic domino logic and differential cascode voltage switch logic (DCVSL) are faster than complementary static CMOS logic. These logic styles are illustrated in Figure 2.6. Complementary CMOS logic suffers because PMOS transistors are roughly 2× slower than NMOS transistors of the same width, which is particularly a problem for NOR gates. With the two PMOS transistors in series in Figure 2.6(a), the PMOS transistors must be sized about 4× larger for equal rise and fall delays, substantially increasing the load on the fanins. The high speed logic styles can be used to reduce the critical path delay, increasing performance. Alternatively, the additional timing slack can be used to achieve lower power at high performance targets. Complementary CMOS logic is lower power than other logic styles when high performance is not required. Hence, low power custom designs primarily use complementary CMOS, with faster logic only on critical paths. ASIC designs are mapped to slower, purely complementary CMOS logic standard cell libraries.

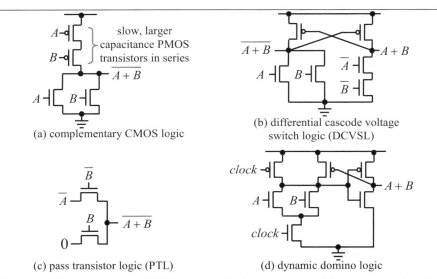

(a) complementary CMOS logic

(b) differential cascode voltage switch logic (DCVSL)

(c) pass transistor logic (PTL)

(d) dynamic domino logic

Figure 2.6 This figure shows NOR2 logic gate implementations in different logic styles. The domino logic output is inverted, so that after precharging the inputs to domino logic gates are low to avoid them being discharged until an input transition to high occurs [71].

The StrongARM used primarily complementary CMOS, with static DCVSL to implement wide NOR gates [62]. In the custom IDCT multiplier, the carry and sum of the full adder cells are both on the critical path [102]. A complementary CMOS gate generated the carry out, and a static DCVSL gate generated the sum. This full adder was 37% faster than a purely complementary CMOS mirror adder.

The StrongARM and XScale used some dynamic logic. Dynamic DCVSL (dual rail domino logic) has twice the activity of single rail domino logic. The Samsung Halla used dynamic DCVSL and is higher power than the complementary CMOS ARM1020E at 400MHz. However, the Halla runs at up to 1.2GHz, while the ARM1020E is limited to 400MHz [60]. Zlatanovici [104] compared 0.13um single rail domino and complementary static CMOS 64-bit adders. Domino could achieve as low as 6.8 FO4 delays at 34pJ/cycle. The fastest static CMOS version was 12.5 FO4 delays, but only 18pJ/cycle.

PTL is a high speed and low energy logic style [7]. In a 0.6um study, a complementary CMOS carry-lookahead 32-bit adder was 20% slower than complementary PTL, but the complementary CMOS adder was 71% lower power [103]. At maximum frequency in 0.25um, a complementary CMOS 3-input XOR ring oscillator had 1.9× delay and 1.3× power compared to versions in PTL and DCVSL [52]. The XScale ALU bypass adder was implemented in PTL. At 1.1V, this was 14% slower than single rail domino, but it has no precharge and lower switching activity [22].

High speed logic styles can increase the speed of combinational logic by 1.5× [18]. We discuss the potential power savings with reduced combinational logic delay calculated from the pipeline model in Section 3.5. We optimistically assumed no extra power consumption for using a high speed logic style on critical paths. At a tight performance constraint, pipelines with only complementary static CMOS combinational logic had up to 2.0× higher energy per operation.

2.5.5.1 What's the problem?

PTL, DCVSL, and dynamic logic libraries are used as in-house aids to custom designers. Standard cell libraries with these logic styles are not available to ASIC designers. All of these logic styles are less robust than complementary CMOS logic, and have higher leakage power.

Differential cascode voltage switch logic is faster than complementary CMOS logic, but is higher energy [7][20]. DCVSL requires both input signal polarities and has higher switching activity than complementary CMOS logic. Static DCVSL has cross-coupled outputs, resulting in longer periods of time with a conducting path from supply to ground and larger short circuit current. The DCVSL inputs and their negations must arrive at the same time to limit the duration of the short circuit current, requiring tight control of the layout to ensure similar signal delays.

Dynamic logic is precharged on every clock cycle, increasing the clock load, activity, and dynamic power. The precharged node may only be discharged once, so glitches are not allowed. Shielding may be required to prevent electromagnetic noise due to capacitive cross-coupling discharging the precharged node. To avoid leakage through the NMOS transistors discharging the node, a weak PMOS transistor is required as a "keeper" [99]. There can be charge sharing between dynamic nodes or on PTL paths.

Pass transistor logic suffers a voltage drop of Vth across the NMOS pass transistor when the input voltage is high [71]. Consequently, the reduced voltage output from PTL may need to be restored to full voltage to improve the noise margin and to avoid large leakage currents in fanouts. The voltage drop can be avoided by using a complementary PMOS transistor in parallel with the NMOS transistor, but this increases the loading on the inputs, reducing the benefit of PTL. Buffering is needed if the fanins and fanouts are not near the PTL gates, and an inverter may be needed to generate a negated input.

Using these logic styles requires careful cell design and layout. A typical EDA flow gives poor control over the final layout, thus use of these logic styles would result in far more yield problems and chip failures.

2.5.5.2   What can we do about it?

The foundry requirement of high yield means that the only standard cell libraries available to ASIC designers will continue to be robust complementary static CMOS logic. Thus an EDA design flow cannot reduce the power gap for logic style.

An alternative is for designers to adopt a semi-custom design flow: high speed custom cells and manual layout can be used for timing critical logic; or custom macros can be used.

2.5.6   Logic design

Logic design refers to the topology and the logic structure used to implement datapath elements such as adders and multipliers. Arithmetic structures have different power and delay trade-offs for different logic styles, technologies, and input probabilities.

2.5.6.1   What's the problem?

Custom designers tend to pay more attention to delay critical datapaths. Specifying logic design requires carefully structured RTL and tight synthesis constraints. For example, we found that flat synthesis optimized out logic that reduced switching activity in multiplier partial products [47], so the scripts were written to maintain the multiplier hierarchy during synthesis. The reduced switching activity reduced the power-delay product by $1.1\times$ for the 64-bit multiplier.

Careful analysis is needed to compare alternate algorithmic implementations for different speed constraints. For example, high-level logic transition analysis showed that a 32-bit carry lookahead adder had about 40% lower power-delay product than carry bypass or carry select adders [13]. There was also a 15% energy difference between 32-bit multipliers. Zlatanovici compared 64-bit domino adders in 0.13um, and found that the radix-4 adders achieved smaller delay and about 25% lower energy than radix-2 [104].

We estimate that incomplete evaluation of logic design alternatives may result in $1.2\times$ higher power for a typical ASIC.

2.5.6.2   What can we do about it?

Synthesis tools can compile to arithmetic modules. The resulting energy and delay is on par with tightly structured RTL. In general, ASIC designers should be able to fully exploit logic design.

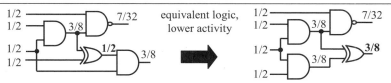

Figure 2.7 This figure illustrates how refactoring logic can reduce the switching activity while giving the same functional result. Switching activities are annotated on the diagram, as propagated from independent inputs that have equal probability of being zero or one.

2.5.7 Technology mapping

In technology mapping a logical netlist is mapped to a standard cell library in a given technology. Different combinations of cells can implement a gate with different activity, capacitance, power and delay. For example to implement an XOR2, an AO22 with inverters may be smaller and lower power, but slower. (An AO22 logic gate computes $ab + cd$, so XOR2 may be implemented by $a\overline{b} + \overline{a}b$.) Refactoring can reduce switching activity (see Figure 2.7). Common sub-expression elimination reduces the number of operations. Balancing path delays and reducing the logic depth decreases glitch activity. High activity nets can be assigned to gate pins with lower input capacitance. [63][77]

2.5.7.1 What's the problem?

While there are commercial tools for power minimization, power minimization subject to delay constraints is still not supported in the initial phase of technology mapping. Minimizing the total cell area minimizes circuit capacitance, but it can increase activity. For a 0.13um 32-bit multiplier after post-synthesis power minimization, the power was 32% higher when using minimum area technology mapping. This was due to more (small) cells being used, increasing activity. We had to use technology mapping combining delay and area minimization targets for different parts of the multiplier. Technology mapping for low power may improve results; without this and other low power technology mapping techniques, ASICs may have 1.4× higher power than custom.

2.5.7.2 What can we do about it?

Power minimization tools do limited remapping and pin reassignment, along with clock gating and gate sizing [84]. These optimizations are applied after technology mapping for minimum delay, or minimum area with delay constraints. EDA tools should support technology mapping for minimum power with delay constraints. This requires switching activity analysis, but it is not otherwise substantially more difficult than targeting minimum area.

For a given delay constraint, technology mapping can reduce the power by 10% to 20%, for a 10% to 20% increase in area [63][77]. Low power encoding for state assignment can also give 10% to 20% power reduction [90]. Logic transformations based on logic controllability and observability, common sub-expression elimination, and technology decomposition can give additional power savings of 10% to 20% [68]. Pin assignment can provide up to 10% dynamic power savings by connecting higher activity inputs to gate input pins with lower capacitance [74].

ASICs should not lag custom power consumption due to technology mapping, if better EDA tool support is provided.

2.5.8   Gate sizing and wire sizing

Wires and transistors should be sized to ensure correct circuit operation, meet timing constraints, and minimize power consumption. ASICs must choose cell sizes from the range of drive strengths provided in the library. ASIC wire widths are usually fixed. Downsizing transistors gives a linear reduction in their capacitance and thus dynamic power, and also gives a linear reduction in leakage. Reducing the wire width gives a linear reduction in wire capacitance but a linear increase in wire resistance, increasing signal delay on the wire.

2.5.8.1   What's the problem?

There is a trade-off between power and delay with gate sizing. To reduce delay, gates on critical paths are upsized, increasing their capacitance. In turn, their fanin gates must be upsized to drive the larger capacitance. This results in oversized gates and buffer insertion on the critical paths. Delay reductions come at the price of increasingly more power and worse energy efficiency.

To balance rise and fall delays, an inverter has PMOS to NMOS width ratio of about 2:1 as a PMOS transistor has about half the drain current of a NMOS transistor of the same width. Accounting for the number of transistors in series, other logic gates also have 2:1 P/N ratio to balance rise and fall delays for an inverter of equivalent drive strength, as illustrated in Figure 2.8. However, to minimize the average of the rise delay and fall delay, the P/N ratio for an inverter should be about 1.5:1 [37]. Reducing the P/N ratio provides a small reduction in delay and a substantial reduction in power consumption, by reducing the capacitance of the larger PMOS transistors. The optimal P/N ratio to minimize the delay is larger for larger loads [73]. In addition, sometimes the rise and fall drive strengths needed are different – for example, the rising output transition from a cell may be on a critical path, but the falling transition may not be critical.

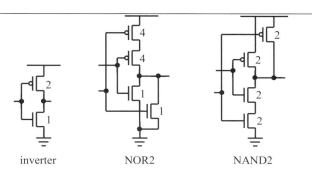

inverter   NOR2   NAND2

Figure 2.8 This figure shows the relative NMOS and PMOS transistor widths for equal rise and fall delays in different logic gates of equivalent drive strength.

The ratio of pullup to pulldown drive strength determines at what input voltage a gate switches from low to high or high to low [99]. Equal rise and fall delays maximize the noise margin for a high or low input. Thus skewing the P/N ratio reduces the noise margin. Ideally, standard cell libraries should provide a range of drive strength skews and lower power cells with reduced P/N ratio, but often only cells with equal rise and fall drive strength are available to ensure high yield.

A design-specific standard cell library developed for the iCORE [73] gave a 20% speed increase by using reduced P/N width ratio, and by using larger transistor widths to increase drive strength instead of buffering. The larger transistor widths required increased cell height, but the net impact on layout area was minimal as they were only used in the most critical paths. However, the design time for this library was about two worker years.

Custom libraries may be finer grained, which avoids oversizing gates, and have skewed drive strengths. Cells in datapath libraries are denser and have smaller input capacitance [18]. Specific cell instances can be optimized. Cells that connect to nearby cells don't need guard-banding. This avoids the need for buffering to handle driving or being driven by long wires.

Wire widths can also be optimized in custom designs. Gong et al. [34] optimized global clock nets on a 1.2um chip. By simultaneously optimizing buffer and wire sizes, they reduced the clock net power by about 63%. This amounts to a 10% to 20% saving in total power.

The basic approach to gate sizing in commercial EDA software has changed little in the past 20 years. These gate sizers like TILOS [29] proceed in a greedy manner, picking the gate with the best power or area versus delay tradeoff to change, and iterating. There are known circuit examples where these approaches perform suboptimally, but it has not been clear how much of a problem this is for typical circuits for real world applications. We found power savings of up to 32.3% versus gate sizing in Design Compiler, which is commonly used in EDA flows for circuit synthesis, and 16.3%

savings on average across the ISCAS'85 benchmarks and three typical datapath circuits (see Section 6.5.3). Gate sizing is an NP-complete problem, but circuit sizes are large and optimization software must have runtimes of $O(n^2)$ or less to be of practical use [81], where n is the number of gates in the circuit. The TILOS-like greedy approaches are relatively fast, being $O(n^2)$, and other approaches that perform better with similar static timing analysis (STA) accuracy have had worse computational complexity.

Some commercial power minimization software has only recently provided the option of minimizing the total power. Previously, the user had to prioritize minimizing either the dynamic power or the leakage power, which can be suboptimal.

We estimate that these limitations in gate sizing and wire sizing for typical ASICs may lead to a power gap of 1.6× versus custom.

2.5.8.2 What can we do about it?

Gate downsizing to reduce power consumption is well supported by power minimization tools. Some commercial tools support clock tree wire sizing, but there are no commercial tools available for sizing other wires. Automated cell creation, characterization and in-place optimization tools are available. Standard cell libraries with finer grained drive strengths and lower power consumption are available, though users may be charged a premium.

We synthesized the base configuration of a Tensilica Xtensa processor in 0.13um. The power/MHz was 42% lower and the area was 20% less at 100MHz than at the maximum clock frequency of 389MHz, due to using smaller gates and less buffers. If delay constraints are not too tight, tools can reduce power by gate downsizing without impacting delay. At 325MHz, Power Compiler was able to reduce the power consumption by 26% and reduce the area by 12% for no delay penalty.

Libraries with fine granularity help to reduce the power by avoiding use of oversized gates. In a 0.13um case study of digital signal processor (DSP) functional macros, using a fine grained library reduced power consumption by 13% (see Chapter 13).

After place and route when wire lengths and capacitive loads are accurately known, in place optimization can remove guard banding where it is unnecessary. ASIC designers have tended to distrust this approach, as the optimized cells without guard banding cannot be safely used at earlier stages in the EDA flow. Skewing the pullup to pulldown drive strength to optimize the different timing arcs through a gate can also improve energy efficiency. A prototype tool flow for in place cell optimization increased circuit speed by 13.5% and reduced power consumption by 18%, giving a 1.4× increase in energy efficiency for the 0.35um 12,000 gate bus controller [25]. 300 optimized cells were generated in addition to the original standard cell library that had 178 cells.

Our linear programming gate sizing approach discussed in 0 takes a global view of the circuit rather than performing greedy "peephole" optimization. We achieved up to 32.3% power savings and on average 16.3% power savings versus gate sizing in Design Compiler for the combinational netlists. Our optimization approach has between $O(n)$ and $O(n^2)$ runtime growth, making it scalable to large circuit sizes.

ASICs may have 1.1× worse power than custom due to gate and wire sizing, as wire sizing tools are not available other than for the clock tree, and some design-specific cell optimizations are not possible without custom cell design, beyond what is possible with automated cell creation.

2.5.9 Voltage scaling

Reducing the supply voltage Vdd quadratically reduces switching power. Short circuit power also decreases with Vdd. Reducing Vdd also reduces leakage. For example, with our 0.13um library leakage decreases by a factor of three as Vdd is decreased from 1.2V to 0.6V. As Vdd decreases, a gate's delay increases. To reduce delay, threshold voltage Vth must also be scaled down. As Vth decreases, leakage increases exponentially. Thus there is a tradeoff between performance, dynamic power and leakage power.

Ideally, we want to operate at as low Vdd as possible, with Vth high enough to ensure little leakage. For example, dynamic scaling of the supply voltage from 3.8V to 1.2V gives a 10× increase in energy efficiency at the price of decreasing performance by a factor of 14 for Burd's 0.6um ARM implementation [11]. Reducing the power consumption in this manner requires timing slack.

Power consumption may be reduced by using multiple supply voltages and multiple threshold voltages. High Vdd and low Vth can be used on critical paths to reduce their delay, while lower Vdd and higher Vth can be used elsewhere to reduce dynamic and leakage power.

2.5.9.1 What's the problem?

Custom designs can achieve at least twice the speed of ASICs with high performance design techniques [18]. At the same performance target as an ASIC, a custom design can reach lower Vdd using the additional timing slack. Compare Vdd of Burd, StrongARM and XScale to other ARMs in Table 2.1. With lower Vdd they save between 40% and 80% dynamic power versus other ARMs in the same technology. This is the primary reason for their higher energy efficiency. To use lower Vdd, ASICs must either settle for lower performance or use high speed techniques, such as deeper pipelining, to maintain performance.

Dynamically adjusting the supply voltage for the desired performance requires a variable voltage regulator and takes time, during which correct signals must be maintained to avoid transitioning into illegal states from which behavior is unknown. To change from 1.2V to 3.8V in Burd's ARM [11] required energy equal to that consumed in 712 cycles of peak operation, and there was a delay of 70us. Increasing or decreasing the supply voltage by 800mV took 50us in the XScale [22].

Several barriers remain to ASICs using low Vdd. Using lower Vdd requires lower Vth to avoid large increases in gate delay. Vth is determined by the process technology. A foundry typically provides two or three libraries with different Vth: high Vth for low power; and low Vth for high speed at the expense of significant leakage power. Most ASIC designers cannot ask a foundry to fine tune Vth for their particular design, even if an intermediate Vth might be preferable to reduce leakage. Vdd can be optimized for ASICs, but typical ASIC libraries are characterized at only two nominal supply voltages – say 1.2V and 0.9V in 0.13um. To use Vdd of 0.6V, the library must be re-characterized. There is also less noise immunity at lower Vdd.

Use of multiple supply voltages either requires that the wells of PMOS transistors in low Vdd gates are reverse biased by connecting them to high Vdd, or spatial isolation between the wells connected to low Vdd and high Vdd. Layout tools must support these spacing constraints. Low voltage swing signals must be restored to full voltage swing with a voltage level converter to avoid large leakage currents when a high Vdd gate is driven by a low Vdd input. Most level converter designs require access to both high Vdd and low Vdd, which complicates layout and may require that they straddle two standard cell rows, additionally the PMOS wells connected to different Vdd must be spatially isolated. Voltage level converters are not available in ASIC libraries. Synthesis and optimization tools must insert level converters where needed, and prevent low Vdd gates driving high Vdd gates in other cases.

If voltage level converters are combined with the flip-flops, the power and delay overheads for voltage level restoration are less. Due to the additional power and delay overheads for asynchronous level converters (those not combined with flip-flops), there have been reservations about whether they provide any practical benefits over only using level converter flip-flops [93]. There has also been concern about their noise immunity [46].

Multi-Vdd circuitry has more issues with capacitive cross-coupling noise due to high voltage swing aggressors on low voltage swing lines. Thus it may be best to isolate high Vdd and low Vdd circuitry into separate voltage islands, rather than using multi-Vdd at a gate level. Multi-Vdd at the gate-level can also require additional voltage rails. Gate level multi-Vdd requires tool support to cluster cells of the same Vdd to achieve reasonable layout density. An additional voltage regulator is needed to generate the lower Vdd.

Using multiple threshold voltages is expensive. Each additional PMOS and NMOS threshold voltage requires another mask to implant a different dopant density, substantially increasing processing costs. A set of masks costs on the order of a million dollars today and an additional Vth level increases the fabrication cost by 3% [69]. Each additional mask increases the difficulty of tightly controlling yield, motivating some manufacturers to limit designs to a single NMOS and single PMOS threshold voltage.

To take full advantage of multiple threshold voltages within gates, standard cells with multi-Vth and skewed transistor widths must be provided. High Vth can be used to reduce leakage while low Vth can be used to reduce dynamic power. For example, using low Vth PMOS transistors and high Vth NMOS transistors in a complementary CMOS NOR gate, as leakage is less through the PMOS transistors that are in series. In gates that have an uneven probability of being high or low, there is more advantage to using high Vth to reduce leakage for the pullup or pulldown network that is more often off. Similarly, for wider transistors with high Vth may be preferable for gates that have low switching activity, while narrower transistors with low Vth is better when there is higher switching activity.

2.5.9.2 What can we do about it?

There are tools to automate characterizing a library at different Vdd operating points. Characterization can take several days or more for a large library. Standard cell library vendors can help by providing more Vdd characterization points. Commercial tools do not adequately support multi-Vdd assignment or layout, but separate voltage islands are possible.

There are voltage level converter designs that only need to connect to high Vdd (see Figure 13.8). Some asynchronous level converters designs have been shown to be robust and have good noise immunity in comparison to typical logic gates at low Vdd [53]. It would help if voltage level converters were added to standard cell libraries.

Foundries often support high and low Vth cells being used on the same chip. Power minimization tools can reduce power by using low Vth cells on the critical path, with high Vth cells elsewhere to reduce leakage. Combining dual Vth with sizing reduced leakage by 3 to 6× for a 5% increase in delay on average versus using only low Vth [78]. From a design standpoint, an advantage of multiple threshold voltages is that changing the threshold voltage allows the delay and power of a logic gate to be changed without changing the cell footprint, and thus not perturbing the layout. As discussed in Chapter 7, multi-threshold voltage optimization is straightforward, providing those cells are provided in the library. Optimization runtime increases at worst linearly with the number of cells in the library.

Geometric programming optimization results on small benchmark circuits suggest that multi-Vdd and multi-Vth may only offer 20% power savings versus optimal choice of single Vdd, single NMOS Vth, and single PMOS Vth [16]. As ASIC designers are limited to Vth values specified by the foundry, there may be more scope for power savings in ASICs when Vth is suboptimal. After scaling Vdd from 1.2V to 0.8V by using a low Vth of 0.08V, we found power savings of up to 26% by using a second higher Vth to reduce leakage with our linear programming optimization approach in Chapter 7, and average power savings were 16%. We found that power savings with gate-level multi-Vdd were generally less than 10%. Using multi-Vdd is more appropriate at a module level, making a good choice of a single supply voltage for the module based on the delay of critical paths.

With 9% timing slack versus the maximum clock frequency, Stok et al. in Chapter 13 reduced power consumption by 31% by scaling from Vdd of 1.2V to Vdd of 1.0V. Usami et al. [94] implemented automated tools to assign dual Vdd and place dual Vdd cells, with substrate biasing for the transistors to operate at low Vth in active mode to increase performance and high Vth in standby mode to reduce leakage. They achieved total power reduction of 58% with only a 5% increase in area. The ARM1176JZ-S [35] synthesizable core supports dynamic voltage scaling, but this requires additional software and hardware support. This demonstrates that ASICs can use such methods with appropriately designed RTL, software, and EDA tool support, reducing the power gap due to voltage scaling alone to 1.0×.

2.5.10 Floorplanning, cell placement and wire routing

The quality of floorplanning of logic blocks and global routing for wires, followed by cell placement and detailed wire routing, have a significant impact on wire lengths. A significant portion of the capacitance switched in a circuit is wiring capacitance. The power consumption due to interconnect is increasing from about 20% in 0.25um to 40% in 0.09um [82]. Wire lengths depend on cell placement and congestion. Larger cells and additional buffers are needed to drive long wires. We estimate that poor floorplanning, cell sizing and cell placement with inaccurate wire load models can result in 1.5× worse power consumption in ASICs compared to custom.

2.5.10.1 What's the problem?

Custom chips are partitioned into small, tightly placed blocks of logic. Custom datapaths are manually floorplanned and then bit slices of layout may be composed. Automatic place and route tools are not good at recognizing layout regularity in datapaths.

We used BACPAC [82] to examine the impact of partitioning. We compared partitioning designs into blocks of 50,000 or 200,000 gates in 0.13um, 0.18um, and 0.25um. Across these technologies, using 200,000 gate blocks increased average wire length by about 42%. This corresponds to a 9% to 17% increase in total power. The delay is also about 20% worse with larger partitions [18]. The net increase in energy per operation is 1.3 to 1.4×.

When sizing gates and inserting buffers, the first pass of synthesis uses wire load models to estimate wire loads. Wire load models have become increasing inaccurate, with wires contributing a larger portion of load capacitance in the deep submicron. A conservative wire load model is required to meet delay constraints, but this results in most gates being over-sized [18], making the power higher.

Physical synthesis iteratively performs placement and cell sizing, to refine the wire length estimates. Cell positions are optimized then wire lengths are estimated with Steiner trees. Steiner tree wire length models used by physical synthesis are inaccurate if a wire route is indirect. There can be too many critical paths to give them all a direct route. Power minimization increases path delay, so more paths are critical, increasing congestion. This may degrade performance. For example for the base configuration of Tensilica's Xtensa processor for a tight performance target of 400MHz clock frequency in 0.13um, we found that the clock frequency was 20% worse after place and route when power minimization was used.

2.5.10.2   What can we do about it?

Physical synthesis, with iteratively refined wire length estimates and cell placement, produces substantially better results than a tool flow using only wire load models. In our experience, physical synthesis can increase speed by 15% to 25%. The cell density (area utilization) increases, reducing wire lengths, and then cells may be downsized, which reduces power by 10% to 20%.

Earlier power minimization tools often ended up increasing the worst critical path delay after layout if the delay constraint was tight. This is less of a problem in today's tools, where power minimization is integrated with physical synthesis. Tool flow integration has also improved, particularly as some of the major CAD software vendors now have complete design flows with tools that perform well throughout the design flow – rather than using for example Synopsys tools for synthesis and Cadence tools for place and route.

An ASIC designer can generate bit slices from carefully coded RTL with tight aspect ratio placement constraints. Bit slices of layout may then be composed. With bit slices, Chang showed a 70% wire length reduction versus automated place-and-route [15], which would give a 1.2 to 1.4× increase in energy efficiency. Stok et al. in Chapter 13 found that bit slicing and some logic optimization, such as constant propagation, improved clock frequency by 22% and reduced power consumption by 20% for seven DSP functional macros implemented in 0.13um, improving the energy efficiency by a factor of 1.5×. Compared to bit slicing using a library of datapath cells, manual placement and routing can still achieve smaller wire lengths [15], leaving a gap of about 1.1×.

2.5.11   Process technology

After the layout is verified, the chip is fabricated in the chosen process technology by a foundry. Within the same nominal technology generation, the active power, leakage power, and speed of a chip differ substantially depending on the process used to fabricate the circuit. Older technologies are slower and are cheaper per mask set. However, newer technologies have more dies per wafer and thus may be cheaper per die for larger production runs. Newly introduced technologies may have lower yield, though these problems are typically ironed out as the technology matures [61].

High performance chips on newer technologies have substantially higher subthreshold leakage power as threshold voltage is scaled down with supply voltage to reduce dynamic power. Gate tunneling leakage is also higher as transistor gate oxide thickness is reduced for the lower input voltage to the transistor gate to retain control of the transistor.

Gate leakage can be reduced if the gate oxide thickness t_{ox} is increased, which requires a high-k gate dielectric permittivity ε_{ox} to maintain the drive current (see Equation (4.1)). For example, Intel will use hafnium oxide in their 45nm process [44], which has dielectric permittivity of about an order magnitude larger than silicon oxide that is used in most of today's processes, enabling Intel to reduce the gate leakage by more than 10×.

The power consumption and power per unit area can be less in deeper submicron technologies if performance is not increased [55]. For example in 65nm, Intel's low power P1265 process reduces leakage 300×, but has 55% lower saturation drain current and hence is about 2.2× slower [48], compared to their higher performance P1264 technology [91]. To reduce leakage they increased oxide thickness from 1.2nm to 1.7nm, increased gate length from 35nm to 55nm, and increased threshold voltage from about 0.4V to 0.5V (at drain-source voltage of 0.05V) [48]. Note that the higher threshold voltage results in a greater delay increase if supply voltage is reduced.

While Intel started selling processors produced in 65nm bulk CMOS technology at the start of 2006, AMD is still producing chips in 90nm silicon-on-insulator (SOI) technology [8][36]. AMD is on track to offer 65nm SOI chips in the last quarter of 2006 [67]. Intel is a technology generation ahead, and has the cost advantage of using cheaper bulk CMOS and more dies per wafer with its smaller technology. However, SOI has better performance per watt than bulk CMOS, so Intel has only a slight advantage in terms of performance and energy efficiency.

In the same nominal technology generation, there are substantial differences between processes. Different technology implementations differ by up to 25% in speed [18], 60% in dynamic power, and an order of magnitude in leakage. We compared several gates in Virtual Silicon's IBM 8SF and UMC L130HS 0.13um libraries. 8SF has about 5% less delay and only 5% of the leakage compared to L130HS, but it has 1.6× higher dynamic power [97]. Our study of two TSMC 0.13um libraries with the base configuration of Tensilica's Xtensa processor showed that TSMC's high V_{th}, low-k library was 20% lower power/MHz, with 66% less leakage power and 14% less dynamic power, than the low V_{th}, low-k library (see Table 2.5).

The power consumption, wire RC delays, and IR drop in the wires can be reduced by use of copper wires and low-k interlayer dielectric insulator. Copper interconnect has 40% lower resistivity than aluminum. Low-k dielectrics of 2.7 to 3.6 electrical permittivity (k) are used in different processes, compared to SiO_2's dielectric constant of 3.9. Using low-k interlayer dielectric insulator reduces interconnect capacitance by up to 25%, reducing dynamic power consumption by up to 12%. High-k transistor gate dielectrics increase the transistor drive strength and thus speed, and can also reduce the gate tunneling leakage by an order of magnitude [59].

Narendra et al. showed that silicon-on-insulator (SOI) was 14% to 28% faster than bulk CMOS for some 0.18um gates. The total power was 30% lower at the same delay, but the leakage was 1.2 to 20× larger [65]. A 0.5um DSP study showed that SOI was 35% lower power at the same delay as bulk CMOS [76]. Double-gated fully depleted SOI is less leaky than bulk CMOS.

The StrongARM caches were 90% of the chip area and were primarily responsible for leakage. A 12% increase in the NMOS channel length reduced worst case leakage 20×. Lengthening transistors in the cache and other devices reduced total leakage by 5× [62]. Transistor capacitance, and thus dynamic power, increases linearly with channel length. Channel length can be varied in ASICs to reduce leakage if such library cells are available.

As a process technology matures, incremental changes can improve yield, improve performance and reduce power consumption. In Intel's 0.25um P856 process the dimensions were shrunk by 5% and, along with other modifications, this gave a speed improvement of 18% in the Pentium II [10]. The 0.18um process for the Intel XScale had a 5% shrink from P858, and other changes to target system-on-chip applications [22]. There was also a 5% linear shrink in Intel's 0.13um P860 process and the effective gate length was reduced from 70nm to 60nm [87]. A 5% shrink reduces transistor capacitance and dynamic power by about 5%. These process improvements are typical of what is available to high volume custom designs.

We estimate that different choices within the same process technology generation may give up to 1.6× difference in power.

Table 2.5 Dynamic and leakage power for two different 0.13um TSMC libraries for Tensilica's Xtensa processor for the base configuration with a clock frequency of 100MHz.

Library	low Vdd, low k-dielectric	low Vdd, low k-dielectric, high Vth
Dynamic power (uW)	6.48	5.66
Leakage power (mW)	0.67	0.25
Total power (mW)	7.15	5.90

2.5.11.1 What's the problem?

Standard cells are characterized in a specific process. The cells must be modified and libraries updated for ASIC customers to take advantage of process improvements. Without such updates, 20% speed increase and greater reductions in power may be unavailable to ASIC customers. Finding the lowest power for an ASIC requires synthesis with several different libraries to compare power at performance targets of interest. The lowest power library and process may be too expensive.

2.5.11.2 What can we do about it?

Generally, it requires little extra work to re-target an ASIC EDA flow to a different library. ASICs can be migrated quickly to different technology generations, and updated for process improvements. In contrast, the design time to migrate custom chips is large. Intel started selling 90nm Pentium 4 chips in February 2004 [36], but a 90nm version of the XScale was only reported in June 2005 [72] and is not currently in production to our knowledge. Meanwhile, ARM has synthesized the more recent Cortex-A8 core for 65nm [4]. ASICs should be able to take full advantage of process improvements, closing the gap for process technology to 1.0×.

2.5.12 Process variation

Chips fabricated in the same process technology vary in power and speed due to process variation, as illustrated in Figure 2.9. Some of the chips fabricated may be too slow, while some are significantly faster. In previous technology generations, the faster chips could be sold at a premium. However, faster chips have more leakage power and greater variation in leakage power [9]. Thus the faster chips may consume too much power, particularly if run at a higher clock frequency where dynamic power is also higher as it increases linearly with clock frequency.

There are a number of sources of process variation, such as optical proximity effects, and wafer defects. The channel length L, transistor width, wire width and wire height have about 25% to 35% variation from nominal at three standard deviations (3σ). Transistor threshold voltage Vth and oxide thickness have about 10% variation at 3σ [66]. Decreased transistor oxide thickness substantially increases gate tunneling leakage, and a decrease in Vth or L can cause a large increase in subthreshold leakage current, though these transistors are faster. Dynamic power scales linearly with transistor and wire dimensions, as capacitances increase.

To ensure high yield accounting for process variation, libraries are usually characterized at two points. To meet the target speed, the process' worst case speed corner is used – typically 125°C, 90% of nominal Vdd, with slow transistors. To prevent excessive power, the active power may be characterized at a worst case power corner, e.g. –40°C, 110% of nominal Vdd, and fast transistors. Leakage is worse at high temperature. Due to Vdd alone, the active power is 50% higher at the worst case power corner than at the worst case speed corner. These process corners are quite conservative and limit a design. The fastest chips fabricated in a typical process may be 60% faster than estimated from the worst case speed corner [18]. Similarly, examining the distribution of power of fabricated 0.3um MPEG4 codecs [85], the worst case power may be 50% to 75% higher than the lowest power chips produced.

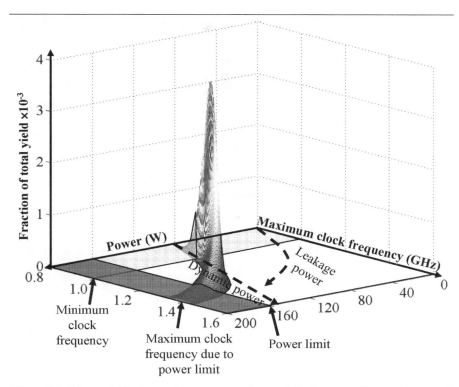

Figure 2.9 This graph illustrates yield versus maximum clock frequency f and total power P at that clock frequency. The minimum frequency is 1.0GHz and the maximum power is 160W. The maximum frequency of about 1.4GHz is determined from the power constraint. 2.3% of the chips are slower than 1.0GHz and 2.2% are faster than 1.4GHz. 10.7% of the chips have power consumption of more than 160W. Data was generated with a normal distribution of $f = N(1.2,0.1)$ and distribution for total power of $P = 100f + e^{-10+10f+N(0,0.4)}$, with dynamic power of $100f$. The leakage distribution is similar to the 0.18um technology in [9].

Table 2.6 This table compares the rated power consumption of chips operating at the same clock frequency that are sold by Intel and AMD today [1][2][3][41][42][45][86]. Higher speed parts can operate at a lower supply voltage, reducing the power consumption. These lower power processors are sold at a premium.

Processor	Model	Codename	Technology (nm)	Frequency (GHz)	Voltage (V)	Power (W)	Power Increase
Athlon 64 X2	4800+	Windsor	90	2.40	1.25	65	
Athlon 64 X2	4800+	Windsor	90	2.40	1.35	89	×1.4
Athlon 64 X2	3800+	Windsor	90	2.00	1.08	35	
Athlon 64 X2	3800+	Windsor	90	2.00	1.25	65	×1.9
Athlon 64 X2	3800+	Windsor	90	2.00	1.35	89	×2.5
Athlon 64	3500+	Orleans	90	2.20	1.25	35	
Athlon 64	3500+	Orleans	90	2.20	1.40	62	×1.8
Turion 64	MT-40	Lancaster	90	2.20	1.20	25	
Turion 64	ML-40	Lancaster	90	2.20	1.35	35	×1.4
Core 2 Duo	T7600	Merom	65	2.33	1.30	35	
Xeon 5100	5148	Woodcrest	65	2.33	1.25	40	×1.1
Xeon 5100	5140	Woodcrest	65	2.33	1.40	65	×1.9
Core 2 Duo	T7200	Merom	65	2.00	1.30	35	
Xeon 5100	5130	Woodcrest	65	2.00	1.40	65	×1.9
Core Duo	L2500	Yonah	65	1.83	1.21	15	
Core Duo	T2400	Yonah	65	1.83	1.33	31	×2.1
Core Duo	L2400	Yonah	65	1.66	1.21	15	
Core Duo	T2300	Yonah	65	1.66	1.33	31	×2.1

Exploiting the variation in power consumption, Intel and AMD have been selling lower power chips at a premium. The power consumption of the cheaper, higher power parts is typically up to about 2× that of the low power chips, as shown in Table 2.6. Note that Intel's Merom (laptop), Conroe (desktop) and Woodcrest (server) chips are essentially the same [40], though voltages, caching strategies and so forth may be changed for lower power but lower performance for the laptop version.

Custom circuitry can be designed to ameliorate process variation in fabricated chips. In the Pentium 4, the clock is distributed across the chip to 47 domain buffers, which each have a 5 bit programmable register to remove skew from the clock signal in that domain to compensate for process variation [54]. A similar scheme was used to reduce clock skew in the 90nm XScale [21]. The body bias can be changed to adjust the transistor threshold voltage, and thus the delay and leakage power. Body bias can be applied at a circuit block level to reduce the standard deviation in clock frequency between dies from 4.1% to 0.21%, improving speed by 15% versus the slower chips without body bias, while limiting the range in leakage power to 3× for a 0.15um test chip [88]. To do this, representative critical path delays and the leakage current must be measured while the bias is varied. Additional power rails are needed to route the alternate NMOS and PMOS body bias voltages from the body bias generator circuitry, resulting in a 3% area overhead [88]. Forward body bias allowed Vdd to be reduced from 1.43V to 1.37V giving a 7% reduction in total power for a 0.13um 5GHz 32-bit integer execution core [95].

2.5.12.1 What's the problem?

For ASIC parts that are sold for only a few dollars per chip, additional testing for power or speed binning is too expensive. Such ASICs are characterized under worst case process conditions to guarantee good yield. Thus ASIC power and speed are limited by the worst case parts. Without binning, there may be a power gap of ×2 versus custom chips that are binned. Custom chips that have the same market niche as ASICs have the same limitation on testing for binning, unless they are sold at a much higher price per chip.

The complicated circuitry and tight control of layout and routing required to compensate for process variation in a fabricated chip is not possible within an ASIC methodology.

2.5.12.2 What can we do about it?

To account for process variation, ASIC power may be characterized after fabrication. Parts may then be advertised with longer battery life. However, post-fabrication characterization of chip samples does not solve the problem if there is a maximum power constraint on a design. In this case, ASICs may be characterized at a less conservative power corner, which requires better characterization of yield for the standard cell library in that process. For typical applications, the power consumption is substantially less than peak power at the worst case power corner. Additional steps may be taken to limit peak power, such as monitoring chip temperature and powering down if it is excessive.

We estimate a power gap of up to 1.3× due to process variation for ASICs in comparison to custom designs that compensate for process variation, from analysis of a 15% increase in custom speed with the pipeline model in Chapter 3.

2.6 SUMMARY

We compared synthesizable and custom ARM processors from 0.6um to 0.13um. We also examined discrete cosine transform cores, as an example of dedicated low power functional units. In these cases, there was a power gap of 3 to 7× between custom and ASIC designs.

We have given a top-down view of the factors contributing to the power gap between ASIC and custom designs. From our analysis, the most significant opportunity for power reduction in ASICs is using microarchitectural techniques to maintain performance while reducing power by voltage scaling.

Reducing the pipeline delay overhead and using pipelining to increase timing slack can enable substantial power savings by reducing the supply voltage and downsizing gates. Multiple threshold voltages may be used to limit leakage while enabling a lower Vdd to be used. Choosing a low power process technology and limiting the impact of process variation reduces power by a large factor.

In summary, at a tight performance constraint for a typical ASIC design, we believe that the power gap can be closed to within 2.6× by using these low power techniques with fine granularity standard cell libraries, careful RTL design and EDA tools targeting low power. The remaining gap is mostly from custom designs having lower pipelining overhead and using high speed logic on critical paths. Using a high speed logic style on critical paths can provide timing slack for significant power savings in custom designs. High speed logic styles are less robust and require careful layout, and thus are not amenable to use in an ASIC EDA methodology.

An example of combining low power and high performance design techniques on DSP functional macros is in Chapter 13. To improve performance and reduce power consumption, they used arithmetic optimizations, logic optimization, a finer grained library, voltage scaling from 1.2V to 1.0V, and bit-slicing. Performance improved from 94MHz to 177MHz and energy efficiency increased from 0.89MHz/mW to 2.78MHz/mW – a factor of 3.1×. This demonstrates the power savings that may be achieved by using low power techniques in ASICs.

The next chapter details our power and delay model that incorporates the major factors that contribute to the power gap between ASIC and custom. It includes pipelining, logic delay, voltage scaling and gate sizing. The logic delay is determined by factors such as the logic style, wire lengths after layout, process technology, and process variation which affects the worse case delay.

2.7   REFERENCES

[1] AMD, AMD Processors for Desktops: AMD Athlon 64 Processor and AMD Sempron Processor, September 2006. http://www.amdcompare.com/us-en/desktop/

[2] AMD, AMD Turion 64 Mobile Technology Model Number, Thermal Design Power, Frequency, and L2 Cache Comparison, September 2006.http://www.amd.com/us-en/Processors/ProductInformation/0,,30_118_12651_12658,00.html

[3] AMD, Processor Pricing, September 2006. http://www.amd.com/us-en/Corporate/VirtualPressRoom/0,,51_104_609,00.html?redir=CPPR01

[4] ARM, ARM Cortex-A8, 2006. http://www.arm.com/products/CPUs/ARM_Cortex-A8.html

[5] ARM, ARM Processor Cores. http://www.armdevzone.com/open.nsf/htmlall/A944EB65693A4EB180256A440051457A/$File/ARM+cores+111-1.pdf

[6] Bhavnagarwala, A., et al., "A Minimum Total Power Methodology for Projecting Limits on CMOS GSI," *IEEE Transactions on VLSI Systems*, vol. 8, no. 3, June 2000, pp. 235-251.

[7] Bisdounis, L., et al., "A comparative study of CMOS circuit design styles for low-power high-speed VLSI circuits," *International Journal of Electronics*, vol. 84, no. 6, 1998, pp. 599-613.

[8] Bohr, M., "Staying Ahead of the Power Curve: Q&A with Intel Senior Fellow Mark T. Bohr," *Technology@Intel Magazine*, August 2006.

[9] Borkar, S. et al., "Parameter variation and impact on Circuits and Microarchitecture," in *Proceedings of the Design Automation Conference*, 2003, pp. 338-342.

[10] Brand, A., et al., "Intel's 0.25 Micron, 2.0 Volts Logic Process Technology," *Intel Technology Journal*, Q3 1998, 8 pp. http://developer.intel.com/technology/itj/q31998/pdf/p856.pdf

[11] Burd, T., et al., "A Dynamic Voltage Scaled Microprocessor System," *IEEE Journal of Solid State Circuits*, vol. 35, no. 11, 2000, pp. 1571-1580.

[12] Burd, T., *Energy-Efficient Processor System Design*, Ph.D. dissertation, Department of Electrical Engineering and Computer Sciences, University of California, Berkeley, CA 2001, 301 pp.

[13] Callaway, T., and Swartzlander, E., "Optimizing Arithmetic Elements for Signal Processing," *VLSI Signal Processing*, 1992, pp. 91-100.

[14] Chandrakasan, A., and Brodersen, R., "Minimizing Power Consumption in Digital CMOS Circuits," in *Proceedings of the IEEE*, vol. 83, no. 4, April 1995, pp. 498-523.

[15] Chang, A., "VLSI Datapath Choices: Cell-Based Versus Full-Custom," S.M. Thesis, Massachusetts Institute of Technology, February 1998, 146 pp. http://cva.stanford.edu/publications/1998/achang_sm_thesis.pdf

[16] Chinnery, D, *Low Power Design Automation*, Ph.D. dissertation, Department of Electrical Engineering and Computer Sciences, University of California, Berkeley, 2006.

[17] Chinnery, D., et al., "Automatic Replacement of Flip-Flops by Latches in ASICs," chapter 7 in *Closing the Gap Between ASIC & Custom: Tools and Techniques for High-Performance ASIC Design*, Kluwer Academic Publishers, 2002, pp. 187-208.

[18] Chinnery, D., and Keutzer, K., *Closing the Gap Between ASIC & Custom: Tools and Techniques for High-Performance ASIC Design*, Kluwer Academic Publishers, 2002, 432 pp.

[19] Chinnery, D., Nikolić, B., and Keutzer, K., "Achieving 550 MHz in an ASIC Methodology," in *Proceedings of the Design Automation Conference*, 2001, pp. 420-425.

[20] Chu, K., and Pulfrey, D., "A Comparison of CMOS Circuit Techniques: Differential Cascode Voltage Switch Logic Versus Conventional Logic," *Journal of Solid-State Circuits*, vol. sc-22, no. 4, August 1987, pp. 528-532.

[21] Clark, L., "The XScale Experience: Combining High Performance with Low Power from 0.18um through 90nm Technologies," presented at the Electrical Engineering and Computer Science Department of the University of Michigan, September 30, 2005. http://www.eecs.umich.edu/vlsi_seminar/f05/Slides/VLSI_LClark.pdf

[22] Clark, L., et al., "An Embedded 32-b Microprocessor Core for Low-Power and High-Performance Applications," *Journal of Solid-State Circuits*, vol. 36, no. 11, November 2001, pp. 1599-1608.

[23] Clark, L., et al., "Standby Power Management for a 0.18um Microprocessor," in *Proceedings of the International Symposium on Low Power Electronics and Design*, 2002, pp. 7-12.

[24] Clark, L., Morrow, M., and Brown, W., "Reverse-Body Bias and Supply Collapse for Low Effective Standby Power," *IEEE Transactions on VLSI Systems*, vol. 12, no. 9, 2004, pp. 947-956.

[25] Cote, M., and Hurat, P. "Faster and Lower Power Cell-Based Designs with Transistor-Level Cell Sizing," chapter 9 in *Closing the Gap Between ASIC & Custom: Tools and Techniques for High-Performance ASIC Design*, Kluwer Academic Publishers, 2002, pp. 225-240.

[26] Dai, W., and Staepelaere, D., "Useful-Skew Clock Synthesis Boosts ASIC Performance," chapter 8 in *Closing the Gap Between ASIC & Custom: Tools and Techniques for High-Performance ASIC Design*, Kluwer Academic Publishers, 2002, pp. 209-223.

[27] Duarte, D., et al., "Evaluating run-time techniques for Leakage Power Reduction," in Proceedings of the Asia and South Pacific Design Automation Conference, 2002, pp. 31-38.

[28] Fanucci, L., and Saponara, S., "Data driven VLSI computation for low power DCT-based video coding," *International Conference on Electronics, Circuits and Systems*, vol.2, 2002, pp. 541-544.

[29] Fishburn, J., and Dunlop, A., "TILOS: A Posynomial Programming Approach to Transistor Sizing," in *Proceedings of the International Conference on Computer-Aided Design*, 1985, pp. 326-328.

[30] Flynn, D., and Keating, M., "Creating Synthesizable ARM Processors with Near Custom Performance," chapter 17 in *Closing the Gap Between ASIC & Custom: Tools and Techniques for High-Performance ASIC Design*, Kluwer Academic Publishers, 2002, pp. 383-407.

[31] Furber, S., *ARM System-on-Chip Architecture*. 2nd Ed. Addison-Wesley, 2000.

[32] Ganswijk, J., Chip Directory: ARM Processor family. http://www.xs4all.nl/~ganswijk/chipdir/fam/arm/

[33] Garg, M., "High Performance Pipelining Method for Static Circuits using Heterogeneous Pipelining Elements," in *Proceedings of the European Solid-State Circuits Conference*, 2003, pp. 185-188.

[34] Gong, J., et al., "Simultaneous buffer and wire sizing for performance and power optimization," *International Symposium on Low Power Electronics and Design*, 1996, pp. 271-276.

[35] Greenhalgh, P., "Power Management Techniques for Soft IP," *Synopsys Users Group European Conference*, May 6, 2004, 12 pp.

[36] Hare, C. 786 Processors Chart. http://users.erols.com/chare/786.htm

[37] Harris, D., "High Speed CMOS VLSI Design – Lecture 2: Logical Effort & Sizing," November 4, 1997.

[38] Harris, D., et al. "The Fanout-of-4 Inverter Delay Metric," unpublished manuscript, 1997, 2 pp. http://odin.ac.hmc.edu/~harris/research/FO4.pdf

[39] Ho, R., Mai, K.W., and Horowitz, M., "The Future of Wires," in *Proceedings of the IEEE*, vol. 89, no. 4, April 2001, pp. 490-504.

[40] Horan, B., "Intel Architecture Update," presented at the IBM EMEA HPC Conference, May 17, 2006.www-5.ibm.com/fr/partenaires/forum/hpc/intel.pdf

[41] Intel, Dual-Core Intel Xeon Processor 5100 Series, Features, September 2006. http://www.intel.com/cd/channel/reseller/asmona/eng/products/server/processors/5100/feature/index.htm

[42] Intel, Intel Core Duo Processor Specifications, September 2006.http://www.intel.com/products/processor/coreduo/specs.htm

[43] Intel, Intel XScale Microarchitecture: Benchmarks. http://developer.intel.com/design/intelxscale/benchmarks.htm

[44] Intel, Meet the World's First 45nm Processor, 2007. http://www.intel.com/technology/silicon/45nm_technology.htm?iid=search

[45] Intel, Processor Number Feature Table, September 2006. http://www.intel.com/products/processor_number/proc_info_table.pdf

[46] Ishihara, F., Sheikh, F., and Nikolić, B., "Level Conversion for Dual-Supply Systems," *IEEE Transactions on VLSI Systems*, vol. 12, no. 2, 2004, pp. 185-195.

[47] Ito, M., Chinnery, D., and Keutzer, K., "Low Power Multiplication Algorithm for Switching Activity Reduction through Operand Decomposition," in *Proceedings of the International Conference on Computer Design*, 2003, pp. 21-26.

[48] Jan, C., et al., "A 65nm Ultra Low Power Logic Platform Technology using Uni-axial Strained Silicon Transistors," *Technical Digest of the International Electron Devices Meeting*, 2005, pp. 60-63.

[49] Keshavarzi, A., et al., "Effectiveness of Reverse Body Bias for Leakage Control in Scaled Dual Vt CMOS ICs," *International Symposium on Low-Power Electronics Design*, 2001, pp. 207-212.

[50] Kim, J., "GHz ARM Processor Design," tutorial at the International System-on-Chip Conference, October 23, 2002.

[51] Kitahara, T., et al., "A clock-gating method for low-power LSI design," in *Proceedings of the Asia and South Pacific Design Automation Conference*, 1998, pp. 307-312.

[52] Kosonocky, S., et al., "Low-Power Circuits and Technology for Wireless Data Systems," *IBM Journal of Research and Development*, vol. 47, no. 2/3, March/May 2003, pp. 283-298.

[53] Kulkarni, S., and Sylvester, D., "Fast and Energy-Efficient Asynchronous Level Converters for Multi-VDD Design," *IEEE Transactions on VLSI Systems*, September 2004, pp. 926-936.

[54] Kurd, N.A, et al., "A Multigigahertz Clocking Scheme for the Pentium® 4 Micro-processor," *IEEE Journal of Solid-State Circuits*, vol. 36, no. 11, November 2001, pp. 1647-1653.

[55] Kuroda, T., "Low-power CMOS design in the era of ubiquitous computers," *OYO BUTURI*, vol. 73, no. 9, 2004, pp. 1184-1187.

[56] Lee, D., et al., "Analysis and Minimization Techniques for Total Leakage Considering Gate Oxide Leakage," *proceedings of the Design Automation Conference*, 2003, pp. 175-180.

[57] Lee, D., et al., "Simultaneous Subthreshold and Gate-Oxide Tunneling Leakage Current Analysis in Nanometer CMOS Design," in *Proceedings of the International Symposium on Quality Electronic Design*, 2003, pp. 287-292.

[58] Lee, D., and Blaauw, D., "Static Leakage Reduction through Simultaneous Threshold Voltage and State Assignment," in *Proceedings of the Design Automation Conference*, 2003, pp. 191-194.

[59] Lee, D., Blaauw, D., and Sylvester, D., "Gate Oxide Leakage Current Analysis and Reduction for VLSI Circuits," *IEEE Transactions on VLSI Systems*, vol. 12, no. 2, 2004, pp. 155-166.

[60] Levy, M., "Samsung Twists ARM Past 1GHz," *Microprocessor Report*, October 16, 2002.

[61] McDonald, C., "The Evolution of Intel's Copy Exactly! Technology Transfer Method," *Intel Technology Journal*, Q4 1998. http://developer.intel.com/technology/ itj/q41998/pdf/copyexactly.pdf

[62] Montanaro, J., et al., "A 160MHz, 32-b, 0.5W, CMOS RISC Microprocessor," *Journal of Solid-State Circuits*, vol. 31, no. 11, 1996, pp. 1703-1714.

[63] Moyer, B., "Low-Power Design for Embedded Processors," in *Proceedings of the IEEE*, vol. 89, no. 11, November 2001.

[64] Mutoh, S., et al., "1-V Power Supply High-Speed Digital Circuit Technology with Multithreshold-Voltage CMOS," *Journal of Solid-State Circuits*, vol. 30, no. 8, 1995, pp. 847-854.

[65] Narendra, S., et al., "Comparative Performance, Leakage Power and Switching Power of Circuits in 150 nm PD-SOI and Bulk Technologies Including Impact of SOI History Effect," *Symposium on VLSI Circuits*, 2001, pp. 217-218.

[66] Nassif, S., "Delay Variability: Sources, Impact and Trends," *International Solid-State Circuits Conference*, 2000.

[67] Ostrander, D., "Logic Technology and Manufacturing," slides presented at AMD's Technology Analyst Day, 2006. http://www.amd.com/us-en/assets/content_type/Down-loadableAssets/DarylOstranderAMDAnalystDay.pdf

[68] Pradhan, D., et al., "Gate-Level Synthesis for Low-Power Using New Transformations," *International Symposium on Low Power Electronics and Design*, 1996, pp. 297-300.

[69] Puri, R., et al., "Pushing ASIC Performance in a Power Envelope," in *Proceedings of the Design Automation Conference*, 2003, pp. 788-793.

[70] Quinn, J., *Processor98: A Study of the MPU, CPU and DSP Markets*, Micrologic Research, 1998.

[71] Rabaey, J.M., *Digital Integrated Circuits*. Prentice-Hall, 1996.

[72] Ricci, F., "A 1.5 GHz 90 nm Embedded Microprocessor Core," *Digest of Technical Papers of the Symposium on VLSI Circuits*, 2005, pp. 12-15.

[73] Richardson, N., et al., "The iCORE™ 520MHz Synthesizable CPU Core," Chapter 16 of *Closing the Gap Between ASIC and Custom*, 2002, pp. 361-381.

[74] Shen, W., Lin, J., and Wang, F., "Transistor Reordering Rules for Power Reduction in CMOS Gates," in *Proceedings of the Asia South Pacific Design Automation Conference*, 1995, pp. 1-6.

[75] Shenoy, N., "Retiming Theory and Practice," *Integration, The VLSI Journal*, vol. 22, no. 1-2, August 1997, pp. 1-21.

[76] Simonen, P., et al., "Comparison of bulk and SOI CMOS Technologies in a DSP Processor Circuit Implementation," *International Conference on Microelectronics*, 2001.

[77] Singh, D., et al., "Power Conscious CAD Tools and Methodologies: a Perspective," in *Proceedings of the IEEE*, vol. 83, no. 4, April 1995, pp. 570-594.

[78] Sirichotiyakul, S., et al., "Stand-by Power Minimization through Simultaneous Threshold Voltage Selection and Circuit Sizing," in *Proceedings of the Design Automation Conference*, 1999, pp. 436-441.

[79] Staszewski, R., Muhammad, K., and Balsara, P., "A 550-MSample/s 8-Tap FIR Digital Filter for Magnetic Recording Read Channels," *IEEE Journal of Solid-State Circuits*, vol. 35, no. 8, 2000, pp. 1205-1210.

[80] Stojanovic, V., and Oklobdzija, V., "Comparative Analysis of Master-Slave Latches and Flip-Flops for High-Performance and Low-Power Systems," *IEEE Journal of Solid-State Circuits*, vol. 34, no. 4, April 1999, pp. 536-548.

[81] Stok, L., et al., "Design Flows," chapter in the *CRC Handbook of EDA for IC Design*, CRC Press, 2006.

[82] Sylvester, D. and Keutzer, K., "Getting to the Bottom of Deep Sub-micron," in *Proceedings of the International Conference on Computer Aided Design*, November 1998, pp. 203-211.

[83] Synopsys, *Design Compiler User Guide*, version U-2003.06, June 2003, 427 pp.

[84] Synopsys, *Power Compiler User Guide*, version 2003.06, 2003.

[85] Takahashi, M., et al., "A 60-mW MPEG4 Video Codec Using Clustered Voltage Scaling with Variable Supply-Voltage Scheme," *Journal of Solid-State Circuits*, vol. 33, no. 11, 1998, pp. 1772-1780.

[86] techPowerUp! CPU Database, August 2006. http://www.techpowerup.com/cpudb/

[87] Thompson, S., et al., "An Enhanced 130 nm Generation Logic Technology Featuring 60 nm Transistors Optimized for High Performance and Low Power at $0.7 - 1.4$ V," *Technical Digest of the International Electron Devices Meeting*, 2001, 4 pp.

[88] Tschanz, J. et al., "Adaptive Body Bias for Reducing the Impacts of Die-to-Die and Within-Die Parameter Variations on Microprocessor Frequency and Leakage," *IEEE International Solid-State Circuits Conference*, 2002.

[89] Tschanz, J., et al., "Dynamic Sleep Transistor and Body Bias for Active Leakage Power Control of Microprocessors," *IEEE Journal of Solid-State Circuits*, vol. 38, no. 11, 2003, pp. 1838-1845.

[90] Tsui, C., et al., "Low Power State Assignment Targeting Two- And Multi-level Logic Implementations," in *Proceedings of the International Conference on Computer-Aided Design*, 1994, pp. 82-87.

[91] Tyagi, S., et al., "An advanced low power, high performance, strained channel 65nm technology," *Technical Digest of the International Electron Devices Meeting*, 2005, pp. 245-247.

[92] Usami, K., et al., "Automated Selective Multi-Threshold Design For Ultra-Low Standby Applications," in *Proceedings of the International Symposium on Low Power Design*, 2002, pp. 202-206.

[93] Usami, K., and Horowitz, M., "Clustered voltage scaling technique for low power design," in *Proceedings of the International Symposium on Low Power Design*, 1995, pp. 3-8.

[94] Usami, K., and Igarashi, M., "Low-Power Design Methodology and Applications Utilizing Dual Supply Voltages," in *Proceedings of the Asia and South Pacific Design Automation Conference*, 2000, pp. 123-128.

[95] Vangal, S., et al., "5GHz 32b Integer-Execution Core in 130nm Dual-V_T CMOS," *Digest of Technical Papers of the IEEE International Solid-State Circuits Conference*, 2002, pp. 334-335, 535.

[96] Veendrick, H., "Short-circuit dissipation of static CMOS circuitry and its impact on the design of buffer circuits," *Journal of Solid-State Circuits*, vol. SC-19, August 1984, pp. 468-473.

[97] Virtual Silicon. http://www.virtual-silicon.com/

[98] Weicker, R., "Dhrystone: A Synthetic Systems Programming Benchmark," *Communications of the ACM*, vol. 27, no. 10, 1984, pp. 1013-1030.

[99] Weste, N., and Eshraghian, K., *Principles of CMOS VLSI Design*, Addison-Wesley, 1992.

[100] Wolfe, A., "Intel Clears Up Post-Tejas Confusion," VARBusiness magazine, May 17, 2004. http://www.varbusiness.com/sections/news/breakingnews.jhtml?articleId=18842588

[101] Xanthopoulos, T., and Chandrakasan, A., "A Low-Power DCT Core Using Adaptive Bitwidth and Arithmetic Activity Exploiting Signal Correlations and Quantization," *Journal of. Solid State Circuits*, vol. 35, no. 5, May 2000, pp. 740-750.

[102] Xanthopoulos, T., and Chandrakasan, A., "A Low-Power IDCT Macrocell for MPEG-2 MP@ML Exploiting Data Distribution Properties for Minimal Activity," *Journal of Solid State Circuits*, vol. 34, May 1999, pp. 693-703.

[103] Zimmerman, R., and Fichtner, W., "Low-Power Logic Styles: CMOS Versus Pass-Transistor Logic," *Journal of Solid-State Circuits*, vol. 32, no. 7, July 1997, 1079-1090.

[104] Zlatanovici, R., and Nikolić, B., "Power-Performance Optimal 64-bit Carry-Lookahead Adders," *European Solid-State Circuits Conference*, 2003, pp. 321-324.

Chapter 3

PIPELINING TO REDUCE THE POWER

David Chinnery, Kurt Keutzer
Department of Electrical Engineering and Computer Sciences
University of California at Berkeley
Berkeley, CA 94720, USA

Algorithmic and architectural choices can reduce the power by an order of magnitude [56]. We assume that ASIC and custom designers make similar algorithmic and architectural choices to find a low power implementation that meets performance requirements for the target application.

Circuit designers explore trade-offs for different microarchitectural features that implement a given architecture for typical applications. The analysis may be detailed using cycle accurate instruction simulators, but low level circuit optimizations are not usually examined until a much later design phase. High level microarchitectural choices have a substantial impact on the performance and power consumption, affecting the design constraints for low level optimizations.

This chapter examines the power gap between ASIC and custom with pipelining and different architectural overheads. Other researchers have proposed high level pipelining models that consider power consumption, but they do not consider gate sizing and voltage scaling. We will augment a pipeline model with a model of power savings from voltage scaling and gate sizing versus timing slack. This enables simultaneous analysis of the power and performance trade-offs for both high-level and low-level circuit optimizations.

Pipelining does not reduce power by itself. Pipelining reduces the critical path delay by inserting registers between combinational logic. Glitches may be prevented from propagating across register boundaries, but logic activity is otherwise unchanged. The clock signal to registers has high activity which contributes to the dynamic power. Pipelining reduces the instructions per clock cycle (IPC), due to high branch misprediction penalties and other hazards, and thus can reduce the energy efficiency. The timing slack from pipelining can be used for voltage scaling and gate downsizing to achieve significant power savings (see Figure 3.1 and Figure 3.2).

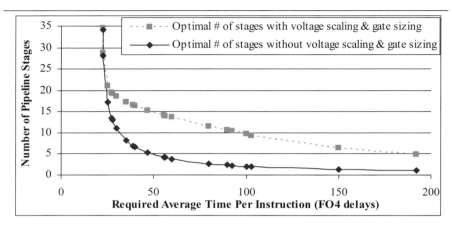

Figure 3.1 The optimal number of pipeline stages to minimize energy/instruction is shown versus the performance constraint. At a tight performance constraint additional stages penalize performance, little timing slack is available, and there is little opportunity for voltage scaling and gate sizing. At more relaxed performance constraints, additional stages provide timing slack for a substantial power reduction with voltage scaling and gate downsizing, as shown in Figure 3.2. These results are for the custom design parameters with our model.

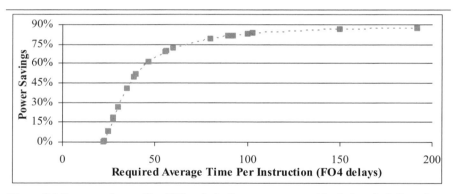

Figure 3.2 Power savings with additional pipeline stages to provide timing slack for voltage scaling and gate sizing versus power consumption without these methods and fewer stages.

From our analysis, pipelining contributes up to a factor of 5.1× to the power gap between ASIC and custom at a tight performance constraint. There is no timing slack at a tight performance constraint for the ASIC where additional pipeline stages will reduce performance, but at this point there is still timing slack for voltage scaling and gate downsizing in a custom design. A custom design may also use additional pipeline stages to further improve performance, as pipeline stage delay overheads are less for custom. The power gap is less as the performance constraint is relaxed, reducing the gap to 4.0× at only 7% lower performance. The gap can be reduced to 1.9× if the pipeline stage delay overhead is reduced.

3.1   INTRODUCTION

Pipelining and parallelism allow the same performance to be achieved at lower clock frequencies. The timing slack can be used to reduce the power by using a lower supply voltage, a higher threshold voltage and reduced gate sizes. Parallel computation trades off area for increased throughput. Pipelining breaks up a datapath into multiple stages with registers between each stage to store the intermediate results. The shorter critical path length from pipelining allows a higher clock frequency. Computation in each pipeline stage can proceed simultaneously if there is no data interdependency, and thus the throughput is higher.

Chandrakasan and Brodersen examined pipelining and parallelism on an 8-bit datapath composed of an adder and comparator in 2.0um technology [10]. If two such datapaths are run in parallel, the clock period can be doubled, and the timing slack can be used to reduce to decrease the supply from 5V to 2.9V. The circuit capacitance more than doubles to 2.15× due to wiring overheads for the parallel datapath and multiplexing of the results. Running the datapath in parallel and reducing the supply reduces the dynamic power by 64%. If instead the datapath is pipelined with registers between the adder and comparator, the supply can be reduced from 5V to 2.9V and the capacitance overhead for the latches is 15%, giving a net power reduction of 61%. The area is more than doubled to 3.4× for the two parallel datapaths, whereas the area for the pipelined datapath is only 1.3× with the additional registers.

For an inverse discrete cosine (IDCT) core in 0.7um process technology with 0.5um channel length, the pipelining power overhead was about 20% of the total power. Without pipelining, the critical path would have been 4× as long and Vdd would have to be increased from 1.32V to 2.2V to achieve the same performance, increasing the total power by 2.2× [70].

Not all microarchitectural techniques for higher performance enable increased energy efficiency. Multiple instructions are executed in parallel execution units in a superscalar architecture, but the additional hardware to determine which instructions can be executed in parallel and reorder the instructions can reduce the energy efficiency [36]. Speculative execution before the outcome of a branch instruction is known wastes energy if the branch is mispredicted. Implementing speculative execution requires branch prediction logic and may require logic to rewind incorrect results. Software hints for branch prediction can reduce the hardware overhead [36].

Very deep pipelines are less energy efficient, as the pipelining over-heads are too large and there is an increased penalty for pipeline hazards. Consequently, Intel is moving from the Pentium 4 NetBurst architecture with 31 stages to the Intel Core architecture with two processor cores that run in parallel, each having a 14 stage pipeline [37]. The Cedar Mill Pentium

4 has about 4.4× the energy/operation of the Yonah Core Duo, despite Yonah having only 2% lower performance on the SPEC CINT2000 benchmark and both being 65nm designs [28].

3.1.1   Power and performance metrics

A typical metric for performance is millions (MIPS) or billions of instructions per second (BIPS) on a benchmark. Commonly used performance benchmarks are the Dhrystone integer benchmark [67], and the integer (SPECint) and floating point (SPECfp) benchmarks from the Standard Performance Evaluation Corporation [58]. Power is measured in watts (W). To account for both power and performance, metrics such as $BIPS^3/W$, $BIPS^2/W$, and $BIPS/W$ are used [9][57]. The inverse of these metrics are also often used. For example, energy per instruction (EPI) corresponds to W/BIPS, and energy-delay product corresponds to $W/BIPS^2$ if we assume that the CPI is fixed.

Minimizing energy or power consumption leads to very large clock periods and low performance being optimal, as dynamic and leakage power can be greatly reduced by using the timing slack to reduce gate sizes, to reduce the supply voltage, and to increase the transistor threshold voltages. Thus metrics placing more emphasis on performance are often used, for example $BIPS^3/W$ and $BIPS^2/W$. More pipeline stages are optimal for metrics with higher weights on performance. Alternatively, the power consumption may be minimized for a specified performance or delay constraint. Changing the microarchitecture may change the delay constraint on the clock period to meet the given performance constraint, for example computing inverse discrete cosine transform serially or in parallel.

3.1.2   Parallel datapath model

Bhavnagarwala et al. developed a model for using parallel datapaths to scale down the supply voltage V_{dd} [8]. To meet the same performance with n datapaths, the clock frequency can be reduced by a factor of $1/n$, and the net switching activity and the dynamic power for the datapaths remain the same if the supply voltage is fixed. If voltages are fixed, the leakage power increases because there are n datapaths leaking rather than one. There is additional routing and multiplexing circuitry for the parallel datapaths, which adds to the dynamic and static power consumption. The expression for the total power that they derive is [8]

$$P_{total} = \frac{1}{2}\alpha C_{datapath}V_{dd}^2 F\left(1 + \frac{C_{overhead}}{C_{datapath}}\right) + P_{static\ for\ datapath}\left(n + \frac{C_{overhead}}{C_{datapath}}\right) \quad (3.1)$$

where $C_{datapath}$ is the total datapath capacitance that switches with activity α, F is the number of operations per second, and $C_{overhead}$ is the capacitance of the additional routing and multiplexing circuitry. They estimate the overhead capacitance to depend quadratically on the number of parallel datapaths [8]:

$$C_{overhead} = (mn^2 + \Gamma)C_{datapath} \qquad (3.2)$$

where m and Γ are fitting constants. From this they estimate that the power savings with parallel datapaths range from 80% power savings with four parallel datapaths for technology with channel length of 0.25um to 15% power savings with two parallel datapaths for technology with channel length of 0.05um. The optimal number of parallel datapaths decreases with technology generation as supply and threshold voltage V_{th} are scaled down, and the ratio of V_{dd}/V_{th} decreases, increasing the performance penalty for lower V_{dd} [8].

The overhead for parallel datapaths is very application dependent, with m in Equation (3.2) having a value from 0.1 to 0.7 depending on the application [8]. Thus the usefulness of parallelism depends greatly on the application. Generally, ASIC and custom designs can make similar use of parallel datapaths, but ASICs have larger wiring overheads with automatic place and route. ASICs suffer higher delay overheads than custom for pipelining – this has a much greater impact on the energy efficiency than ASIC overheads for parallel datapaths, so the remainder of this chapter focuses on the power gap between ASIC and custom with pipelining.

3.1.3   Pipeline model

Pipeline delay models suggest that deeply pipelined designs with logic depth of as low as 8 FO4 delays per stage are optimal for performance [38]. For integer and floating point SPEC 2000 benchmarks, Srinivasan et al. found that the optimal pipeline stage delay was 10 FO4 delays to maximize BIPS, 18 FO4 delays for the BIPS^3/W metric, and 23 FO4 delays for BIPS^2/W [57]. They assumed an unpipelined combinational logic delay of 110 FO4 delays and 3 FO4 timing delay overhead.

Harstein and Puzak did similar analysis following the work of Srinivasan et al. They assumed an unpipelined combinational logic delay of 140 FO4 delays and 2.5 FO4 timing delay overhead [32]. The optimal pipeline stage delay was 22.5 FO4 delays for the BIPS^3/W metric, which is close to the result from Srinivasan et al. given the difference in unpipelined delays. In their models, the pipeline stage delay T is given by [32]

$$T = \frac{t_{comb\ total}}{n} + t_{timing\ overhead} \qquad (3.3)$$

where $t_{comb\ total}$ is the unpipelined delay, n is the number of pipeline stages, and $t_{timing\ overhead}$ is the timing overhead for the registers and clocking. Their performance metric, average time per instruction, can be written for a scalar architecture as [32]

$$T / \text{instruction} = T(1 + \gamma n) \tag{3.4}$$

where γ is the increase in cycles per instruction (CPI) per pipeline stage due to pipeline hazards, and it is assumed that on average an instruction would complete execution every cycle in the absence of hazards. To determine the power for the registers, they use the expression from Srinivasan et al. [32][57],

$$P_{timing} = \left(\frac{1}{T} \alpha_{clock\ gating} E_{dynamic} + P_{leakage} \right) N_L n^{\eta} \tag{3.5}$$

where $\alpha_{clock\ gating}$ is the fraction of time the pipeline is not clock gated; $E_{dynamic}$ and $P_{leakage}$ are respectively the dynamic switching energy and the leakage power for a latch; N_L is the number of latches if there is only a single pipeline stage; n is the number of pipeline stages; and η is the latch growth factor with the number of pipeline stages.

We augment Harstein and Puzak's model by allowing timing slack to be used for voltage scaling and gate sizing to reduce the dynamic power and leakage power for the combinational logic and the registers. In addition, we assume different ASIC and custom values for $t_{timing\ overhead}$ and include pipeline imbalance in the pipeline stage delay for ASICs.

Harstein and Puzak assume that $\alpha_{clock\ gating}$ is $1/(1 + \gamma n)$ [32], with dynamic power consumption for pipeline hazards avoided by shutting off the clock to stalled pipeline stages. This is a reasonable assumption if there is no speculative execution. We will make the same assumption for the value of $\alpha_{clock\ gating}$. We do not consider power gating or reverse body biasing to reduce the leakage during a pipeline stall. For these leakage reduction techniques, the delay and power overhead to raise and lower the voltage are only justified when the circuitry will be unused for at least tens of clock cycles [14].

With an unpipelined combinational logic delay of 180 FO4 delays and 3 FO4 timing delay overhead for custom, we find that a clock period of 8 FO4 delays is optimal to maximize performance, 21 FO4 delays to maximize BIPS^3/W, and 59 FO4 delays to maximize BIPS^2/W. When power is included in the metric, the optimal clock period is significantly larger than that determined by Srinivasan et al., because we allow timing slack to be used to reduce power by voltage scaling and gate sizing. The optimal clock period for a typical ASIC is 2× to 4× larger than custom due to the 20 FO4 delay pipeline stage overhead.

The largest power gap between ASIC and custom is when it is difficult for the ASIC to meet the performance constraint. At the maximum performance for a typical ASIC of 56 FO4 delays on average per instruction, the ASIC power is 5.1× that of custom. As the performance constraint is relaxed, the power gap decreases to 4.0× at only 7% lower performance. For very low performance requirements, the energy efficiency of ASIC and custom microarchitectures is essentially the same.

The delay overhead is the most important factor for the power gap between ASIC and custom with pipelining. If the pipeline stage delay overhead can be reduced from 20 FO4 delays to 5 FO4 delays, the power gap between ASIC and custom is only up to 1.9×.

In Section 3.2, we discuss the overheads for pipelining in ASIC and custom designs. Using our geometric programming optimization results for dynamic power and leakage power with voltage scaling and gate sizing [11], we augment Harstein and Puzak's model with power reduction versus timing slack in Section 3.3. Then in Section 3.4, this augmented model of pipeline power and delay is used to estimate the power gap between ASIC and custom due to pipelining focusing on the impact of the required performance and the pipeline stage delay overhead. The effect of other factors in the pipeline model on the power gap is considered in Section 3.5. Glitching and additional power overheads affect the minimum energy per operation as discussed in Section 3.6. The results are summarized in Section 3.7.

3.2 PIPELINING OVERHEADS

There are several pipelining overheads that we need to consider when comparing pipelining for ASIC and custom designs. There is timing overhead for the registers that store the combinational logic outputs of each stage, and the power consumption for the registers and clock signal. The delay of combinational logic in different pipeline stages may be imbalanced. The penalty for pipeline hazards that delay the next instruction being executed increases with the number of pipeline stages. These overheads are typically less for carefully designed custom circuits compared to ASICs.

Pipeline hazards include data dependency, branch misprediction, cache misses, and so forth. For example, the Willamette Pentium 4 with 20 pipeline stages has 10% to 20% less instructions per cycle than the Pentium III which has only 10 pipeline stages [40].

Adding the timing overhead and pipeline imbalance, the *pipelining delay overhead* is typically about 30% of the clock period for ASICs and 20% of the clock period for custom designs [13]; however, custom designs usually have a much smaller clock period than ASICs. The pipelining delay overhead may be 30% for a custom design with many pipeline stages such as the Pentium 4. When we compared the microarchitectural impact on ASIC and custom speeds, we estimated the pipelining delay overhead in FO4 delays for a variety of custom and ASIC processors as shown in Table 3.1 and Table 3.2 [13]. The pipelining delay overhead ranges from as low as about 2 FO4 delays in some custom designs to 20 FO4 delays in the ASIC processors.

Table 3.1 Characteristics of ASICs and super-pipelined Pentium 4 processors assuming 30% pipelining delay overhead [2][3][4][5][23][45][49][52][59][61][62][63][64][65][71].

Custom PCs	Frequency (MHz)	Technology (nm)	Effective Channel Length (nm)	Voltage (V)	Integer Pipeline Stages	FO4 delays/stage	30% Pipelining Overhead (FO4 delays)	Unpipelined Clock Period (FO4 delays)	Pipelining Overhead % of Unpipelined Clock Period	Clock Frequency Increase by Pipelining
Pentium 4 (Willamette)	2000	180	100	1.75	20	10.0	3.0	143	2.1%	×14.3
Pentium 4 (Gallatin)	3466	130	60	1.60	31	9.6	2.9	212	1.4%	×22.0
ASICs										
Xtensa T1020 (Base)	250	180	130	1.80	5	61.5	18.5	234	7.9%	×3.8
Lexra LX4380	266	180	130	1.80	7	57.8	17.4	301	5.8%	×5.2
iCORE	520	180	150	1.80	8	25.6	7.7	151	5.1%	×5.9
ARM 926EJ-S	200	180	130	1.80	5	64.1	19.2	244	7.9%	×5.9
ARM 1026EJ-S	540	90	50	1.00	6	61.7	18.5	278	6.7%	×4.5
ARM 1136J-S	400	130	80	1.20	8	52.1	15.6	307	5.1%	×5.9
ARM Cortex-A8	800	65	40	1.20	13	62.5	18.8	588	3.2%	×9.4

Table 3.2 Custom design characteristics assuming 20% timing overhead [19][23][24][27] [29][30][31][35][39][41][48][49][51][55][60][66].

Custom Processors	Frequency (MHz)	Technology (nm)	Effective Channel Length (nm)	Voltage (V)	Integer Pipeline Stages	FO4 delays/stage	20% Pipelining Overhead (FO4 delays)	Unpipelined Clock Period (FO4 delays)	Pipelining Overhead % of Unpipelined Clock Period	Clock Frequency Increase by Pipelining
Alpha 21264	600	350	250	2.20	7	13.3	2.7	77	3.4%	×5.8
IBM Power PC	1000	250	150	1.80	4	13.3	2.7	45	5.9%	×3.4
Custom PCs										
Athlon XP (Palomino)	1733	180	100	1.75	10	11.5	2.3	95	2.4%	×8.2
Athlon 64 (Clawhammer)	2600	130	80	1.50	12	9.6	1.9	94	2.0%	×9.8
Pentium III (Coppermine)	1130	180	100	1.75	10	17.7	3.5	145	2.4%	×8.2
Core 2 Extreme (Conroe)	2930	65	35	1.34	14	19.5	3.9	222	1.8%	×11.4
Custom ARMs										
StrongARM	215	350	250	2.00	5	37.2	7.4	156	4.8%	×4.2
XScale	800	180	135	1.80	7	18.5	3.7	107	3.4%	×5.8
Halla (ARM 1020E)	1200	130	80	1.10	6	20.8	4.2	104	4.0%	×5.0

The total delay for the logic without pipelining was calculated from the number of pipeline stages, estimated pipelining overhead, and the FO4 delay for the process and operating conditions. For Table 3.1 and Table 3.2, the FO4 delay was calculated assuming worst case operating conditions and typical process conditions using Equation (2.2), except for the ARM cores in Table 3.1 where the process conditions were worst case and Equation (2.3) was used. The estimated FO4 delay for the custom processes may be more than the real FO4 delay in fabricated silicon for these chips, because of speed-binning, unreported process improvements, and better than worse case operating conditions. As a result, the custom FO4 delays/stage may be underestimated in Table 3.1 and Table 3.2.

The following subsections estimate timing overhead; pipeline imbalance; instructions per cycle with number of pipeline stages for ASIC and custom designs; and discuss the power overhead for pipelining. We will look at a pipeline model incorporating these factors in Section 3.3.

3.2.1 Timing overhead per pipeline stage for ASIC and custom designs

The timing overhead specifies the delay for registers and synchronization of the clock signal to the registers. It includes the setup time during which the input to the register must be stable before the clock signal arrives; the delay for a signal to propagate from a register's input to output; clock skew accounting for the clock signal arriving at different registers at different times; and clock jitter in the arrival time of the periodic clock signal.

The timing overhead for an ASIC may be as much as 10 FO4 delays, but can be reduced to 5 FO4 delays if latches are used instead of D-type flip-flops. We have used Design Compiler scripts to automate replacement of flip-flops by latches, achieving 5% to 20% speed increase in the Xtensa processor [12]. In comparison, the custom timing overhead can be as low as 2.6 FO4 delays as detailed in Table 3.3.

3.2.2 Pipeline imbalance in ASIC and custom designs

The pipeline imbalance for an ASIC with flip-flops can range from 10 FO4 delays down to 2.6 FO4 delays in a carefully balanced design. Unbalanced critical path delays in different pipeline stages can be addressed in several ways. ASICs may use automatic retiming of the register positions to balance critical path delays in different stages. Slack passing by using transparent latches or by useful clock skew is commonly used in custom designs. Useful clock skew tailors the arrival time of the clock signal to different registers by adjusting buffers in the clock tree, and can be used in ASIC designs [17]. Pipeline imbalance with different design techniques is summarized in Table 3.4.

From our experiments replacing flip-flops by latches in the 5-stage Tensilica Xtensa processor [12], we estimate that a typical ASIC may have imbalance of 15% of the clock period. The base configuration of the Xtensa has a maximum stage delay (clock period) of about 67 FO4 delays, of which 15% is 10 FO4 delays.

The imbalance between pipeline stages for a well balanced ASIC can be as low as 10% of the clock period. For example, the 8-stage iCORE has about 10% imbalance in the critical sequential loop through IF1, IF2, ID1, ID2, and OF1 back to IF1 through the branch target repair loop [52]. The iCORE has about 26 FO4 delays per pipeline stage, thus the imbalance is about 2.6 FO4 delays.

Table 3.3 Comparison of ASIC and custom timing overheads, assuming balanced pipeline stages [13]. Alpha 21164 [7], Alpha 21264 [29] and Pentium 4 [42] setup times were estimated from known setup times for latches and pulse-triggered flip-flops. The pulse-triggered latches in the Pentium 4 are effectively used as flip-flops rather than as transparent latches. Timing overhead for flip-flops was calculated from $t_{CQ} + t_{su} + t_{sk} + t_j$. As there are two latches, positive and negative-edge triggered, per clock cycle, the timing overhead for latches was calculated from $2t_{DQ} + t_{j\ multicycle}$, where multi-cycle jitter $t_{j\ multi-cycle}$ of 1.0 FO4 delays was assumed for ASICs.

Contributions to ASIC and custom timing overhead in FO4 delays	ASICs		Custom		
	D-type flip-flops	Good Latches	Pass transistor latches in Alpha 21164	Edge-triggered flip-flops in Alpha 21264	Clock-pulsed latches in Pentium 4
Clock-to-Q delay t_{CQ}	4.0			2.0	2.0
2× D-to-Q latch propagation delay ($2 \times t_{DQ}$)		4.0	2.6		
Flip-flop setup time t_{su}	2.0			0.0	0.0
Edge jitter t_j				0.1	0.7
Clock skew t_{sk}				0.7	0.3
Budget for clock skew and edge jitter $t_{sk} + t_j$	4.0	1.0			
Timing overhead per clock cycle	10.0	5.0	2.6	2.8	3.0

Table 3.4 Summary of pipeline imbalance for ASIC and custom designs

	Pipeline Imbalance (FO4 delays)
Typical ASIC with flip-flops	10.0
Carefully balanced ASIC with flip-flops	2.6
Optimal design with flip-flops, no slack passing	1.0
Slack passing via latches or cycle stealing	0.0

Automated flip-flop retiming won't typically achieve a pipeline imbalance of a single gate delay. Retiming is based on assumptions such as fixed register delay, fixed register setup time, and fixed gate delays. In reality, this depends on the drive strength and type of flip-flop chosen, and the gate loads which are changed by retiming. Additionally, the combinational gates and registers will usually be resized after retiming. Reducing the clock period by retiming may be limited by input or output delay constraints, such as reading from and writing to the cache [33]. These issues limit the optimality of auto-mated retiming.

Slack passing is not limited by the delay of a particular stage. Custom designers also have tighter control of gate delays, register positions, and better knowledge of wire loads that depend on layout – which is not known for retiming in the synthesis stage of an ASIC EDA methodology. Thus custom designs may be able to balance stages, whereas ASICs typically suffer some pipeline imbalance.

With useful clock skew or transparent latches, slack passing between stages can eliminate pipeline imbalance. From the iCORE and Xtensa exam-ples, a 10% to 15% reduction in clock period can be achieved by slack passing for ASICs with imbalanced pipeline stages.

Table 3.5 Cycles per instruction (CPI) for various processors [16][18][34][43][52][54][72].

Processor	# of Pipeline Stages	IPC	CPI	Increase in CPI/stage, γ
ARM7TDMI	3	0.53	1.90	30.0%
ARM9TDMI	5	0.67	1.50	10.0%
ARM810	5	0.71	1.40	8.0%
DEC StrongARM	5	0.61	1.63	12.7%
Intel XScale	7	0.56	1.78	11.2%
STMicroelectronics iCORE	8	0.70	1.43	5.4%
Pentium 4 (Willamette)	20	not known		3.0%
Pentium 4 (Cedar Mill)	31	0.54	1.84	2.7%

3.2.3 Instructions per cycle versus number of pipeline stages

Instructions per cycle (IPC) and its reciprocal *cycles per instruction* (CPI) are measures of how quickly instructions are executed after accounting for pipeline stalls due to hazards. Reductions in IPC can be caused by cache misses, waiting for data from another instruction that is executing, branch misprediction, and so forth. The CPI for a number of processors is summarized in Table 3.5.

The CPI is very application dependent, as some applications have more branches and other hazards. For the Cedar Mill Pentium 4 with 31 pipeline stages, the CPI ranges from 0.64 to 7.87 for different benchmarks in the SPEC CINT2000 benchmark set. The geometric mean for Cedar Mill for the SPEC

CINT2000 benchmark set was 1.84 [18]. The 1.5GHz Willamette Pentium 4 with 20 pipeline stages is 15% to 20% faster for integer applications than a 1.0GHz Pentium III with 10 pipeline stages [34], which corresponds to a 20% to 23% worse IPC.

For a variety of benchmarks, the IPC for the five stage DEC StrongARM ranges from 0.30 to 0.83 [72], and the IPC ranges from 0.38 to 0.82 for the seven stage Intel XScale [16]. The geometric means of the IPC values were 0.61 and 0.56 respectively.

The IPC for the three stage ARM7TDMI was 0.5, whereas the ARM9TDMI with five pipeline stages had an IPC of 0.7 [54]. The higher IPC for the five stage ARM810 and ARM9 pipelines was achieved by adding static branch prediction, single cycle load, single cycle store, and doubling the memory bandwidth [43]. The eight stage STMicroelectronics iCORE also achieved an IPC of 0.7 by microarchitectural optimizations in an ASIC EDA methodology [52]. The ARM810 used standard cells for the control logic, but was otherwise full custom [43]. The ARM7TDMI and ARM9TDMI were full custom, but synthesizable versions of these processors were also created [21].

Without additional microarchitectural features such as data forwarding and improved branch prediction to maintain high IPC, the CPI increases approximately linearly with the number of pipeline stages as more pipeline stages are stalled when a hazard is encountered [32]. Assuming that an unpipelined design has an IPC of close to 1.0, which is somewhat optimistic as there may be cache misses and off-chip memory will take more than a cycle to read, the CPI increase per pipeline stage ranges from 30.0% to 2.7%.

3.2.4   Power overheads for pipelining

The majority of the power overhead for pipelining is power consumption in the clock tree and registers. The registers and clock tree can consume from 18% to 36% of the total power, as shown in Table 3.6. Clock gating was used in these processors to reduce the power consumed by the registers and clock tree.

Branch prediction, data forwarding and other microarchitectural techniques to maintain a high IPC with deeper pipelines also take some power. We assume that these additional power overheads are small relative to the clock tree and register power, and do not explicitly include them in the pipeline power model in the same manner as [32] and [57].

The percentage of power consumed by registers and the clock tree depends on the application. For example, the clock tree accounts for 18% of the XScale for the DSP FIR benchmark, but it is 23% for the Dhrystone MIPS 2.1 benchmark [15].

Table 3.6 Register and clock tree power consumption in various processors. The StrongARM [47] and XScale [15] are custom embedded processors. The Alpha 21264 [26] and Itanium [1] are custom desk top processors. The MCORE [25] is a semi-custom processor, and the 16-bit CompactRISC ASIC [46] was synthesized.

Processor	# of Pipeline Stages	Registers and Clock Tree Power as % of Total Power
16-bit CompactRISC	3	34%
MCORE	4	36%
StrongARM	5	25%
XScale (DSP FIR filter)	7	18%
XScale (Dhrystone MIPS)	7	23%
Alpha 21264	7	32%
Itanium	8	33%

In Motorola's 0.36um 1.8V MCORE embedded processor with a four stage pipeline, the datapath and clock tree each contribute 36% of the total power, and control logic contributes the other 28% of the total power. If the custom datapath was instead synthesized, the datapath's power would have been 40% higher [25].

Few breakdowns of power data for synthesized processors are available. We expect register and clock tree power to consume a similar portion of the total power in ASICs. In a synthesized 0.18um 1.8V National Semi-conductor 16-bit CompactRISC processor with a three stage pipeline, the register file consumed 34% of the processor core's total dynamic power [46].

We now examine a model that incorporates these pipelining overheads.

3.3 PIPELINING POWER AND DELAY MODEL

To build the pipeline power and delay model, we first calculate the minimum pipeline stage delay T_{min}. Given some upper limit on the clock period, the actual clock period T can be anywhere between the upper limit and the minimum. We will discuss a simple experimental fit to determine the reduced power from voltage scaling and gate downsizing with timing slack $(T - T_{min})$. We can then find the optimal number of pipeline stages and optimal amount of timing slack to use for power reduction in order to mini-mize the power.

3.3.1 Pipeline stage delay

The number of pipeline stages n in a processor varies widely. For example the ARM7 architecture has a three stage pipeline comprising inst-ruction fetch, instruction decode, and execute [22]; whereas the Cedar Mill Pentium 4 has 31 stages [37]. The total unpipelined delay $t_{comb\ total}$ can range from about 50 to 300 FO4 delays, as was estimated earlier in Table 3.1 and Table 3.2.

If a pipeline with n stages is ideally balanced, the combinational delay per pipeline stage is

$$t_{comb} = \frac{t_{comb\ total}}{n} \qquad (3.6)$$

The maximum delay of a pipeline stage, which limits the minimum clock period T_{min}, is

$$T_{min} = \frac{t_{comb\ total}}{n} + t_{imbalance} + t_{timing\ overhead} \qquad (3.7)$$

where $t_{imbalance}$ accounts for the pipeline stages being unbalanced, and $t_{timing\ overhead}$ is the timing overhead.

The clock period T that is used must be at least T_{min}, but may be larger to provide slack for power reduction by voltage scaling and gate downsizing.

3.3.2   Utilizing slack for voltage scaling and downsizing to reduce power

We can reduce the dynamic energy per clock cycle and reduce the leakage power by reducing the supply voltage V_{dd}, increasing the threshold voltage V_{th}, and reducing the gate size. The impact of gate size, supply voltage and threshold voltage on the leakage power and dynamic power is detailed in Chapter 4. In this section, we are just interested in how the leakage power and dynamic power decrease as timing slack is used to downsize gates and scale the voltages.

Increasing a gate's size reduces the delay, but increases the load on fanins, requiring them to also be upsized. Consequently, at a tight delay constraint there is substantially higher power consumption with many gates having been upsized. To meet a tight delay constraint, the supply voltage will also be higher and the threshold voltage may be lower to reduce the critical path delay. The rapidly increasing power consumption as T approaches the minimum delay T_{min} results in the classic "banana" curve shape shown for dynamic power and leakage power in Figure 3.3. From a tight delay constraint, a small amount of timing slack can be used to significantly reduce the energy consumed per clock cycle.

The power versus delay curves with gate sizing and voltage scaling for the dynamic power and leakage power versus clock period are fit well by hyperbolic functions of the form

$$a + \frac{b}{\left(\dfrac{T}{T_{min}} + c \right)^d} \qquad (3.8)$$

where a, b, c and d are experimentally fitted constants. We require that $a \geq 0$, so that the power does not become negative as T becomes large.

As the leakage depends exponentially on the threshold voltage, and the threshold voltage can be increased with increasing clock period, the accuracy of the fit to leakage power can be improved by including an exponential term:

$$a + \frac{be^{\lambda T/T_{\min}}}{\left(\dfrac{T}{T_{\min}} + c\right)^d} \tag{3.9}$$

where exponent λ is also an experimentally fitted constant.

Accurate fits for the dynamic and leakage power for a benchmark in 0.13um technology are shown in Figure 3.3. The relative root mean square error is 0.5% for the dynamic power fit and 3.1% for the leakage power fit, where the relative root mean square (RMS) error is given by

$$\text{Relative Root Mean Square Error} = \sqrt{\frac{1}{N}\sum_{i=1}^{N}\left(\frac{y_{data,i} - y_{fit,i}}{y_{data,i}}\right)^2} \tag{3.10}$$

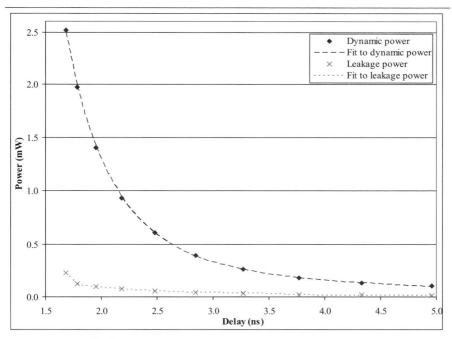

Figure 3.3 Curve fits for the dynamic and leakage power for ISCAS'85 benchmark c880. The total power was minimized by choosing optimal gate sizes, single supply voltage, single NMOS threshold voltage and single PMOS threshold voltage by geometric program optimization [11]. The allowed range for the supply voltage was 1.3V to 0.6V. The allowed ranges for the threshold voltages were ±0.13V from the nominal threshold voltage.

The leakage power contributes from 5.9% to 11.8% of the total power. The maximum error in the total power is 0.7%. The fits are

$$P_{dynamic}(T) = 0.0422 + \frac{0.758}{\left(\dfrac{T}{T_{min}} - 0.352\right)^{2.704}} \qquad (3.11)$$

$$P_{leakage}(T) = \frac{0.212e^{-0.971T/T_{min}}}{\left(\dfrac{T}{T_{min}} - 0.997\right)^{0.182}} \qquad (3.12)$$

where fitting coefficients have been shown to three significant figures, and T_{min} for c880 was 1.69ns.

With pipelining allowing higher clock frequency, the switching activity and hence dynamic power increases proportionally to the clock frequency, that is as $1/T$. The dynamic power fit implicitly includes the dependence of switching activity on T.

Glitching caused by spurious transitions from signals propagating through the logic at different speeds also affects the switching activity. Glitching depends approximately linearly on the logic depth [57], so pipelining reduces glitching by reducing the logic depth. Glitching only has a small impact on the power gap between ASIC and custom, so we do not consider it at this stage. Section 3.6.1 discusses the impact of glitching.

We must also account for the power consumption of the registers and clock tree, as the number of registers varies with the pipeline depth.

3.3.3   Power consumption of the registers and clock tree

Deeper pipelines typically require more registers per pipeline stage, because balancing the stage delays may require the additional registers to be placed at cut points where there are more edges. To take into account the register and the clock tree power, Harstein and Puzak [32] use a power model from Srinivasan et al. [57], which has the form

$$P_{timing} = \left(\frac{1}{T}\alpha_{clock\ gating}E_{dynamic} + P_{leakage}\right)\beta n^{\eta} \qquad (3.13)$$

where $\alpha_{clock\ gating}$ is the fraction of time the pipeline is not clock gated; βn^{η} is the additional fraction of power due to the registers and the clock tree; and η is a scaling factor – the "latch growth factor". η takes into account how many additional registers are required as the pipeline depth increases. If $\eta = 1$, the number of registers for per pipeline stage does not vary with the depth.

Srinivasan et al. show that the latch growth factor η is about 1.7 for a floating point unit, and 1.9 for a Booth recoder and Wallace tree multiplier. Throughout most of their analysis they assume a value for η of 1.1 [57].

Harstein and Puzak also assume an η value of 1.1 [32]. For a fixed circuit with neither gate sizing nor voltage scaling, as the dynamic power consumption is proportional to the switching activity and hence the clock frequency, the dynamic energy per clock cycle is independent of the clock frequency. Thus, they assume that the register and clock tree power is the only significant change in the total power with the number of pipeline stages n, as the dynamic energy for the combinational logic is fixed.

Voltage scaling and gate sizing can be used to reduce the power for the registers and clock tree, and we assume that their power scales in the same manner versus timing slack as the combinational logic. Including the combinational logic's power consumption in Equation (3.13) gives

$$P_{total} = \left(\frac{1}{T} \alpha_{clock\ gating} E_{dynamic} + P_{leakage} \right)(1 + \beta n^\eta) \qquad (3.14)$$

Allowing timing slack $(T - T_{min})$ for gate sizing and voltage scaling to help reduce the power, using the dynamic and leakage power models from Section 3.3.2 normalized to their value at T_{min},

$$P_{total}(T) = \frac{1}{T_{min}} \left(\begin{array}{c} \alpha_{clock\ gating} \dfrac{P_{dynamic}(T)}{P_{dynamic}(T_{min})}(1 - k_{leakage}) \\[2ex] + k_{leakage} \dfrac{P_{leakage}(T)}{P_{leakage}(T_{min})} \end{array} \right)(1 + \beta n^\eta) \qquad (3.15)$$

where $k_{leakage}$ is the fraction of total power due to leakage at T_{min}. Our dynamic and leakage power models account for the decrease in switching activity for $T > T_{min}$, but we still require the $1/T_{min}$ factor to account for switching activity varying with T_{min}. The $1/T_{min}$ factor also affects leakage, because the fraction of total power due to leakage at T_{min} is determined versus the dynamic power without clock gating ($\alpha_{clock\ gating} = 1$).

As in the earlier models, we will assume that a value for η of 1.1 is a reasonable estimate for the integer pipeline of a processor. Assuming a given value for η, we can calculate the value of β in Equation (3.15) from the number of pipeline stages and the percentage of register and clock tree power in a processor. From the register and clock tree power data for different processors in Table 3.6 and assuming η of 1.1, β ranges from 0.026 to 0.15. We use a value for β of 0.05, which is typical for most of the processors of five to eight pipeline stages.

We use Harstein and Puzak's model for the clock gating factor [32],

$$\alpha_{clock\ gating}(n) = 1/(1 + \gamma n) \qquad (3.16)$$

where clock gating enables avoiding dynamic power consumption due to pipeline hazards by shutting off the clock to stalled pipeline stages. This is a reasonable assumption if there is no speculative execution.

We now look at using these models to choose the optimal number of pipeline stages and allocation of slack for voltage scaling and gate sizing.

3.3.4   The pipeline power and delay model for optimization

We now have the pipeline stage delay and can calculate the power reduction from the timing slack used for voltage scaling and gate sizing. The number of clock cycles per instruction must be accounted for. Assuming one instruction would be executed per cycle if there were no hazards, and that the penalty for pipeline hazards increases linearly with the number of stages as discussed in Section 3.2.3, the average time per instruction is [32]

$$T / \text{instruction} = T(1 + \gamma n) \tag{3.17}$$

where γ is the increase in CPI per pipeline stage due to hazards.

Typical metrics that we wish to optimize include maximizing the performance, minimizing the energy per operation, and minimizing the power for a given performance constraint. The numerical solution of these optimization problems will usually give a non-integer value for the number of pipeline stages n, though in a real circuit n must be integral.

3.3.4.1   Maximum performance: minimum *T*/instruction

The maximum performance is found by minimizing the average time per instruction in Equation (3.17), where T is given by Equation (3.7). Setting the derivative with respect to n to zero, the solution is

$$n = \sqrt{\frac{t_{comb\ total}}{\gamma(t_{imbalance} + t_{timing\ overhead})}}$$

$$T = t_{imbalance} + t_{timing\ overhead} + \sqrt{\gamma t_{comb\ total}(t_{imbalance} + t_{timing\ overhead})} \tag{3.18}$$

$$\min T / instruction = t_{imbalance} + t_{timing\ overhead} + \gamma t_{comb\ total}$$
$$+ 2\sqrt{\gamma t_{comb\ total}(t_{imbalance} + t_{timing\ overhead})}$$

3.3.4.2   Maximum BIPS^m/W: minimum $P_{total}(T/\text{instruction})^m$

Minimizing the energy per operation is equivalent to maximizing BIPS/W (instructions per second per unit of power). The minimum energy per operation is found when there is substantial timing slack to reduce the dynamic and leakage power, and the pipelining power overheads are minimized. Consequently, a single pipeline stage is optimal to minimize the energy per operation, as this minimizes the power overhead for registers. More than one pipeline stage is optimal to minimize energy per operation when glitching and additional power overheads are accounted for – see Section 3.6 for further discussion.

The solution for minimum energy per operation is not particularly interesting from a circuit design viewpoint as most applications require higher performance than where the minimum energy per operation occurs. Thus the performance is usually more heavily weighted in the objective. The more general optimization problem is to maximize BIPS^m/W by finding the optimal number of pipeline stages n and optimal clock period T in

minimize $\quad P_{total}(T(1+\gamma n))^m$

subject to $\quad T_{min} = \dfrac{t_{comb\ total}}{n} + t_{imbalance} + t_{timing\ overhead}$ $\hfill$ (3.19)

$\quad\quad\quad\quad T \geq T_{min}$

$\quad\quad\quad\quad n \geq 1$

$$P_{total} = \frac{1}{T_{min}} \left(\begin{array}{c} \dfrac{1}{(1+\gamma n)} \dfrac{P_{dynamic}(T)}{P_{dynamic}(T_{min})}(1-k_{leakage}) \\[2ex] +k_{leakage} \dfrac{P_{leakage}(T)}{P_{leakage}(T_{min})} \end{array} \right) (1+\beta n^\eta)$$

where the dynamic and leakage power are fitted as in equations (3.11) and (3.12). m is typically 0, 1, 2, or 3. m of 0 minimizes power, regardless of performance. $m=1$ minimizes the energy per operation. The energy-delay product is given by $m=2$. Values for m of 2 or more emphasize minimizing delay over minimizing power. The optimization problem in Equation (3.19) can be solved easily with the Newton-Raphson gradient descent algorithm.

3.3.4.3 Minimum power P_{total} at performance $T_{required\ per\ instruction}$

Many applications require a specific performance. To find the minimum power for a given required performance, or average time per instruction $T_{required\ per\ instruction}$, we solve for n and T in

minimize $\quad P_{total}$

subject to $\quad T_{min} = \dfrac{t_{comb\ total}}{n} + t_{imbalance} + t_{timing\ overhead}$

$\quad\quad\quad\quad T \geq T_{min}$

$\quad\quad\quad\quad n \geq 1$

$\quad\quad\quad\quad T(1+\gamma n) = T_{required\ per\ instruction}$

$$P_{total} = \frac{1}{T_{min}} \left(\begin{array}{c} \dfrac{1}{(1+\gamma n)} \dfrac{P_{dynamic}(T)}{P_{dynamic}(T_{min})}(1-k_{leakage}) \\[2ex] +k_{leakage} \dfrac{P_{leakage}(T)}{P_{leakage}(T_{min})} \end{array} \right) (1+\beta n^\eta)$$

$\hfill$ (3.20)

Table 3.7 This table lists parameters and variables used for pipeline modeling. The default values assumed for the parameters are listed. The excellent ASIC corresponds to an ASIC using latches, or faster pulsed flip-flops where needed and useful clock skew, to reduce the register delay and impact of clock skew, and to address imbalance in pipeline stage delays.

Parameter for Design Style Overhead	Represents	Typical ASIC	Excellent ASIC	Custom
$t_{imbalance}$ (FO4 delays)	unbalanced pipeline stage delay overhead	10	0	0
$t_{timing\ overhead}$ (FO4 delays)	timing overhead per pipeline stage	10	5	3

Parameter	Represents	Value
$k_{leakage}$	fraction of total power from leakage at T_{min}	0.1
m	exponent for delay per instruction in the objective	varies
$T_{required\ per\ instruction}$ (FO4 delays)	required delay per instruction	varies
$t_{comb\ total}$ (FO4 delays)	total unpipelined combinational delay	180
β	coefficient for power due to registers and the clock tree	0.05
γ	increase in clock cycles per instruction (CPI) with pipeline stages due to hazards	0.05
η	latch growth factor for increase in number of registers with pipeline depth	1.1

Optimization Variable	Represents
n	number of pipeline stages
T (FO4 delays)	clock period

Dependent Variable	Represents
$P_{dynamic}$	dynamic power
$P_{leakage}$	leakage power
P_{timing}	power consumption of registers and clock
P_{total}	total power consumption
t_{comb} (FO4 delays)	combinational delay per pipeline stage
T_{min} (FO4 delays)	minimum clock period
$\alpha_{clock\ gating}$	fraction of time pipeline is not clock gated

We now use the solutions of Equation (3.20) to compare the minimum power for ASIC and custom designs for a given performance constraint.

3.4   ASIC VERSUS CUSTOM PIPELINING

We will now estimate the gap between ASIC and custom designs due to microarchitecture using the model in Section 3.3 with default values for parameters listed in Table 3.7, which correspond to an integer pipeline in a high performance ASIC processor and the custom equivalent. For other applications, different parameter values should be considered – the impact of varying the parameters is discussed later in Section 3.5.

Pipeline stage delay overhead has the greatest impact on the results, so we initially focus on the differences due to this. The total pipelining delay overhead for a typical ASIC is 20 FO4 delays due to slow D-type flip-flops and imbalanced pipeline stages, compared to 3 FO4 delays for custom. If an ASIC design uses latches, or faster pulsed flip-flops where needed and useful clock skew, then the pipelining delay overhead may be as low as 5 FO4 delays.

We wish to determine the impact of microarchitecture in the absence of other factors such as slower logic style, so we assume an unpipelined combinational delay of 180 FO4 delays which is reasonable for the custom processors in Table 3.2, though less than we estimated for the ASICs in Table 3.1. To reduce the clock period to 40 FO4 delays, at least nine pipeline stages must be used for our typical ASIC that has a pipeline stage delay overhead of 20 FO4 delays; whereas a custom design with a pipeline stage delay overhead of 3 FO4 delays needs only five pipeline stages. As a large portion of the ASIC's clock period is devoted to the pipelining delay overhead, reducing the delay overhead is very important to improve ASIC performance.

As was assumed by Srinivasan et al. [57] and Harstein and Puzak [32], we assume a value of 1.1 for the latch growth factor η for an integer pipeline. From this and the clock tree and register power in Table 3.6, we estimate the coefficient β for the clock and register power to be 0.05, which is typical for most of the processors of five to eight pipeline stages [11].

We assume that ASIC and custom designers can take the same advantage of data forwarding, branch prediction and other techniques to reduce the CPI penalty for deeper pipelines. A CPI penalty per stage of 0.05/stage for both ASIC and custom is assumed.

The power-delay curve fits from geometric programming optimization of ISCAS'85 benchmark c880 will be used to estimate the power savings that can be achieved by voltage scaling and gate sizing. The dynamic power and leakage power are normalized as in Equation (3.15), and the minimum delay is set to the minimum stage delay from pipelining in Equation (3.7). When Vdd and Vth are chosen to minimize the total power consumption, leakage may be 8% to 21% of the total power consumption as discussed in Section 4.6.1. We assume that leakage is 10% of the total power at the minimum delay, as may be typical for high performance circuits in 0.13um.

We will first examine the maximum performance that can be achieved by ASICs and custom, and then look at metrics that include power consumption as well as performance. Then we will compare the power gap between ASIC and custom at maximum performance and relaxed performance constraints. The results for different metrics are summarized in Table 3.8.

Table 3.8 This table compares the minimum of various metrics for ASIC and custom. Below the normalized comparison versus custom are listed the optimal number of pipeline stages n, clock period T, delay per instruction T/instruction, et al. to optimize each metric. Note that the numerical solution below for the optimization problem has a non-integer value for the number of pipeline stages n, though in a real circuit n must be integral.

	Normalized versus custom			
	T/instruction	$P\,(T$/instruction$)^3$	$P\,(T$/instruction$)^2$	Energy/operation
Typical ASIC	2.5	3.5	1.7	1.0
Excellent ASIC	1.2	1.3	1.1	1.0

Minimum T/instruction (maximizing BIPS)

	n	T_{min}	T (FO4 delays)	T/instruction (FO4 delays)	Power P	Energy / Operation
Typical ASIC	13.3	33.5	33.5	55.8	0.0356	1.987
Excellent ASIC	26.5	11.8	11.8	27.4	0.1174	3.220
Custom	34.2	8.3	8.3	22.4	0.1795	4.020

Minimum $P\,(T$/instruction$)^3$ (maximizing BIPS^3/W)

	n	T_{min}	T (FO4 delays)	T/instruction (FO4 delays)	Power P	Energy / Operation	$P\,(T$/instruction$)^3$
Typical ASIC	8.5	41.1	64.6	92.2	0.0054	0.497	4225
Excellent ASIC	14.2	17.7	27.3	46.7	0.0143	0.668	1457
Custom	16.6	13.9	21.3	38.9	0.0192	0.747	1129

Minimum $P\,(T$/instruction$)^2$ (maximizing BIPS^2/W)

	n	T_{min}	T (FO4 delays)	T/instruction (FO4 delays)	Power P	Energy / Operation	$P\,(T$/instruction$)^2$
Typical ASIC	6.5	47.8	139.2	184.3	0.0010	0.175	32.3
Excellent ASIC	9.6	23.7	69.6	103.1	0.0020	0.204	21.0
Custom	10.6	20.0	58.6	89.6	0.0024	0.214	19.2

Minimum energy/operation $P\,(T$/instruction) (maximizing BIPS/W)

	n	T_{min}	T (FO4 delays)	T/instruction (FO4 delays)	Power P	Energy / Operation
Typical ASIC	1.0	200.0	892.2	936.8	0.0001	0.105
Excellent ASIC	1.0	185.0	825.3	866.5	0.0001	0.105
Custom	1.0	183.0	816.4	857.2	0.0001	0.105

3.4.1 Maximum performance (minimum delay/instruction)

The minimum delay per instruction is 22.4 FO4s for custom, 27.4 FO4s for an excellent ASIC, and 55.8 FO4s for a typical ASIC (see Table 3.8). The corresponding clock period is 8.3 FO4s for custom, 11.8 FO4s for an excellent ASIC, and 33.5 FO4s for a typical ASIC. These model results are comparable to the custom 3.466GHz 0.13um Gallatin Pentium 4 with clock period of 9.6 FO4s and the high performance, synthesized 520MHz 0.18um iCORE with clock period of 25.6 FO4s (from Table 3.1).

The typical ASIC with 20 FO4 stage delay overhead is about 2.5× slower than custom with 3 FO4 stage delay overhead, whereas the excellent ASIC with only 5 FO4 stage delay overhead closes the performance gap to 1.2×. These results correspond fairly well with our earlier analysis which estimated a performance gap due to microarchitecture and timing overhead of 2.6× for a typical ASIC to 1.4× for an excellent ASIC [13].

Maximizing performance leads to deep pipelines being optimal, from 34.2 pipeline stages for custom to 13.3 stages for a typical ASIC. For a real world comparison, Intel's Prescott and Cedar Mill Pentium 4 custom processors have 31 integer pipeline stages [37] from Intel's pushing to the extreme higher performance and higher clock frequency to compete with AMD; while ARM's Cortex-A8 synthesizable processor has thirteen pipeline stages [6].

3.4.2   Maximum BIPS^m/W with voltage scaling and gate sizing

The optimal clock period increases and the optimal number of pipeline stages decreases when power is included in the optimization objective. We can maximize metrics of the form BIPS^m/W with Equation (3.19), minimizing $P(T/\text{instruction})^m$. Results are listed in Table 3.8, except for minimizing power for which an infinite clock period is optimal to avoid dynamic power.

A single pipeline stage is optimal to avoid the pipelining power overhead for more registers when minimizing the power ($m = 0$) or energy per operation ($m = 1$). More than one pipeline stage is optimal to minimize energy per operation when glitching and additional power overheads are accounted for – see Section 3.6 for further discussion.

The optimal clock period of 816 FO4 delays for custom to 892 FO4 delays for ASIC to minimize the energy/operation corresponds to a clock frequency of 31MHz and 28MHz respectively in 0.13um technology with 0.08um channel length. Some applications such as the discrete cosine transform (DCT) and its inverse (IDCT) [20][69][70] can be performed at such low clock frequencies via parallel datapaths, achieving low energy per operation. The DCT and IDCT cores do have more than one pipeline stage – pipelining is used to implement the algorithm and allow clock gating of units that are not in use to reduce the dynamic power.

Many applications require higher performance, so metrics placing a greater weight on delay are commonly used. Comparing the inverse of BIPS^3/W, a typical ASIC has 3.5× the $P(T/\text{instruction})^3$ of custom, while an excellent ASIC is only 1.3× worse. The gap is larger for BIPS^3/W as the performance gap is multiplied. For both ASIC and custom, the ratio of T/T_{min} is about 1.5 to maximize BIPS^3/W providing a significant amount of slack for voltage scaling and gate downsizing.

The optimal number of pipeline stages to maximize BIPS^3/W is roughly half the number of stages to maximize performance, as inclusion of power consumption in the objective substantially penalizes very deep pipelines for their additional registers. To maximize BIPS^3/W for custom, the optimal number of pipeline stages from the model is 16.6 and the optimal clock period is 21.3 FO4 delays, which is comparable to Intel's Conroe Core 2 Extreme with 14 pipeline stages and a clock period of 19.5 FO4 delays. Conroe's predecessor, the Yonah Core Duo, was specifically designed to be more energy efficient than the Pentium 4 models with 31 integer pipeline stages that maximized performance and have 4.4× more energy per operation [28]. To maximize BIPS^3/W for a typical ASIC, the optimal clock period of 64.6 FO4 delays from the model is very similar to the clock period of a number of the ASICs in Table 3.1, though the number of pipeline stages for these ASICs ranges from 5 to 13.

We now look at the minimum power for a given performance constraint.

3.4.3   Minimum power for a given performance constraint

A fixed performance constraint can be used instead of weighting performance in the objective. The minimum power consumption to satisfy the performance constraint can be determined by solving Equation (3.20).

The power gap between a typical ASIC and custom ranges from 5.1× at the maximum performance for the typical ASIC down to 1.0× depending on the performance constraint, as shown in Figure 3.4. If the pipeline stage delay overhead is reduced from 20 FO4 delays to 5 FO4 delays, the power gap is at most 1.9× at the maximum performance for the excellent ASIC. The corresponding optimal number of pipeline stages is shown in Figure 3.5. The optimal clock period for both ASIC and custom is very similar and varies almost linearly with the performance constraint [11].

When the required average time per instruction is large, a typical ASIC has more pipeline stages than custom to get similar timing slack for power reduction. As the required performance increases, fewer stages are optimal for an ASIC due to the larger pipeline stage delay overhead. As the maximum performance is approached, the number of pipeline stages increases rapidly and the timing slack for power reduction approaches zero with the ratio of T/T_{min} approaching one (see Figure 3.6).

At a more relaxed constraint of 80 FO4 delays per instruction, the power gap between a typical ASIC and custom is only 2.8×, and the excellent ASIC is only 1.1× worse. At this point, the optimal clock period for the typical ASIC of 55 FO4 delays corresponds to about 450MHz in 0.13um with channel length of 0.08um. Typical low power embedded processors in 0.13um are slower than this, indicating that the power gap between ASIC and custom is not that large in their lower performance market niche.

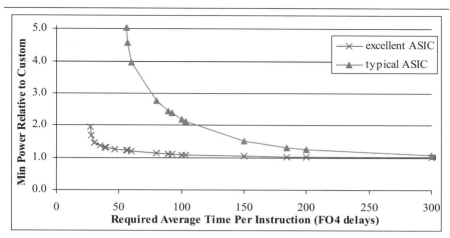

Figure 3.4 Minimum power relative to custom for the parameters listed in Table 3.7.

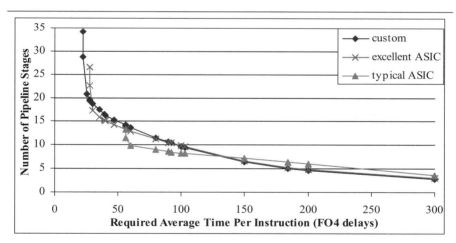

Figure 3.5 Optimal number of pipeline stages *n* to minimize power.

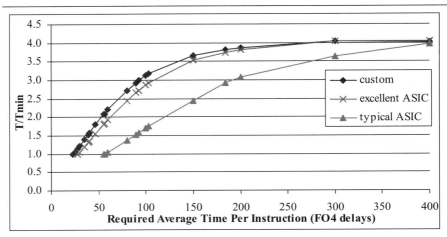

Figure 3.6 The ratio of the clock period to the minimum pipeline stage delay T/T_{min} approaches one as the performance constraint becomes tight. Consequently, there is substantially less timing slack $(T - T_{min})$ for power reduction by voltage scaling and gate sizing, and the dynamic power and the leakage power increase significantly.

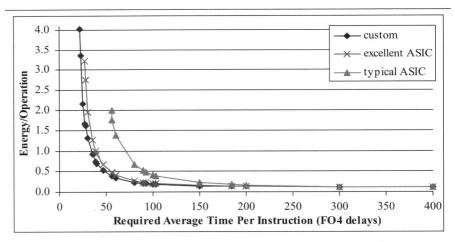

Figure 3.7 The minimum energy per operation at different performance constraints.

The minimum energy per operation for custom at maximum performance is 39× larger than the minimum energy per operation of 0.105 (see Figure 3.7). The energy/operation for the combinational logic is 12× larger, but the power for the registers and clock tree has grown from 5% of the total power with a single pipeline stage to 71% of the total power with 34.6 pipeline stages. The number of registers has increased by a factor of $34.6^{1.1} = 49.3$.

3.5   OTHER FACTORS AFFECTING
THE POWER GAP

Our discussion has focused on the impact of the required performance and the pipeline stage delay overhead on the power gap. The difference in pipeline delay overhead is the most significant cause of the power gap between ASIC and custom with pipelining.

We have examined the significance of other model parameters: glitching in complementary CMOS logic; increased CPI penalty per pipeline stage; no clock gating to avoid dynamic power during pipeline stalls; higher leakage power; and increased clock tree and register power [11]. While these factors can have a significant impact on the power consumption, they increase the power gap between ASIC and custom by at most 20%, assuming the same parameter values for both. The two other factors that have a larger impact on the gap are how much savings can be achieved with voltage scaling and gate sizing, and use of high performance logic styles in custom designs.

The power savings with voltage scaling and gate sizing depend on how steep the power-delay curve is, which depends on the range of allowable supply and threshold voltages in the process technology and the range of gate sizes in the library. The power gap due to pipelining is only 1.6× with no voltage scaling nor gate sizing. Allowing timing slack to be utilized for gate sizing can increase the power gap by 1.6× and voltage scaling can increase the power gap by 3.2× [11].

Custom designers can take advantage of domino logic or other high performance logic styles to reduce the combinational logic delay. For example, dynamic domino logic used for the 1.0GHz IBM PowerPC was 50% to 100% faster than static combinational logic with the same functionality [50]. For our model, this corresponds to reducing the unpipelined combinational logic delay from 180 to 120 FO4 delays. If the combinational logic delay is reduced to 120 FO4 delays for custom, the power gap is 7.9× between custom and a typical ASIC at maximum performance for the typical ASIC, and the power gap is 3.9× between custom and the excellent ASIC at maximum performance for the excellent ASIC. Thus the impact of *logic style* on the power gap is 1.6× for the typical ASIC and 2.0× for the excellent ASIC, but we have not accounted for the additional power that may be required for high performance logic styles.

3.6   OTHER FACTORS AFFECTING THE MINIMUM
ENERGY PER OPERATION

We concluded in Section 3.4.2 that a single pipeline stage was optimal to minimize energy/operation, but this is no longer the case when glitching and other power overheads are included in the pipeline power model.

Other causes of power consumption include the memory; off-chip communication; video display and other peripheral devices. These other power overheads do not affect the minimum power of a processor for a given performance constraint, and thus do not affect the power gap between ASIC and custom designs. However, the additional power does affect the energy per operation [11].

We will illustrate how other power overheads may be incorporated in the pipeline model by adding in glitching.

3.6.1 Glitching in complementary static CMOS logic

Different logic styles have different switching activity and some logic styles suffer from spurious signal transitions, glitches, propagating through the logic. Glitching increases the switching activity in complementary CMOS logic, but by construction glitches may not occur in dynamic logic. Glitches typically cause 15% to 20% of the switching activity in complementary CMOS logic [56].

Glitches do not propagate through edge-triggered flip-flops, providing the setup time is not violated, nor through level-sensitive latches when they are opaque. Thus pipeline registers reduce switching activity due to glitches.

Based on experimental data from a dynamic circuit timing simulator, Srinivasan et al. modeled glitching's contribution to the dynamic power of the pipeline's combinational logic as depending linearly on the logic depth [57]. A generated glitch was assumed to have a high probability of propagating through the combinational logic. While the glitching power may be fit reasonably well by a linear model over the range of pipeline depths considered by Srinivasan et al., glitching power data for pipelined 32-bit [53] and 64-bit [68] FPGA multipliers has sublinear growth with logic depth [11].

The growth of glitching with logic depth depends on a number of factors. Glitches from a gate's output may propagate through the fanout logic gates. The glitch may not propagate if it is not the controlling input of a fanout gate. If the delay of paths through the logic are unbalanced, there is more glitching [44]. Some functional blocks have more glitching than others. For example, in an inverse discrete cosine transform (IDCT) core about 37% of the power consumption in the accumulators was due to glitches, whereas glitches accounted for only 14% of the power for the chip as a whole [102].

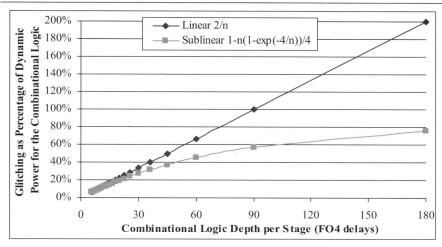

Figure 3.8 Glitching overhead estimated by the linear and sublinear glitching models in equations (3.22) and (3.23) respectively versus logic depth.

3.6.1.1   Glitching power model

To account for glitching in complementary CMOS logic, as glitching only affects the dynamic power for the combinational logic, we use [11]

$$P_{total}\left(T\right)=\frac{(1+\beta n^{\eta})}{T_{min}}\left(k_{leakage}\frac{P_{leakage}(T)}{P_{leakage}(T_{min})}\right)$$
$$+\frac{(1+\alpha_{glitching}(n)+\beta n^{\eta})}{T_{min}}\left(\frac{1}{(1+\gamma n)}\frac{P_{dynamic}(T)}{P_{dynamic}(T_{min})}(1-k_{leakage})\right) \quad (3.21)$$

where $\alpha_{glitching}(n)$ is the model for glitching as a fraction of the dynamic power due to non-spurious transitions.

In the vein of Srinivasan et al. [57], we consider a linear model for glitching with logic depth, which is inversely proportional to the number of pipeline stages,

$$\alpha_{glitching}(n)=\frac{2}{n} \quad (3.22)$$

Based on fits to glitching in 32-bit and 64-bit FPGA multipliers, we also consider a model for glitching growing sublinearly with logic depth,

$$\alpha_{glitching}(n)=1-\frac{n}{4}(1-e^{-4/n}) \quad (3.23)$$

The dynamic power overhead for the combinational logic estimated by these glitching models is shown in Figure 3.8.

The sublinear model in Equation (3.23) provides similar results to the linear model in Equation (3.22) for deeper pipelines, with less than 10% lower glitching for $n = 14$ and more pipeline stages. For shallower pipelines, the glitching estimated by the sublinear model is substantially less than that from the linear model: for a single pipeline stage, the glitching overhead is 75% from the sublinear model, and 200% from the linear model.

3.6.1.2   Impact of glitching

The maximum performance for a typical ASIC was achieved with 13 pipeline stages, by which point glitching accounts for only 15% of the dynamic power for the combinational logic and only 6% of the total power. Moreover, the optimal number of pipeline stages for custom and ASIC designs are similar (this was shown in Figure 3.5), resulting in similar contributions from glitching. Consequently, glitching increases the power gap between ASIC and custom by at most 5%.

For a single stage pipeline, glitches contribute a large percentage of the total power, and thus have a significant impact on the minimum energy per operation. More pipeline stages are optimal to prevent glitches propagating. For the linear glitching model in Equation (3.22), 5.3 pipeline stages minimizes the energy per operation. With the sublinear glitching model in Equation (3.23), 3.9 stages is optimal. The minimum energy/operation is 0.17 and 0.16 respectively, about 60% more than without glitching.

3.7   SUMMARY

ASICs have a substantially higher pipelining delay overhead than custom circuits, which reduces the benefit of additional pipeline stages and substantially reduces the timing slack available for power reduction. With pipelining to provide timing slack for power reduction, a typical ASIC with a 20 FO4 pipeline stage delay overhead may have 5.1× the power of a custom processor with only 3 FO4 delay overhead at a tight performance constraint. The power gap is less at more relaxed performance constraints, reducing to 4.0× at only 7% lower performance. The delay overhead in an ASIC can be reduced by using latches or faster pulsed flip-flops on critical paths with useful clock skew, instead of slower D-type flip-flops that don't allow slack passing between unbalanced pipeline stages. If the ASIC pipeline stage delay overhead can be reduced to 5 FO4 delays, the gap is only 1.9×.

The difference in pipeline delay overhead is the most significant cause of the power gap between ASIC and custom with pipelining. The impact of other factors such as glitching in complementary CMOS logic, increased CPI penalty per pipeline stage, no clock gating to avoid dynamic power during pipeline stalls, higher leakage power, and increased clock tree and register power is at most 20%.

Only custom designs can make use of high performance logic styles such as dynamic domino logic. If the custom design reduces the combinational logic delay with a high performance logic style, the ASIC may have up to 2.0× larger power gap at maximum performance, ignoring additional power consumption of high performance logic styles. We attribute this factor to logic style rather than microarchitecture.

Inclusion of voltage scaling and gate sizing in the pipeline model has a substantial impact on the power consumption. It is important to consider high level circuit techniques to provide timing slack along with these low level circuit techniques that can reduce power if there is timing slack. The improvements with gate sizing and voltage scaling depend greatly on the steepness of the power-delay curves, which depend on the range of allowable supply and threshold voltages in the process technology and the range of gate sizes in the library.

Pipeline model parameters can be estimated from the particular micro-architecture being considered for a design. For good estimates, the dynamic and leakage power with gate sizing and/or voltage scaling must be fit over a range of delay targets for representative circuit benchmarks in the target process technology for a design.

3.8 REFERENCES

[1] Anderson, F., Wells, J., and Berta, E., "The Core Clock System on the Next Generation Itanium™ Microprocessor," *Digest of Technical Papers of the IEEE International Solid-State Circuits Conference*, 2002, pp. 146-147, 453.
[2] ARM, ARM926EJ-S, 2006. http://www.arm.com/products/CPUs/ARM926EJ-S.html
[3] ARM, ARM1026EJ-S, 2006. http://www.arm.com/products/CPUs/ARM1026EJS.html
[4] ARM, ARM1136J(F)S, 2006. http://www.arm.com/products/CPUs/ARM1136JF-S.html
[5] ARM, ARM Cortex-A8, 2006. http://www.arm.com/products/CPUs/ARM_Cortex-A8.html
[6] ARM, ARM Cortex-A8, 2005.http://www.arm.com/products/CPUs/ARM_Cortex-A8.html
[7] Benschneider, B.J., et al., "A 300-MHz 64-b Quad-Issue CMOS RISC Microprocessor," *IEEE Journal of Solid-State Circuits*, vol. 30, no. 11, November 1995, pp. 1203-1214.
[8] Bhavnagarwala, A., et al., "A Minimum Total Power Methodology for Projecting Limits on CMOS GSI," *IEEE Transactions on VLSI Systems*, vol. 8, no. 3, June 2000, pp. 235-251.
[9] Brooks, D., et al., "Power-Aware Microarchitecture: Design and Modeling Challenges for Next-Generation Microprocessors," *IEEE Micro*, vol. 20, no. 6, 2000, pp. 26-44.
[10] Chandrakasan, A., and Brodersen, R., "Minimizing Power Consumption in Digital CMOS Circuits," in *Proceedings of the IEEE*, vol. 83, no. 4, April 1995, pp. 498-523.
[11] Chinnery, D, *Low Power Design Automation*, Ph.D. dissertation, Department of Electrical Engineering and Computer Sciences, University of California, Berkeley, 2006.
[12] Chinnery, D., et al., "Automatic Replacement of Flip-Flops by Latches in ASICs," chapter 7 in *Closing the Gap Between ASIC & Custom: Tools and Techniques for High-Performance ASIC Design*, Kluwer Academic Publishers, 2002, pp. 187-208.
[13] Chinnery, D., and Keutzer, K., *Closing the Gap Between ASIC & Custom: Tools and Techniques for High-Performance ASIC Design*, Kluwer Academic Publishers, 2002, 432 pp.
[14] Clark, L., "The XScale Experience: Combining High Performance with Low Power from 0.18um through 90nm Technologies," presented at the Electrical Engineering and

Computer Science Department of the University of Michigan, September 30, 2005. http://www.eecs.umich.edu/vlsi_seminar/f05/Slides/VLSI_LClark.pdf

[15] Clark, L., et al., "An Embedded 32-b Microprocessor Core for Low-Power and High-Performance Applications," *Journal of Solid-State Circuits*, vol. 36, no. 11, November 2001, pp. 1599-1608.

[16] Contreras, G., et al., "XTREM: A Power Simulator for the Intel XScale Core," in *Proceedings of the ACM Conference on Languages, Compilers, and Tools for Embedded Systems*, 2004, 11 pp.

[17] Dai, W., and Staepelaere, D., "Useful-Skew Clock Synthesis Boosts ASIC Performance," chapter 8 in *Closing the Gap Between ASIC & Custom: Tools and Techniques for High-Performance ASIC Design*, Kluwer Academic Publishers, 2002, pp. 209-223.

[18] Davies, B., et al., "iPART: An Automated Phase Analysis and Recognition Tool," Intel Research Tech Report IR-TR-2004-1, 2004, pp. 12.

[19] De Gelas, J. AMD's Roadmap. February 28, 2000. http://www.aceshardware.com/Spades/read.php?article_id=119

[20] Fanucci, L., and Saponara, S., "Low-Power VLSI Architectures for 3D Discrete Cosine Transform (DCT)," in *Proceedings of the International Midwest Symposium on Circuits and Systems*, 2003, pp. 1567-1570.

[21] Flynn, D., and Keating, M., "Creating Synthesizable ARM Processors with Near Custom Performance," chapter 17 in *Closing the Gap Between ASIC & Custom: Tools and Techniques for High-Performance ASIC Design*, Kluwer Academic Publishers, 2002, pp. 383-407.

[22] Furber, S., *ARM System-on-Chip Architecture*. 2nd Ed. Addison-Wesley, 2000.

[23] Ghani, T., et al., "100 nm Gate Length High Performance/Low Power CMOS Transistor Structure," *Technical digest of the International Electron Devices Meeting*, 1999, pp. 415-418.

[24] Golden, M., et al., "A Seventh-Generation x86 Microprocessor," *IEEE Journal of Solid-State Circuits*, vol. 34, no. 11, November 1999, pp. 1466-1477.

[25] Gonzalez, D., "Micro-RISC architecture for the wireless market," *IEEE Micro*, vol. 19, no. 4, 1999, pp. 30-37.

[26] Gowan, M., Biro, L., and Jackson, D., "Power Considerations in the Design of the Alpha 21264 Microprocessor," in *Proceedings of the Design Automation Conference*, 1998, pp. 726-731.

[27] Greenlaw, D., et al., "Taking SOI Substrates and Low-k Dielectrics into High-Volume Microprocessor Production," *Technical Digest of the International Electron Devices Meeting*, 2003, 4 pp.

[28] Grochowski, E., and Annavaram, M., "Energy per Instruction Trends in Intel Microprocessors," *Technology@Intel Magazine*, March 2006, 8 pp.

[29] Gronowski, P., et al., "High-Performance Microprocessor Design," *IEEE Journal of Solid-State Circuits*, vol. 33, no. 5, May 1998, pp. 676-686.

[30] Hare, C. 586/686 Processors Chart. http://users.erols.com/chare/586.htm

[31] Hare, C. 786 Processors Chart. http://users.erols.com/chare/786.htm

[32] Harstein, A., and Puzak, T., "Optimum Power/Performance Pipeline Depth," in *Proceedings of the 36th International Symposium on Microarchitecture*, 2003, pp. 117-126.

[33] Hauck, C., and Cheng, C. "VLSI Implementation of a Portable 266MHz 32-Bit RISC Core," *Microprocessor Report*, November 2001, 5 pp.

[34] Hinton, G., et al., "A 0.18-um CMOS IA-32 Processor With a 4-GHz Integer Execution Unit," *IEEE Journal of Solid-State Circuits*, vol. 36, no. 11, November 2001, pp. 1617-1627.

[35] Hinton, G., et al., "The Microarchitecture of the Pentium 4 Processor," *Intel Technical Journal*, Q1 2001, pp. 13.

[36] Hofstee, H., "Power Efficient Processor Architecture and the Cell Processor," in *Proceedings of the Symposium on High-Performance Computer Architecture*, 2005, pp. 258-262.

[37] Horan, B., "Intel Architecture Update," presented at the IBM EMEA HPC Conference, May 17, 2006.www-5.ibm.com/fr/partenaires/forum/hpc/intel.pdf

[38] Hrishikesh, M., et al., "The Optimal Logic Depth Per Pipeline Stage is 6 to 8 FO4 Inverter Delays," in *Proceedings of the Annual International Symposium on Computer Architecture*, May 2002, pp. 14-24.

[39] Intel, Intel Unveils World's Best Processor, July 27, 2006. http://www.intel.com/pressroom/archive/releases/20060727comp.htm

[40] Intel, Inside the NetBurst Micro-Architecture of the Intel Pentium 4 Processor, Revision 1.0, 2000. http://developer.intel.com/pentium4/download/netburst.pdf

[41] Keltcher, C., et al., "The AMD Opteron Processor for Multiprocessor Servers," *IEEE Micro*, vol. 23, no. 2, 2003, pp. 66-76.

[42] Kurd, N.A, et al., "A Multigigahertz Clocking Scheme for the Pentium® 4 Microprocessor," *IEEE Journal of Solid-State Circuits*, vol. 36, no. 11, November 2001, pp. 1647-1653.

[43] Larri, G., "ARM810: Dancing to the Beat of a Different Drum," presented at Hot Chips, 1996.

[44] Leitjen, J., Meerbergen, J., and Jess, J., "Analysis and Reduction of Glitches in Synchronous Networks," in *Proceedings of the European Design and Test Conference*, 1995, pp. 398-403.

[45] Lexra, Lexra LX4380 Product Brief, 2002, http://www.lexra.com/LX4380_PB.pdf

[46] Mahnke, T., "Low Power ASIC Design Using Voltage Scaling at the Logic Level," Ph.D. dissertation, Department of Electronics and Information Technology, Technical University of Munich, May 2003, pp. 204.

[47] Montanaro, J., et al., "A 160MHz, 32-b, 0.5W, CMOS RISC Microprocessor," *Journal of Solid-State Circuits*, vol. 31, no. 11, 1996, pp. 1703-1714.

[48] MTEK Computer Consulting, AMD CPU Roster, January 2002. http://www.cpuscorecard.com/cpuprices/head_amd.htm

[49] MTEK Computer Consulting, Intel CPU Roster, January 2002. http://www.cpuscorecard.com/cpuprices/head_intel.htm

[50] Nowka, K., and Galambos, T., "Circuit Design Techniques for a Gigahertz Integer Microprocessor," in *Proceedings of the International Conference on Computer Design*, 1998, pp. 11-16.

[51] Perera, A.H., et al., "A versatile 0.13um CMOS Platform Technology supporting High Performance and Low Power Applications," *Technical Digest of the International Electron Devices Meeting*, 2000, pp. 571-574.

[52] Richardson, N., et al., "The iCORE™ 520MHz Synthesizable CPU Core," Chapter 16 of *Closing the Gap Between ASIC and Custom*, 2002, pp. 361-381.

[53] Rollins, N., and Wirthlin, M., "Reducing Energy in FPGA Multipliers Through Glitch Reduction," presented at the International Conference on Military and Aerospace Programmable Logic Devices, September 2005, 10 pp.

[54] Segars, S., "The ARM9 Family – High Performance Microprocessors for Embedded Applications," in *Proceedings of the International Conference on Computer Design*, 1998, pp. 230-235.

[55] Silberman, J., et al., "A 1.0-GHz Single-Issue 64-Bit PowerPC Integer Processor," *IEEE Journal of Solid-State Circuits*, vol. 33, no.11, November 1998. pp. 1600-1608.

[56] Singh, D., et al., "Power Conscious CAD Tools and Methodologies: a Perspective," in *Proceedings of the IEEE*, vol. 83, no. 4, April 1995, pp. 570-594.

[57] Srinivasan, V., et al., "Optimizing pipelines for power and performance," in *Proceedings of the International Symposium on Microarchitecture*, 2002, pp. 333-344.

[58] Standard Performance Evaluation Corporation, SPEC's Benchmarks and Published Results, 2006. http://www.spec.org/benchmarks.html

[59] STMicroelectronics, "STMicroelectronics 0.25μ, 0.18μ & 0.12 CMOS," slides presented at the annual Circuits Multi-Projets users meeting, January 9, 2002. http://cmp.imag.fr/Forms/Slides2002/061_STM_Process.pdf

[60] techPowerUp! CPU Database, August 2006. http://www.techpowerup.com/cpudb/

[61] Tensilica, Xtensa Microprocessor – Overview Handbook – A Summary of the Xtensa Data Sheet for Xtensa T1020 Processor Cores. August 2000.

[62] Thompson, S., et al., "An Enhanced 130 nm Generation Logic Technology Featuring 60 nm Transistors Optimized for High Performance and Low Power at 0.7 – 1.4 V," *Technical Digest of the International Electron Devices Meeting*, 2001, 4 pp.

[63] TSMC, 0.13 Micron CMOS Process Technology, March 2002.

[64] TSMC, 0.18 Micron CMOS Process Technology, March 2002.

[65] TSMC, TSMC Unveils Nexsys 90-Nanometer Process Technology, August 2006. http://www.tsmc.com/english/technology/t0113.htm

[66] Tyagi, S., et al., "An advanced low power, high performance, strained channel 65nm technology," *Technical Digest of the International Electron Devices Meeting*, 2005, pp. 245-247.

[67] Weicker, R., "Dhrystone: A Synthetic Systems Programming Benchmark," *Communications of the ACM*, vol. 27, no. 10, 1984, pp. 1013-1030.

[68] Wilton, S., Ang, S., and Luk, W., "The Impact of Pipelining on Energy per Operation in Field-Programmable Gate Arrays," in *Proceedings of the International Conference on Field Programmable Logic and Applications*, 2004, pp. 719-728.

[69] Xanthopoulos, T., and Chandrakasan, A., "A Low-Power DCT Core Using Adaptive Bitwidth and Arithmetic Activity Exploiting Signal Correlations and Quantization," *Journal of Solid State Circuits*, vol. 35, no. 5, May 2000, pp. 740-750.

[70] Xanthopoulos, T., and Chandrakasan, A., "A Low-Power IDCT Macrocell for MPEG-2 MP@ML Exploiting Data Distribution Properties for Minimal Activity," *Journal of Solid State Circuits*, vol. 34, May 1999, pp. 693-703.

[71] Yang, F., et al., "A 65nm Node Strained SOI Technology with Slim Spacer," *Technical Digest of the International Electron Devices Meeting*, 2003, pp. 627-630.

[72] Zhuang, X., Zhang, T., and Pande, S., "Hardware-managed Register Allocation for Embedded Processors," in *Proceedings of the ACM Conference on Languages, Compilers, and Tools for Embedded Systems*, 2004, pp. 10.

Chapter 4

VOLTAGE SCALING

David Chinnery, Kurt Keutzer
Department of Electrical Engineering and Computer Sciences
University of California at Berkeley
Berkeley, CA 94720, USA

4.1   INTRODUCTION

Scaling the supply voltage (Vdd) and threshold voltages (Vth) to an optimal point for a design can provide substantial power savings, particularly at a relaxed performance constraint. We will examine how Vdd, Vth and gate size affect the circuit delay, dynamic power and leakage power with analytical models. We compare these models to empirical fits for a 0.13um library characterized at different Vdd and Vth values. These models help us examine the trade-off between power and delay, and determine which power reduction techniques can provide the most benefit in different situations.

In this chapter, we focus on use of a single supply voltage (Vdd) and a single threshold voltage (Vth). In Chapter 7, we will examine use of multiple supply and multiple threshold voltages in comparison to using a single Vdd and single Vth. Throughout this chapter, we assume that the NMOS and PMOS threshold voltages are of about the same magnitude, $V_{thn} = -V_{thp}$, and will generally refer to this value as the threshold voltage.

Dynamic power is due to switching capacitances and short circuit power. Switching power consumption occurs when logic switches from 0 to 1 and 1 to 0, and capacitances are charged and discharged. A short circuit current from supply to ground occurs in a gate when both the pull-up PMOS network of transistors and pull-down NMOS network of transistors are conducting. Signal glitches propagating also cause switching and short circuit power.

Leakage power occurs when logic is idle, whether the circuit is in standby or simply not switching. Leakage power is primarily due to subthreshold leakage and gate leakage.

Table 4.1 Delay and total power consumption for ISCAS'85 benchmark c7552, using PowerArc characterized 0.13um libraries with 1.2V supply voltage, and threshold voltage of 0.23V or 0.08V. Using the netlist delay minimized at 0.23V is sufficient to provide a delay minimized netlist when using the 0.08V library instead, but the power is about 10% higher.

Netlist	Delay & Power with			Total
	Vdd (V)	Vth (V)	Delay (ns)	Power (mW)
Delay minimized at Vdd=1.2V/Vth=0.23V	1.2	0.23	0.847	27.9
Delay minimized at Vdd=1.2V/Vth=0.23V	1.2	0.08	0.695	47.9
Delay minimized at Vdd=1.2V/Vth=0.08V	1.2	0.08	0.695	43.6

Delay and power data for different Vdd and Vth values was analyzed on ISCAS'85 benchmarks [2] with libraries characterized in PowerArc for STMicroelectronics' 0.13um HCMOS9D process. The comparison was at the minimum delay achievable with the particular Vdd and Vth. The power and delay data were normalized to Vdd=1.2V/Vth=0.23V data, and then averaged across the netlists. There was little variation in the normalized data at the same Vdd and same Vth across the netlists [6]. To provide a range of Vth values for the characterization with PowerArc, the zero bias threshold voltage parameter vth0 [1] in the SPICE technology files was adjusted.

The netlists were delay minimized in Design Compiler with the Vdd=1.2V and Vth=0.23V library. To minimize the delay, it was not necessary to resize the circuits when using different Vdd and Vth values, as illustrated in Table 4.1. The wire loads were 3+2×#fanouts fF, and output port loads were 3fF excluding the load of the wire to the port.

The delay dependence on Vdd and Vth is discussed in Section 4.2. We then examine switching power, short circuit power and leakage power in sections 4.3, 4.4 and 4.5. The net power consumption and how to choose Vdd and Vth to minimize it for a given delay constraint is detailed in Section 4.6. These trade-offs are summarized in Section 4.7.

4.2   DELAY

The delay for an input transition to cause a transition at a gate's output is due to the time it takes to charge or discharge the load capacitance and the gate internal capacitances. There are also wire RC delays, but these contribute less than 1% of the critical path delay for our small combinational benchmarks where wires are not that long.

A simple model for the delay can be derived from the saturation drain current through a transistor. The saturation current dependence on input voltage can be modeled with the Sakurai-Newton alpha-power law [12]

$$I_{saturation\ drain\ current} = c\mu \frac{\varepsilon_{ox}}{t_{ox}} \frac{W}{L_{eff}} (V_{in} - V_{th})^{\alpha} \qquad (4.1)$$

where c is a constant; μ is the charge carrier mobility; ε_{ox} is the electric permittivity; t_{ox} is the gate oxide thickness; L_{eff} is the effective transistor channel length; W is the transistor gate width (size); V_{in} is the driving voltage to the NMOS transistor gate; and α is the velocity saturation index which depends on the technology, and is between 1 and 2. A value for α of about 1.3 is typical for today's technologies.

Using the saturation drain current from Equation (4.1), the delay d for charging a capacitance C from 0V to V_{dd} may be approximated as [14]

$$d = CV_{dd} / I_{saturation\ drain\ current} = k \frac{CV_{dd}}{W(V_{dd} - V_{th})^{\alpha}} \qquad (4.2)$$

where k is a constant, and W is the width of the transistor through which current is flowing to charge the capacitor. From Equation (4.2), the delay scaling with Vdd and Vth is [14]

$$\text{Delay scaling factor from } V_{dd1} \& V_{th1} \text{ to } V_{dd2} \& V_{th2} = \frac{V_{dd2}}{V_{dd1}} \frac{(V_{dd1} - V_{th1})^{\alpha}}{(V_{dd2} - V_{th2})^{\alpha}} \qquad (4.3)$$

A gate's delay is reduced as its size increases, but increases as the size of fanout gates and thus their capacitance increases. A gate also has internal capacitances which contribute an additional "parasitic delay" as it is termed in "logical effort" delay models [14][15]. The delay also increases as the supply voltage Vdd is reduced and as the threshold voltage Vth is increased.

Excluding the driving gate from analysis by using a fixed input voltage ramp, as a gate is upsized to reduce the delay, the reduction in gate delay is less for larger gate sizes due to the parasitic delay. The dynamic power for charging and discharging the internal capacitances increases linearly with gate size. This results in the classic power-delay "banana" curve shown in Figure 4.1. Including the driving gate in analysis results in a delay increase for larger gate sizes due to the load on the driving gate, as shown in Figure 4.2. Thus if a gate is upsized, its fanins may also need to be upsized.

By comparing the impact of gate size, threshold voltage, and supply voltage on delay and these power terms, we see that there are significant delay-power trade-offs that must be carefully analyzed. To minimize power, it is not clear that reducing supply voltage to reduce switching power, or increasing threshold voltage to reduce leakage power, is more important than reducing gate size. All of these choices increase the circuit delay, except for gate downsizing, which may increase delay or reduce delay as the reduced load on preceding gates reduces their delay. Optimization approaches which favor one technique, typically Vdd, over the others will be suboptimal in situations where the power-delay sensitivity to this technique is less. As noted by Brodersen et al. [3], reducing the threshold voltage or increasing the supply voltage, subject to process constraints, to provide slack for gate downsizing can sometimes give better overall power savings.

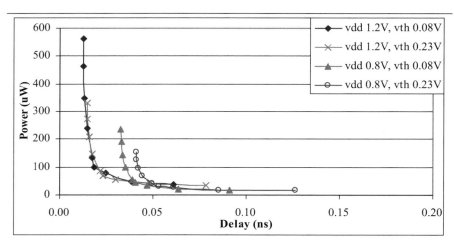

Figure 4.1 Power versus delay for inverter cells from the 0.13um PowerArc characterized libraries. The input signal was a ramp with 0.1ns slew. Each point on a curve is for a different gate size, and larger gate sizes have larger power consumption. The load capacitance was 8fF, and the switching frequency was 4GHz, where dynamic power dominates leakage power.

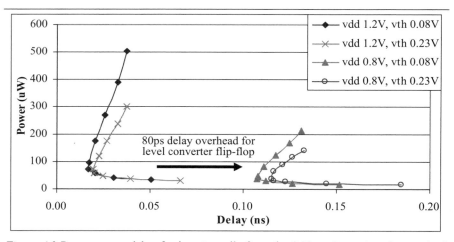

Figure 4.2 Power versus delay for inverter cells from the 0.13um PowerArc characterized libraries. The driving gate was included in the power and delay analysis. In addition for the 0.8V Vdd cells, there is a level converter flip-flop delay overhead of 80ps, but no power overhead, for voltage level restoration to drive the output at 1.2V. Each point on a curve is for a different gate size, and larger gate sizes have larger power consumption. The load capacitance was 8fF, and the switching frequency was 4GHz.

Table 4.2 This table lists the average error magnitude for the delay fits using the analytical model in Equation (4.3) and the empirical fit in Equation (4.4). Vth was 0.08V, 0.12V, 0.14V and 0.23V.

Delay Fit	Mean error magnitude vs. Vdd 0.8V & 1.2V delay data	Mean error magnitude vs. Vdd 0.5V, 0.8V & 1.2V delay data
Analytical model, α=1.30	7.3%	17.0%
Analytical model, α=1.66	1.5%	6.0%
Analytical model, α=1.83	4.8%	4.3%
Empircal fit, α=1.10	0.6%	0.8%

4.2.1   Empirical fit to 0.13um delay data

We will now discuss fitting the delay to 0.13um data at a range of Vdd and Vth values. Several cells were characterized incorrectly with Vdd of 0.5V, so we present delay fits with and without the 0.5V Vdd library.

The derivation of the analytical delay model in equations (4.2) and (4.3) ignores signal slew and other factors. Assuming α of 1.3 for today's technologies for the saturation drain current may be reasonable, but does not give a good fit with Equation (4.3) for the delay. The delay is underestimated by up to 19.4% at Vdd of 0.8V and by up to 52.3% at Vdd of 0.5V.

A least squares fit of Equation (4.3) to the 1.2V and 0.8V Vdd delay data gives a value for α of 1.66. The α=1.66 fit underestimates the 0.5V Vdd delays by up to 24.4%. Including the 0.5V Vdd data in the least squares fit, gives a value for α of 1.83, but the fit errors are still up to 6.6%.

Thus delay scaling with Equation (4.3) in [14] and other papers is at best somewhat inaccurate if α is fitted correctly, and otherwise wrong when a value for α of 1.2 to 1.3 is assumed typically.

From analysis of the derivation in the Sakurai-Newton alpha power law delay model [12] and experimenting with different fits, the best fit to normalized delay was given by

$$d = 0.587 \frac{V_{dd}}{(V_{dd} - V_{th})^{\alpha+1}} + 0.241(V_{dd} - V_{th})^{\alpha} \qquad (4.4)$$

with a value for α of 1.101. This fit had an average error magnitude of 0.8% and the maximum error was only 2.1%. The accuracy of the fits is summarized in Table 4.2.

As the supply voltage is reduced, if the threshold voltage is kept constant, the delay begins to increase rapidly. Whereas if both Vdd and Vth are scaled down, then the delay does not increase as much. This is illustrated in the graph of the normalized delay given by Equation (4.4) in Figure 4.3. For example at Vdd=0.8V/Vth=0.08V, the delay is only 10% worse than at Vdd=1.2V/Vth=0.23V.

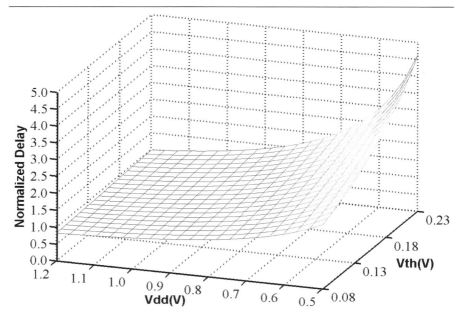

Figure 4.3 Graph of the fit in Equation (4.4) for the minimum delay versus supply voltage and threshold voltage, normalized versus the circuit delays with Vdd=1.2V/Vth=0.23V.

4.3   SWITCHING POWER

As digital logic performs computations, logic transitions occur between 0 and 1, charging circuit capacitances to a high voltage, or discharging them back to a low voltage. The switching power for charging and discharging a capacitance C to a voltage V_{dd} and back to 0V with switching frequency f is

$$P_{switching} = \frac{1}{2}CV_{dd}^{2}f \qquad (4.5)$$

Capacitances in the circuit include internal "parasitic" capacitances within each gate, capacitances of the input pins of each gate, and wire capacitances. In standard cell library characterization, switching power for the internal gate capacitances is included with the short circuit power in the "internal power" of a gate. A gate's internal capacitances and input pin capacitances depend on the transistor width (i.e. the gate size).

With the quadratic dependence of switching power on supply voltage V_{dd} in Equation (4.5), reducing the supply voltage is often seen as the most effective way to reduce dynamic power. However at a tight delay constraint, reducing gate sizes can provide a greater power reduction: as each gate loads its fanins, if a gate is upsized to reduce delay, then its fanins must in turn be upsized to prevent their delay increasing. Consequently, the gate sizes and power increase rapidly as delay is reduced towards a tight delay constraint. Conversely, if timing slack is available near a tight delay constraint, reducing a gate's size allows fanins to be reduced in size, and substantial power savings may be achieved.

4.3.1 Empirical fit to 0.13um input pin capacitance

In a combinational complementary CMOS circuit, the switching power is the power consumed charging and discharging the wire capacitances, output port capacitances and gate input pin capacitances. The power for (dis)charging gate internal capacitances is included in the internal power (see Section 4.4.1). Gate input pin capacitances vary with Vdd and Vth.

By setting the wire loads and output port loads to zero, we determine that wire loads contribute 37% and the output ports contribute 1% of the switching power, excluding switching of gate internal capacitances, on average across the ISCAS'85 benchmarks with the Vdd=1.2V/Vth=0.23V library. The delay is also 25% more with wire loads versus no wire loads. We measured the switching power at the original clock frequency to factor out the delay change.

Given the switching power due to the output port and wire loads, the remainder of the switching power is due to the gate input pin capacitances. We can then determine how gate input pin capacitance C_{in} varies with Vth and Vdd, by dividing by the V_{dd}^2 term in Equation (4.5). A least squares fit with a first order Taylor series gives

$$\text{Normalized input pin capacitance } C_{in} = 0.957 + 0.200 V_{dd} - 0.859 V_{th} \quad (4.6)$$

where C_{in} was normalized to 1.0 at Vdd=1.2V and Vth=0.23V. This fit has an average error magnitude of 1.0% and maximum error of 2.6%.

The increase in gate capacitance with decreases in Vth has been identified previously [13][17], but the dependence of C_{in} on Vdd has not generally been discussed in multi-Vdd optimization research. However, the reduction in Vth with increased Vdd due to drain induced barrier lowering (DIBL) [11] is well known from a process standpoint. In our 0.13um data, there was up to a 22% C_{in} increase if Vdd was increased from 0.5V to 1.2V, and up to a 20% increase in C_{in} if Vth was reduced from 0.23V to 0.08V [6].

4.4   SHORT CIRCUIT POWER

Short circuit power is dissipated when there is a current from supply to ground in a gate when both the pull-up PMOS network of transistors and pull-down NMOS network of transistors are conducting. In 1984, Veendrick [16] derived the short circuit current for an inverter without load. He assumed that the saturation drain current has the form in Equation (4.1) with α of 2, resulting in average short circuit current of [16]

$$I_{short\ circuit} = c\mu \frac{\varepsilon_{ox}}{t_{ox}} \frac{W}{L_{eff}} \frac{1}{V_{dd}} (V_{dd} - 2V_{th})^3 s_{in} \tag{4.7}$$

where c is a constant, and s_{in} is the input slew.

A value for α of about 1.3 is typical for today's technologies to model the saturation drain current with Equation (4.1). Following a similar derivation to Veendrick, the short circuit power without load in terms of the optimization variables of direct interest to us is

$$P_{short\ circuit} = cfW(V_{dd} - 2V_{th})^{\alpha+1} \tag{4.8}$$

where c is a constant and f is the switching frequency.

Consequently, the short circuit power is between quadratically and cubically dependent on the supply voltage, depending on the value of α. The short circuit power is linearly dependent on the gate size, and increases as the threshold voltage is reduced.

The impact of slew is not included in Equation (4.8). As noted by Veendrick [16], the short circuit power contributes only a minor portion of the dynamic power when the input slews to the gate and the gate's output slew are similar. This minimizes the duration when both PMOS and NMOS transistor chains are conducting. A circuit's short circuit power is minimized if the input slews to a gate and its output slews are equal, and the short circuit power increases rapidly as input slew increases relative to output slew [4]. As input slews are usually similar to a gate's output slew, short circuit power typically contributes less than 10% of the dynamic power [5]. However, this can be an underestimate for low threshold voltages.

4.4.1   Empirical fit to 0.13um internal power data

The internal power has two components: the short circuit power, and the switching power for internal "parasitic" capacitances. For a fit to the internal power, we expect terms of the form $(V_{dd} - 2V_{th})^\alpha$ for the short circuit power from Equation (4.8) and a V_{dd}^2 term for the switching power of the gate internal capacitances from Equation (4.5). A good fit was provided by

$$\begin{aligned} P_{internal} &= P_{switching\ internal\ capacitances} + P_{short\ circuit} \\ &= f\left(0.413V_{dd}^2 + 0.958(V_{dd} - 2V_{th})^{3.183} + 0.039(V_{dd} - 2V_{th})^{0.162}\right) \end{aligned} \tag{4.9}$$

The fit has an average error magnitude of 0.4% and maximum error of 1.0%.

The first V_{dd}^2 term in Equation (4.9) accounts for the switching of internal capacitances. Multiplying the V_{dd}^2 term by a first order Taylor series in Vdd and Vth to consider the dependence of internal capacitance on Vth and Vdd, in the manner of Equation (4.6), does not improve the fit substantially. This suggests that gate internal capacitance is independent of Vth and Vdd.

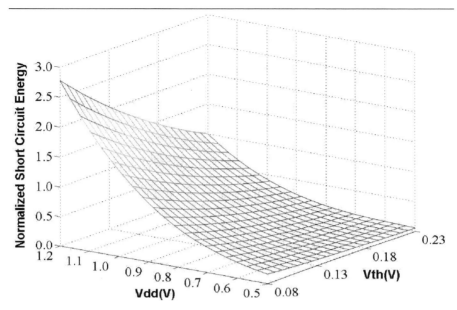

Figure 4.4 Graph of the short circuit energy ($V_{dd} - 2V_{th}$) terms in Equation (4.9), normalized versus the short circuit energy at Vdd=1.2V/Vth=0.23V.

The second term in Equation (4.9) accounts for the majority of the short circuit power, while the last term provides a slight correction to the short circuit power. The fitted exponents for the short circuit power terms do not correspond to what we might expect from Equation (4.8), if α is about 1.3, but Equation (4.8) does not include the effect of input slew and load capacitance which introduce higher order terms.

The short circuit energy is graphed in Figure 4.4. As the threshold voltage is reduced, the short circuit power grows significantly. The short circuit energy approaches zero as V_{dd} approaches $2V_{th}$, as predicted by Veendrick [16], because at $V_{dd} = 2V_{th}$ NMOS and PMOS transistors cannot be "on" (i.e. $V_{GS} > V_{th}$) simultaneously, so there is no short circuit current. The short circuit power reduces substantially as Vdd decreases, by about a factor of 10× as Vdd is reduced from 1.2V to 0.6V at Vth=0.08V. The short circuit power is reduced by this large factor due to the smaller period of time when both pull-up and pull-down transistor chains are conducting.

4.5   LEAKAGE POWER

In deep submicron process technologies with low threshold voltages, the dominant sources of leakage power are subthreshold leakage and gate leakage. Subthreshold leakage occurs when the transistor gate-source voltage V_{GS} is below the threshold voltage V_{th}, and the minority carrier concentration varies across the MOSFET channel causing a diffusion current between drain and source. Gate leakage is due to a high electric field across the thin transistor gate oxide, which results in tunneling of electrons through the transistor gate oxide [11].

An analytical model for the subthreshold leakage current is [7]

$$I_{subthreshold\ leakage} = c\mu \frac{\varepsilon_{ox}}{t_{ox}} \frac{W}{L_{eff}} \left(\frac{kT}{q}\right)^2 e^{q(V_{GS}-V_{th0}-\gamma V_b+\eta V_{DS})/mkT} \left(1 - e^{-qV_{DS}/kT}\right) \quad (4.10)$$

where c is a constant; k is Boltzmann's constant; q is the charge of an electron; T is the temperature; V_{DS} is the transistor drain to source voltage; V_{GS} is the transistor gate to source voltage; V_{th0} is the zero bias threshold voltage, V_b is the body bias voltage; γ is the linearized body effect coefficient; m is the subthreshold swing coefficient; and η is the drain-induced barrier lowering (DIBL) coefficient.

The subthreshold leakage increases rapidly as the temperature increases. At 25°C, The "ideal" subthreshold slope is $\ln(10)kT/q$ = 60mV/decade if m is 1 in Equation (4.10), though in real processes the subthreshold slope is worse (more) than this [10].

Simplifying the expression in Equation (4.10) to the optimization variables of interest to us, the subthreshold leakage power for an NMOS transistor in an inverter is approximately

$$P_{subthreshold\ leakage} = c_1 V_{dd} W e^{c_2(-V_{th0}+\eta V_{dd})} \quad (4.11)$$

where c_1 and c_2 are constants; the input voltage V_{GS} = 0V when it is "off" and leaking; $V_{DS} = V_{dd}$; and V_b = 0V assuming there is no body bias applied. The subthreshold leakage increases exponentially as the threshold voltage V_{th} is reduced. The subthreshold leakage increases linearly with gate size (transistor width W) and with the supply voltage V_{dd}. However, Vdd increases leakage further due to the drain-induced barrier lowering term.

Gate oxide tunneling leakage current I_{ox} can be modeled as [8]

$$I_{ox} = aL_{eff}e^{(bV_{GS}-cT_{ox}^{-2.5})} + aL_{eff}e^{(bV_{GD}-cT_{ox}^{-2.5})} \quad (4.12)$$

where a, b and c are constants; L_{eff} is the effective channel length; V_{GS} is the gate to source voltage; V_{GD} is the gate to drain voltage; and t_{ox} is the transistor gate oxide thickness. The gate leakage increases exponentially as the gate oxide thickness is reduced, and is significant in process technologies below 90nm, in some cases exceeding the subthreshold leakage.

Both subthreshold leakage and gate leakage vary significantly depending on the input state to the logic gate. Subthreshold leakage is largest when the leakage current path from Vdd to ground has only one transistor that is off. This stack effect can be used to reduce the subthreshold leakage, for example with sleep transistors for power gating as discussed in Chapter 10.

The gate tunneling leakage is largest when $V_{GS} = V_{GD} = V_{dd}$, and decreases rapidly when V_{GS} and V_{GD} are reduced [8].

4.5.1   Empirical fit to 0.13um leakage power data

In STMicroelectronics' 0.13um HCMOS9D process, subthreshold leakage is by far the most significant component of leakage power, but gate leakage is also included in the technology models. The leakage power increases exponentially as the threshold voltage is reduced. For our 0.13um data, leakage increases by about 56× as Vth is reduced from 0.23V to 0.08V, and increases by about 3× as Vdd is increased from 0.6V to 1.2V [6].

Fitting the analytical model for subthreshold leakage in Equation (4.11) to the leakage data normalized to the value at Vdd=1.2V/Vth=0.23V, we get

$$P_{leakage} = 158V_{dd}e^{-26.9V_{th}+0.770V_{dd}} \qquad (4.13)$$

This fit has an average error magnitude of 2.6% and a maximum error of 6.0%. This is quite a good fit considering that the function is exponential.

From the fitted V_{th} coefficient in Equation (4.13), we can determine the subthreshold leakage slope at 25°C

$$\text{Subthreshold leakage slope} = \frac{\ln 10}{26.9} = 86\text{mV/decade} \qquad (4.14)$$

The subthreshold slope of 86mV/decade is about what we expect.

4.6   0.13um DATA FOR TOTAL POWER

From the empirical fits and the analysis of the components of switching energy and internal energy, we can determine the individual contributions to the total energy, as shown in Figure 4.5. The leakage power was normalized to about 1% of the total power at Vdd=1.2V/Vth=0.23V at the minimum delay. At high Vdd and high Vth, the majority of the energy is due to switching of capacitances and short circuit current. If Vdd is scaled down to reduce the dynamic power, Vth must be reduced to avoid excessive delay, but then the leakage power becomes large.

The majority of the dynamic power is due to the switching power for the gate internal capacitances, gate input pin capacitances and wire loads. About 42% of the switching power is for (dis)charging the gate input pin capacitances, 32% is due to the gate internal capacitances, and 26% is due to the wire loads. These percentages vary up to ±3% with Vdd and Vth, as gate input pin capacitance varies with Vdd and Vth. Switching of the output ports contributes less than 1% of the switching power, as we assumed only a small output port load of 3fF. However in circuits with buses or chip outputs, these capacitances can be much larger and then I/O, receiving/sending input/output data, contributes a significant portion of the chip power.

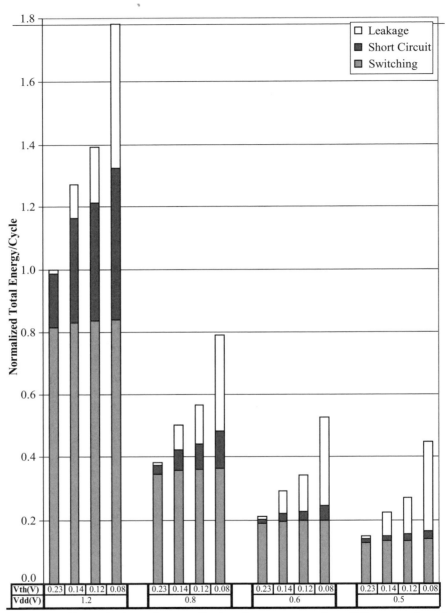

Figure 4.5 This graph shows the normalized total energy per cycle at different supply and threshold voltages, with the breakdown into leakage, short circuit and switching energy.

Leakage power is very significant at low threshold voltages. If threshold voltages are reduced too much, the majority of the power consumption is due to leakage, as leakage increases exponentially with reduction in Vth.

Short circuit power can be quite significant at high Vdd and low Vth – it contributes up to 27.2% of the total power at Vdd=1.2V/Vth=0.08V. Typically, circuit designers say that short circuit power contributes around 10% of total power, and a common assumption in optimization research to ignore the impact of short circuit power. However, our analysis indicates that it is important to include short circuit power when considering lower Vth.

While the power savings by reducing Vdd and increasing Vth can be huge, the accompanying delay increase must also be considered. For example, there is a 20× power reduction going from Vdd=1.2V/Vth=0.14V to Vdd=0.6V/Vth=0.23V, but the delay increases by 3.3×. Assuming no delay constraint, the appropriate metric to use is the energy. There is only a 6× energy reduction from Vdd=1.2V/Vth=0.14V to Vdd=0.6V/Vth=0.23V, as can be seen in Figure 4.5. Note that this comparison is somewhat simplistic as we haven't considered power minimization with gate sizing yet. Instead of reducing Vdd and increasing Vth, the timing slack could be used for gate downsizing to reduce gate internal capacitances and gate input pin capacitances. Thus the actual benefits of Vdd=0.6V/Vth=0.23V may be significantly less.

4.6.1 Optimal Vdd and Vth to minimize the total power

We can now determine the optimal Vdd and Vth to minimize power consumption when meeting a given delay constraint from the delay and power fits for the 0.13um technology. Combining equations (4.4), (4.6), (4.9) and (4.13), we minimize the total power P_{total} with

minimize P_{total}

subject to $T = T_{max}$

$$T = \frac{0.587V_{dd}}{(V_{dd} - V_{th})^{2.101}} + 0.241(V_{dd} - V_{th})^{1.101}$$

$$P_{leakage} = 158V_{dd}e^{-26.7V_{th}+0.770V_{dd}}$$

$$E_{internal} = 0.413V_{dd}^2 + 0.958(V_{dd} - 2V_{th})^{3.183}$$
$$+ 0.039(V_{dd} - V_{th})^{0.162}$$

$$E_{switching} = \begin{pmatrix} 0.62(0.957 + 0.200V_{dd} - 0.859V_{th}) \\ +0.37 + 0.01 \end{pmatrix} V_{dd}^2$$

$$\qquad\qquad (4.15)$$

$$P_{total} = 0.011P_{leakage} + \frac{1}{T}(0.430E_{internal} + 0.559E_{switching})$$

where T is the critical path delay; T_{max} is the delay constraint; $P_{leakage}$ is the leakage power; $E_{internal}$ is the energy/cycle from short circuit currents and switching of internal capacitances; and $E_{switching}$ is the switching energy of transistor gates (62%), wire loads (37%) and output port loads (1%).

While the optimal Vdd and Vth depend on the process technology, conventional wisdom based on theoretical analysis by Nose and Sakurai [9] suggests that leakage should contribute 30% of total power to minimize the total power consumption, independent of the process technology, switching activity and delay constraint. They ignored short circuit current, the dependence of transistor gate capacitance on Vdd and Vth, and the impact of drain-induced barrier lowering on leakage. The largest inaccuracy in their analysis was scaling delay with Equation (4.2) and α of 1.3, which underestimates delay by up to 50% at low Vdd (see Section 4.2.1).

In contrast to Nose and Sakurai's result, to minimize the total power for our 0.13um data, we find that leakage contributes from 8% to 21% of the total power depending on the delay constraint and how much leakage there is versus dynamic power. To model the effect of different activities and process technologies, thus the amount of leakage, we considered three scenarios for leakage power with 0.1×, 1× and 10× the leakage in Equation (4.15). For each order of magnitude increase in the weight on leakage, the optimal Vth averages 0.095V higher and the optimal Vdd averages 0.17V higher, reducing leakage by a factor of 9.0× which mostly cancels out the 10× weight increase, in exchange for a 40% increase in dynamic power on average (see Figure 4.6). The total power consumption is also 40% higher.

Leakage contributes more power at a tight delay constraint, as shown in Figure 4.7, because a lower Vth must be used to meet the delay constraint. As the delay constraint is relaxed, the timing slack enables an exponential reduction in the leakage power by increasing Vth. The minimum contribution from leakage occurs at a delay constraint of about 0.9. When the delay constraint is relaxed further, Vdd is reduced faster than the increase in Vth (see Figure 4.8), because of the exponential dependence of leakage power on Vth. The contribution of leakage slowly increases as the delay constraint is increased beyond 0.9, because the switching activity decreases as $1/T$, reducing dynamic power but not leakage. In the scenario where there is more leakage, at a tight delay constraint the percentage of leakage is larger because Vth must be lower, but at more relaxed delay constraints the percentage of leakage is less because Vdd is higher and there is more dynamic power.

In sequential circuitry, the optimal portion of total power from leakage may be higher as dynamic power in idle units can be avoided by clock gating, but power gating and other leakage reduction methods can only be used in standby mode.

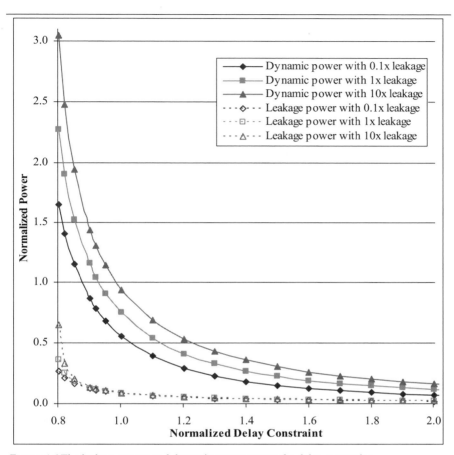

Figure 4.6 The leakage power and dynamic power versus the delay constraint.

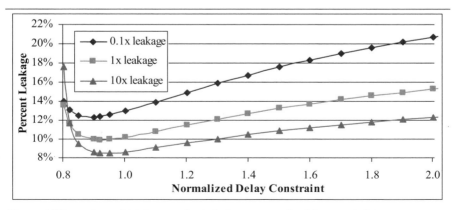

Figure 4.7 Percentage of total power due to leakage versus the delay constraint.

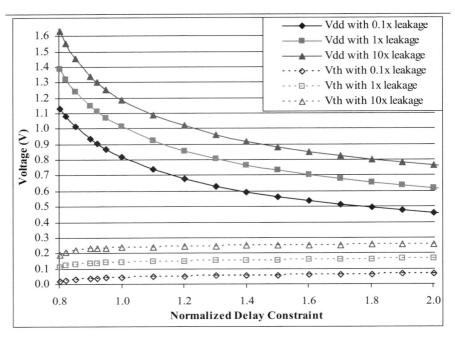

Figure 4.8 Optimal Vth and Vdd to minimize the total power versus the delay constraint.

4.7   SUMMARY

Having examined the power and delay trade-offs, let us summarize the power minimization approaches. *Gate downsizing* reduces the gate internal capacitances and gate input pin capacitances, thus reducing switching power. Reducing the gate size also gives an approximately linear reduction in leakage and short circuit power, due to the higher transistor resistances.

Reducing the supply voltage provides a quadratic reduction in switching power, and also provides substantial reductions in short circuit power and leakage power. *Increasing the threshold voltage* exponentially reduces the leakage power and also reduces the short circuit power. In the 0.13um techno-logy, the leakage at Vth of 0.08V is about 56× the leakage at 0.23V Vth, and the leakage at Vdd of 1.2V is about 3× the leakage at 0.6V.

Both an *increase in Vdd* and a *decrease in Vth* increase the gate input pin capacitance C_{in}, which increases the delay and the switching power. C_{in} increases by 22% if Vdd is increased from 0.5V to 1.2V at Vth of 0.23V, and it increases by 20% if Vth is reduced from 0.23V to 0.08V at Vdd of 0.5V. The dependence of C_{in} on Vdd has not generally been mentioned in other multi-Vdd optimization research, though it can be as significant as the effect of Vth on C_{in}. Other optimization research has noted the impact of Vth on C_{in} [13][17]. In contrast, we found that gate internal capacitances did not depend significantly on Vth or Vdd. Thus reducing supply voltage and increasing threshold voltage provide some additional power reduction by reducing the gate input capacitance.

Except for gate downsizing in some situations, these power minimization approaches come with a delay penalty. If Vth is scaled with Vdd then delay is inversely proportional to Vdd. However, Vth scaling is limited by the exponential increase in leakage power as Vth is reduced. Thus the delay may increase substantially when Vdd is reduced. To avoid the delay penalty for low Vdd and high Vth, we can use high Vdd and low Vth on critical paths, and use low Vdd and high Vth elsewhere to reduce the power. Chapters 7 and 8 will examine the power savings that can be achieved with use of multiple supply voltages and multiple threshold voltages.

Optimization researchers often exclude wire loads and short circuit power to simplify analysis; however, we found that wire loads can contribute 24% of the total power, and short circuit power can account for up to 27% of the total power at high Vdd and low Vth. The wire loads also increase the critical path delay, by 25% on average with the Vdd=1.2V/Vth=0.23V library. Typically, circuit designers say that short circuit power contributes around 10% of total power. However, our analysis indicates that it is important to include short circuit power when considering lower Vth.

From the empirical fits for the delay and power, the optimal Vdd and Vth to minimize the total power consumption can be determined. For example, the optimal Vdd is 1.0V and the optimal Vth is 0.14V for a delay constraint of 1.0; and the optimal Vdd is 0.86V and the optimal Vth is 0.15V for a delay constraint of 1.2, where delays have been normalized to the delay with Vdd of 1.2V and Vth of 0.23V in the 0.13um process technology.

The analysis for the optimal Vdd and Vth does not consider that additional timing slack may be used for gate downsizing. In Chapter 7, we consider selection of Vdd and Vth with gate sizing, but Vdd is limited to 0.6V, 0.8V or 1.2V, and Vth is limited to 0.08V, 0.14V or 0.23V. Without gate sizing, our analysis would predict that Vdd of 1.2V and Vth of 0.23V are the best choice for delay of 1.0, and that Vdd of 0.8V and Vth of 0.08V are the best choice for delay of 1.2. The optimal Vdd is still 1.2V at a delay constraint of 1.0, but the optimal Vth is lower, 0.14V (see Table 7.3), providing timing slack of 12% of the clock period for gate downsizing, compared to no timing slack with Vth of 0.23V. At a delay constraint of 1.2, the optimal Vdd and Vth remain respectively 0.8V and 0.08V (see Table 7.7 with 0.8V input drivers) as there is sufficient timing slack, 8% of the clock period, for gate downsizing.

Conventional wisdom based on theoretical analysis by Nose and Sakurai [9] suggests that leakage should contribute 30% of total power when Vdd and Vth are chosen optimally to minimize the total power consumption, independent of the process technology, switching activity and delay constraint. Choosing Vdd and Vth optimally to minimize the total power with the empirical fits to 0.13um data, we found that leakage contributes from 8% to 21% of the total power depending on the delay constraint and how much leakage there is, thus depending on the process technology and switching activity. However, the possible Vdd and Vth values depend on the particular process technology and available standard cell libraries. For example for the delay constraint of 1.2 with the best library choice with Vdd of 0.8V and Vth of 0.08V, leakage contributed on average 40% of the total power.

4.8   REFERENCES

[1] Avant!, *Star-Hspice Manual*, 1998, 1714 pp.

[2] Brglez, F., and Fujiwara, H., "A neutral netlist of 10 combinational benchmark circuits and a target translator in Fortran," in *Proceedings of the International Symposium Circuits and Systems*, 1985, pp. 695-698.

[3] Brodersen, R., et al., "Methods for True Power Minimization," in *Proceedings of the International Conference on Computer-Aided Design*, 2002, pp. 35-42.

[4] Burd, T. "Low-Power CMOS Library Design Methodology," M.S. Report, University of California, Berkeley, UCB/ERL M94/89, 1994, 78 pp.

[5] Chandrakasan, A., and Brodersen, R., "Minimizing Power Consumption in Digital CMOS Circuits," in *Proceedings of the IEEE*, vol. 83, no. 4, April 1995, pp. 498-523.

[6] Chinnery, D., *Low Power Design Automation*, Ph.D. dissertation, Department of Electrical Engineering and Computer Sciences, University of California, Berkeley, 2006.

[7] De, V., et al., "Techniques for Leakage Power Reduction," chapter in *Design of High-Performance Microprocessor Circuits*, IEEE Press, 2001, pp. 48-52.

[8] Lee, D., Blaauw, D., and Sylvester, D., "Gate Oxide Leakage Current Analysis and Reduction for VLSI Circuits," IEEE Transactions on VLSI Systems, vol. 12, no. 2, 2004, pp. 155-166.

[9] Nose, K., and Sakurai, T., "Optimization of V_{DD} and V_{TH} for Low-Power and High-Speed Applications," in *Proceedings of the Asia South Pacific Design Automation Conference*, 2000, pp. 469-474.

[10] Rabaey, J.M., *Digital Integrated Circuits*. Prentice-Hall, 1996.

[11] Roy, K., Mukhopadhyay, S., and Mahmoodi-Meimand, H., "Leakage Current Mechanisms and Leakage Reduction Techniques in Deep-Submicrometer CMOS Circuits," in *Proceedings of the IEEE*, vol. 91, no. 2, 2003, pp. 305-327.

[12] Sakurai, T., and Newton, R., "Delay Analysis of Series-Connected MOSFET Circuits," *Journal of Solid-State Circuits*, vol. 26, no. 2, February 1991, pp. 122-131.

[13] Sirichotiyakul, S., et al., "Stand-by Power Minimization through Simultaneous Threshold Voltage Selection and Circuit Sizing," in *Proceedings of the Design Automation Conference*, 1999, pp. 436-441.

[14] Stojanovic, V., et al., "Energy-Delay Tradeoffs in Combinational Logic Using Gate Sizing and Supply Voltage Optimization," in *Proceedings of the European Solid-State Circuits Conference*, 2002, pp. 211-214.

[15] Sutherland, I., Sproull, R., and Harris, D., Logical Effort: Designing Fast CMOS *Circuits*, Morgan Kaufmann, 1999.

[16] Veendrick, H., "Short-circuit dissipation of static CMOS circuitry and its impact on the design of buffer circuits," *Journal of Solid-State Circuits*, vol. SC-19, Aug. 1984, pp. 468-473.

[17] Wang, Q., and Vrudhula, S., "Algorithms for Minimizing Standby Power in Deep Submicrometer, Dual-Vt CMOS Circuits," IEEE Transactions on Computer-Aided *Design of Integrated Circuits and Systems*, vol. 21, no. 3, 2002, pp. 306-318.

Chapter 5

METHODOLOGY TO OPTIMIZE ENERGY OF COMPUTATION FOR SOCS

Jagesh Sanghavi, Eliot Gerstner
Tensilica
3255-6 Scott Boulevard
Santa Clara, CA 95054
sanghavi@tensilica.com, gerstner@tensilica.com

We present a novel energy optimization methodology based on processor customization. Unlike previous approaches focused either on behavioral-level optimization with approximate consideration for underlying hardware, or register transfer level (RTL), or gate-level power optimization with limited microarchitectural trade-offs, the new approach compiles cycle count reducing instruction extension description to synthesizable hardware and accurately estimates dynamic power at the register transfer level. For a sample set of digital signal processing (DSP) applications, we see energy reductions exceeding a factor of 10× compared to fixed instruction set processors.

5.1   INTRODUCTION

Power is an important design consideration for a range of battery-operated consumer electronic devices such as PDAs (personal digital assistants), cell phones, and digital cameras. To increase battery life during active use, the real metric to minimize is the energy of computation, i.e., the area under the power curve as a function of time. Secondly, these devices have bursty computation requirements during which a specific signal processing task is performed by a functional unit. Hence, it is important to effectively reduce the power dissipated by a functional unit when it is in the idle state. Finally, these devices must be programmable to cope with evolving standards requirements and provide feature evolution on the same hardware platform.

The dissipated power consists of three components: switching power, short-circuit power, and leakage power. The switching component is power dissipated by charging and discharging circuit nodes. The short-circuit component is due to short-circuit currents when both P-channel and N-channel transistors are partially on during output signal transition. The leakage power is primarily due to gate leakage and subthreshold leakage. Although leakage power dissipation has received a lot of attention, the issue may be mitigated by process technology advances (multiple threshold voltages and high dielectric constant gate oxide), ASIC design methodology changes [14], and non-uniform scaling. Voltage scaling has been an effective technique to reduce the dynamic power (sum of switching and short-circuit power) with every new process technology due to quadratic dependence of power on the supply voltage. However, as the device geometries shrink further to 65nm and 45nm transistor gate lengths, the energy minimization will need to decrease its reliance on voltage scaling and will need to rely more on architectural and microarchitectural explorations, effective clock gating, and design methodology employing power rail shut-off techniques.

The impact of architectural and microarchitectural changes on power requires accurate estimation of the dynamic power. At the minimum, the dynamic power that depends on switching activity in the circuit requires an RTL description to estimate the power with reasonable accuracy [15]. However, it is extremely difficult to explore major architectural and microarchitectural changes while designing at the register transfer level. With time-to-market schedule constraints for a reasonably complex design, it is possible only to perform very limited design explorations. The previous work has focused on behavioral-level power optimization [1] with approximate consideration for the underlying hardware implementation and concomitantly inaccurate power estimates [7].

Extensible processors [5][19] have been proposed as a solution to dramatically improve the application performance. Application-specific processors can be extended by adding custom instructions to efficiently implement algorithmic kernels. By customizing the processor for a specific application or class of applications, extensible processors are able to drastically reduce the cycle count for a range of application benchmarks [4].

In this chapter, we propose a new methodology to optimize the energy of computation based on customizing an extensible processor. The new approach compiles the instruction extension description into synthesizable hardware and uses RTL power estimation to accurately focus on the dynamic power dissipation. The new approach reduces the area under the power curve over time by dramatically reducing the cycle count and shuts off the power to instruction extension units when they are idle.

The rest of the chapter is organized as follows. We define the problem and motivate the solution approach in Section 5.2. We present the energy minimization methodology in Section 5.3. We present the experiment results on a set of case studies in Section 5.4. Finally, we conclude the paper and provide directions for future research in Section 5.5.

5.2   PROBLEM DEFINITION AND SOLUTION APPROACH

The objective is to minimize the energy of computation required to perform a specific processing task in a system context in the presence of possible area and speed constraints. Although it is possible to optimize power using the appropriate algorithm [10], the goal of this paper is to define a methodology that minimizes the energy of computation for a specific C/C++ description of the application.

It is important to be able to accurately estimate dynamic power to optimize it [3][8]. In contrast to approaches that create a power model based on an estimate of switching capacitance in response to input transitions [13] or that estimate power based on parameters such as number and type of operations and number of edges in control/data flow specification [1][17], we measure the power on the generated RTL. RTL power estimation provides a better estimate of dynamic power by performing a quick logic synthesis and potentially a quick physical synthesis to estimate wire capacitances.

Instead of solving the energy optimization problem in general at a behavioral or unconstrained architectural level, we focus our attention on the use of an extensible processor platform. As transistor densities increase in accordance with Moore's law, an extensible processor may emerge as the fundamental building block that provides the next level of abstraction above RTL. We present a set of transformations that reduce the energy of computation in the context of extensible processors.

In the system-on-chip context, energy optimization using an extensible processor compares favorably to a hardwired solution that implements the specific functionality. The energy dissipated by the datapath logic of a hardwired solution will be comparable to an execution unit added in the extensible processor. The advantage of hardwired logic decreases further when the energy requirement for address generation logic and control logic are taken into consideration. In the system context, a hardwired solution typically shares memory with the control processor, hence, one must also take into account the energy costs of additional ports on the memory or contention management logic.

When viewed from the larger system perspective, the overhead of running the processor pipeline to execute specific functionality is outweighed by the design, verification, and integration complexities of the hardwired solution.

5.3   OPTIMIZATION METHODOLOGY

The energy optimization methodology is shown in Figure 5.1. The methodology consists of the following four steps:

1. Extraction of software kernels and design of custom instructions

2. Hardware generation and software tool generation for custom instructions

3. Creating hardware switching activity for the application software kernels

4. Estimating RTL power using generated hardware and switching activity

We describe below each of these steps in detail.

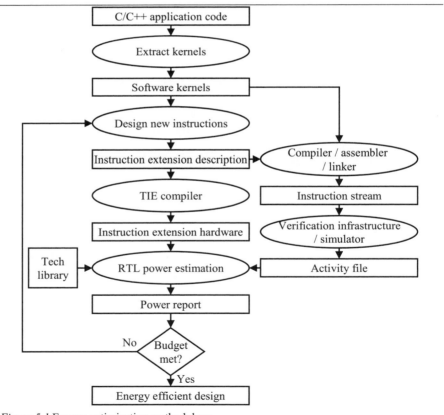

Figure 5.1 Energy optimization methodology

5.3.1 Instruction Set Extension

The designer starts with a C/C++ description of the application. Based on intuition, prior experience, and software profiling tools, the designer extracts the computation kernels that dissipate a significant fraction of the total energy.

Once these software kernels are identified, the designer defines new instructions that reduce the system energy. This aspect of design is based on the following transformation guidelines:

- **Data localization:** When a large amount of computation is performed on a data set of moderate size, it is important to ensure that entire data set is resident in the processor register file. This ensures the loads and stores from local memories are minimized. Not only does this reduce the energy of computation for the processor, but it also reduces the data memory energy dissipation.

- **Combining basic instructions:** By combining multiple instructions into a custom instruction that either uses the same hardware or adds a marginal amount of additional hardware, the power per instruction will grow only slightly. However, the cycle count reduction leads to reduction in the total energy.

- **Parallelization:** The common parallelization techniques used are Single Instruction Multiple Data (SIMD) and multiple operations per instruction word. Additional hardware leads to increased capacitance, so the power per custom instruction that uses parallelization techniques will increase. However, the same computation using the unaugmented hardware would take proportionately longer to compute, where the instruction stream lengthening would be determined by SIMD vector length or the number of operations performed by the instruction word.

For the above transformations, it is important to ensure that the area increase is within the specified budget. The extensible processor must support freely intermixed variable length instructions; any restriction that requires instructions to be the same width will be untenable due to an increase in the instruction memory cost and resulting instruction-fetch cost. Secondly, the transformations above rely on aggressive clock gating that shuts off the clock to the associated execution unit and register file during inactive phases. In fact, it is important to ensure that the execution unit is active only when the custom instruction is being executed in the pipeline and is switched off when a custom instruction retires or when custom instructions in the processor pipeline are killed due to an exception.

5.3.2   Hardware and Software Tool Generation

The set of custom instructions are described in the Tensilica Instruction Extension (TIE) language [19] that is used to extend the Xtensa processor [5]. A TIE Compiler is used to generate the synthesizable hardware that implements execution units that correspond to the custom instructions. In addition to the hardware for custom instructions, the TIE Compiler generates appropriate components of the software tool chain; namely the compiler, assembler, and linker.

The following features of the TIE Compiler are important in achieving the goal of energy minimization:

- **Clock gating:** The clock to the execution unit is gated; the execution-unit logic is activated only during the time when a custom instruction implemented by the execution unit is executing in the processor pipeline.

- **Register file:** The TIE language enables the designer to specify an arbitrary size register file that can be used by the custom instructions.

- **SIMD instructions:** The TIE language has the capability to implement SIMD instructions.

- **Multiple operation instructions:** TIE can describe wide instructions of varying length which can be used to implement multiple operations in parallel. The varying length instructions can be freely intermixed without penalty.

5.3.3   Hardware Switching Activity Generation

The extracted software kernels are compiled into an instruction stream using the software tool chain. The instruction stream is converted into a memory image and simulated using the verification infrastructure. The hardware switching activity information is determined from the simulation.

5.3.4   RTL Power Analysis

The RTL power analysis tool takes in the activity file, RTL description, and standard-cell library information to perform a quick logic synthesis. In power analysis, the wire capacitances are estimated using a wire load model (with capacitance estimated from the number of fanouts) provided with the standard cell library, which was selected as it gave the closest correlation between the RTL power estimate and post-layout power analysis. Capacitance estimates may be improved by performing a quick placement. The RTL power estimation includes clock power based on the type of clock tree. In our experience, the RTL power estimate is within 15% to 20% of the power computed for the post-layout netlist with 2.5-D extraction[1].

5.4   EXPERIMENTAL RESULTS

We present four case studies that demonstrate the reduction of energy of computation. These are dot product computation, Advanced Encryption Standard (AES) encryption computation [9], Fast Fourier Transform (FFT) computation, and Viterbi decoder computation. The kernels presented here are representative signal processing routines commonly found in consumer applications.

The RTL power analysis is performed using Sequence's PowerTheater Analyst tool [15], with an Artisan standard cell library for TSMC's 0.13um low threshold voltage, FSG process[2]. For each of the examples, the design is simulated at the register transfer level and power is estimated assuming 100MHz operation at the typical process corner and typical operating corner (25°C, 1.0V supply). Energy is calculated from power × number of cycles × cycle time, where the cycle time is 10ns at 100MHz.

5.4.1   Dot Product

This example consists of computing a dot product of two vectors, each with 2,048 16-bit entries, as shown in Equation (5.1).

$$\mathbf{a} \cdot \mathbf{b} = \sum_{i=1}^{2048} a_i b_i, \text{ where } \mathbf{a} = \begin{bmatrix} a_1 \\ a_2 \\ \vdots \\ a_{2048} \end{bmatrix}, \text{ and } \mathbf{b} = \begin{bmatrix} b_1 \\ b_2 \\ \vdots \\ b_{2048} \end{bmatrix} \quad (5.1)$$

---

[1] 2.5-D extraction is performed by Silicon Ensemble on a placed and routed netlist to determine the cross-coupling capacitance from looking at nearest neighbors on the same layer (the first two-dimensional parasitic extraction), then all wires that cross over or cross under (the second 2-D extraction). These capacitances are summed together to get the lumped cross-coupling capacitance to the victim net.

[2] FSG is fluorinated silica glass, with a dielectric constant k of 3.6, which is typically used in 0.13um processes. Its dielectric constant is lower than that of pure SiO_2, which has k of 4.0 and used to be the standard dielectric that was used in earlier process technologies. The lower k value for the interlayer dielectric (ILD) reduces the wiring capacitance.

Table 5.1 Cycle count, power, and energy for dot product of two vectors with 2,048 16-bit entries on different configurations. The configurations are the Xtensa with a multiplier in the case of MaxMUL, with extensions of two multiply-accumulate blocks in the case of MAC-2, and four multiply-accumulate blocks in the case of MAC-4.

Configuration	Area (mm^2)	Number of Cycles	Power (mW)	Energy (uJ)
MaxMUL	0.906	11,909	27.8	3.31
MAC-2	1.064	7,426	25.1	1.86
MAC-4	1.263	5,896	26.5	1.56

The measurement results for the dot product computation are shown in Table 5.1. The table compares the energy of computation without the TIE and with TIE instruction extensions. The first column shows the configuration name. The second column is the area after logic synthesis. The third column shows the cycle count to execute the software corresponding to the dot product kernel. The fourth column shows the average power reported by the power estimation tool. And, the final column shows the calculated energy. Please note that the cycle count also includes a "reset" code sequence and other software overhead that is common among MaxMUL, MAC-2, and MAC-4 configurations.

The MaxMUL configuration implements a multiplier but does not support a multiply-accumulate instruction. The MAC-2 and MAC-4 configurations extend the MaxMUL configuration by adding two and four multiply-accumulate (MAC) units, each implementing 16×16 multiplication with 32-bit addition. The area increase is about 10,000 to 12,000 gates per MAC unit, assuming a NAND2 drive-strength-2 gate to be about 8um^2 in 0.13um technology.

The number of cycles decreases by about 4,500 for the MAC-2 configuration. The reduction in the instruction count is due to merged loads and merged computation operations. For the MAC-2 configuration, two loads are merged into one instruction; two multiplications and two addition operations are performed in one instruction compared to four instructions required for MaxMUL. A 64-bit register is used, with one MAC putting results in the upper 32-bits of the register, and the second MAC unit putting its results in the lower 32-bits. The average power reported for the MAC-2 configuration is lower than MaxMUL, because the "reset" code sequence and other software overheads that dissipate much less power (than the load, multiply, and add instructions) constitute a large fraction of the total cycle count for the MAC-2 configuration. The energy of computation is reduced from 3.31uJ to 1.86uJ.

The addition of two more MAC units to give the MAC-4 configuration causes the cycle count to drop by 1,500 due to halving of loads and MAC instructions compared to the MAC-2 configuration. The energy decreases marginally from 1.86uJ to 1.56uJ.

Table 5.2 Cycle count, power, and energy for AES encryption on 55 16-byte blocks of plaintext. The configurations are the Small Xtensa (SX), and SX extended with TIE code to implement AES in hardware (SX+AES).

Configuration	Area (mm^2)	Number of Cycles	Power (mW)	Energy (uJ)
SX	0.367	283,004	21.6	61.13
SX+AES	0.822	2,768	26.9	0.74

5.4.2   Advanced Encryption Standard (AES) Encryption

The advanced encryption standard (AES) works on one 128-bit (16-byte) block of plaintext at a time, and uses a 128-bit encryption key [9]. The 128-bit encryption key is expanded to ten additional 128-bit keys, which together make up the "key schedule" of the algorithm. These keys can be generated before the core of the algorithm, or on the fly as necessary.

The 16-byte block of plaintext is conceptually stored and operated upon in a 4×4 byte state array. There are four encryption steps performed on this 4×4 byte array: SubBytes, ShiftRows, MixColumns, and AddRoundKey. This latter AddRoundKey step involves one of the eleven keys from the key schedule. These four encryption steps are performed eleven times on a given 128-bit block of plaintext to arrive at the AES-encrypted ciphertext block.

Implemented as standard C, the inner loop of four encryption steps plus key expansion takes hundreds of cycles. However, using TIE, and taking advantage of the ability to declare 128-bit wide buses and a custom register file, this can be reduced to a single TIE instruction that takes one cycle to compute. This single TIE instruction is called 11 times for a given plaintext block, and there are also 3 cycles of load/store operations, giving a final count of 14 cycles per 16-byte plaintext block.

The power, cycle count, and energy of computation for AES are shown in Table 5.2. AES encryption is performed on 55 blocks of 16 bytes each. The total number of blocks is reduced from 1,663 to make disk space reasonable for gathering data for a Small Xtensa (SX) configuration. However, this results in a very small cycle count for the SX configuration augmented with AES TIE. As each block requires only 14 cycles for the SX+AES configuration, most of the 2,768 cycles is "reset" code. To report the power more accurately, the power for the SX+AES configuration is from the full 1,663 block run, which takes about 38,000 cycles.

For the AES example, the area increase due to custom instructions exceeds the size of the original processor configuration (124%). The cycle count decreases quite drastically. The average power does go up as expected, due to more computations per cycle. However, the increase in the computation power is offset by reduction in the load and store power. The cycle reduction is achieved by the use of wide loads, wide stores, a user-defined register file that localizes the data, and key manipulation instructions that are very effectively performed in the extension unit hardware. Due to drastic reduction in the cycle count, the energy consumption of 0.74uJ is a small fraction of the 61.1uJ energy required without instruction extensions.

For the SX+AES configuration, the amount of simulation cycles for the "reset" code is a significant fraction of the total cycles. The power dissipated as a function of time for the SX+AES configuration is shown in Figure 5.2. The figure demonstrates the effectiveness of clock gating that shuts off the clock to the AES execution unit during the "reset" code execution, reducing power by a factor of 2 during reset.

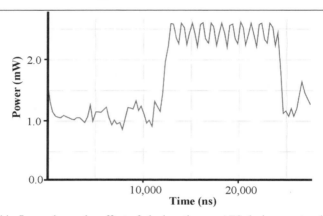

Figure 5.2 This figure shows the effect of clock gating on AES during reset code execution, where the power is halved due to clock gating.

5.4.3   Fast Fourier Transform

The fast Fourier transform (FFT) covers a family of techniques for computing the discrete Fourier transform (DFT). The discrete Fourier transform of a sequence x of N points is given by

$$X[k] = \sum_{n=0}^{N-1} x[n]e^{-2\pi ikn/N} \qquad (5.2)$$

where k is from 0 to $N-1$. The inverse discrete Fourier transform is given by

$$x[n] = \frac{1}{N} \sum_{k=0}^{N-1} X[k]e^{2\pi ikn/N} \qquad (5.3)$$

We chose to implement an N-point radix-2 decimation-in-frequency FFT algorithm for complex input values [11]. Radix-2 refers to the FFT approach where a transform of length N is broken up into two transforms of length N/2, which are then subdivided in a similar manner, down to the complex butterfly computation shown in Figure 5.3.

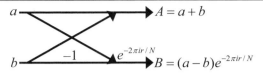

$A = a + b$

-1 $e^{-2\pi i r / N}$

$B = (a - b)e^{-2\pi i r / N}$

Figure 5.3 The butterfly computation for the radix-2 decimation-in-frequency FFT, where *r* is an integer and depends on the particular stage of the computation.

Table 5.3 Cycle count, power, and energy for fast Fourier transform. The comparisons are between C code on an Xtensa with a multiplier (MUL); C code on an Xtensa with a multiplier and extension with two multipliers for the radix-2 butterfly operation (MUL-BFLY); and hand-coded assembly on MUL-BFLY.

Configuration	Area (mm²)	Number of Cycles	Power (mW)	Energy (uJ)
MUL	0.421	325,506	17.4	56.64
MUL-BFLY	0.588	37,676	20.4	7.57
assembly coded	0.588	13,836	18.3	2.53

There are many loops in this algorithm, all of which follow the basic sequence of load-compute-store operations on their respective data. Thus the first concern is how to best perform loop optimizations such that the compute portion of one iteration can be done while waiting to perform the load or store of an adjacent loop iteration. The Xtensa architecture allows for at most one load or store per processor cycle, however the computation portion can be written to occur in parallel, limited only by the amount of compute hardware available. The width of the processor interface (PIF) determines the throughput of the load/store instructions, and with a PIF width of 128 bits, a throughput of two sets of complex butterfly inputs in two cycles is achievable; the throughput of stores is the same. Through this method of loop unrolling and optimization, about one complex butterfly computation can be achieved per cycle.

The cycle count, power, and energy of computation for a 256-point complex FFT are shown in Table 5.3. The table compares the following: a standard C routine running on an Xtensa configuration with a multiplier (MUL); a C routine on an Xtensa configuration with a multiplier and with two multipliers that are used to implement the radix-2 butterfly operation (MUL-BFLY); and hand-coded assembly for the MUL-BFLY configuration.

Custom instructions fold multiple load instructions into a single load instruction, fold multiple store instructions into a single store instruction, and implement radix-2 butterfly operation. This significantly reduces the number of cycles in computation kernels. With hand-coded assembly for the MUL-BFLY (assembly coded) configuration, it is possible to reduce the cycle count further by clever register allocation and inner-loop code reorganization. The additional hardware cost to implement the radix-2 butterfly is about 21,000 gates. The reduction in the energy of computation is from 57uJ for the original approach to less than 3uJ.

5.4.4   Viterbi Decoder

Viterbi decoding is used to determine the maximum likelihood sequence, given a sequence of transmitted data which has some noise. The transmitted data is encoded with a convolution of the current input with a set number of earlier input bits and a masking polynomial. This can be represented by a state machine where only some state transitions are possible. This enables detection and correction of errors at the receiver, by excluding sequences of bits that could not have occurred or have low probability – i.e. by finding the maximum likelihood sequence.

The core part of the Viterbi decoder that determines the throughput is the add-compare-select routine:

$$p_{i,k}{}^{n-1} = s_i{}^{n-1} + b_{i,k}, \; p_{j,k}{}^{n-1} = s_j{}^{n-1} + b_{j,k} \; \text{(add)} \tag{5.4}$$

$$s_k{}^{n} = \min(p_{i,k}{}^{n-1}, p_{j,k}{}^{n-1}) \; \text{(compare and select)} \tag{5.5}$$

where $s_k{}^{n}$ is a state metric (a value representing the likelihood of being in state k at time n); $b_{i,k}$ is a branch metric (a value representing the likelihood of transition from state i to state k); and $p_{i,k}{}^{n-1}$ is a path metric, accounting for the probability of being in state i at time $n-1$ ($s_i{}^{n-1}$) then transitioning to state k at time n. A simple two-state trellis butterfly computation for the Viterbi algorithm is shown in Figure 5.4.

The add-compare-select routine is the main target of TIE optimization for the Viterbi decoder. The Viterbi decoder implemented is for the GSM wireless communication standard, which has 16 states in every trellis column and requires eight butterfly operations to decode a single bit. The inner add-compare-select loop requires about 42 cycles on the base Xtensa. Two TIE instructions are created to optimize this process, one 128-bit wide load instruction into a state register, and one instruction that calculates the shortest arc into each state, stores these in state accumulators, and writes out 16 binary encoded bytes which designates the most-likely arcs going into each subsequent state. With the TIE extension, the inner loop executes 64 butterfly computations in 10 clock cycles – a speedup by about 270× for the inner loop.

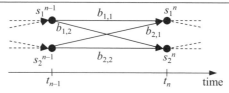

Figure 5.4 The two-state trellis butterfly computation for the Viterbi algorithm.

Table 5.4 Cycle count, power, and energy for Viterbi decoder

Configuration	Area (mm^2)	Number of Cycles	Power (mW)	Energy (uJ)
Viterbi	0.522	279,537	23.5	65.69
Viterbi+TIE	0.595	7,632	26.2	2.00

The data for the Viterbi decoder is shown in Table 5.4. The data consisted of six trials of 192 frames each, with two integer entries per frame. The power measurement for the Viterbi+TIE configuration is on ten trials, because cycle count for six trials is too low to make reset code negligible. The increase in hardware cost is less than 10,000 gates, and the cycle count reduces from 280,000 to less than 8,000 cycles.

5.5 SUMMARY

We present a new approach to energy optimization at microarchitectural and architectural-level using extensible processors. Extensible processors offer an attractive alternative to hardwired logic as the potential energy increase from the use of a processor platform is more than compensated by the complexity of designing, verifying, and integrating hardwired logic. We present a methodology to optimize the energy that consists of designing custom instructions; generating hardware and software tools; generating hardware switching activity; and measuring power at the RT level. We obtain a drastic reduction in the energy of computation compared to fixed-instruction-set processors.

Several approaches have been presented to automatically generate custom instructions for extensible processors [2][6][12][16]. For future work, it is promising to take the energy minimization objectives into account while automatically generating custom instructions from the application code. Also, it is interesting to look at ways to make the software compiler energy aware [18].

5.6 REFERENCES

[1] Chandrakasan, A., et al., "Optimizing power using transformations," *IEEE Transactions on Computer-Aided Design*, vol. 14, no. 1, 1995.

[2] Cheung, N., Henkel, J., and Parameswaran, S., "Inside: Instruction selection/identification and design exploration for extensible processor," *proceedings of the International Conference on Computer-Aided Design*, 2003.

[3] Devadas, S., and Malik, S., "A survey of optimization techniques targeting low power VLSI circuits," *proceedings of the Design Automation Conference*, 1995.

[4] Embedded Microprocessor Benchmark Consortium, http://www.eembc.org/

[5] Gonzalez, R., "Xtensa: A configurable and extensible processor," *IEEE Micro*, March 2000.

[6] Goodwin, D., and Petkov, D., "Automatic generation of application specific processors," *proceedings of the international conference on Compilers, Architecture and Synthesis for Embedded Systems*, 2003, pp. 137-147.

[7] Landman, P., and Rabaey, J., "Power estimation for high level synthesis," proceedings of the European Design Automation Conference, 1993, pp. 361-366.

[8]   Najm, F., "A survey of power estimation techniques in VLSI circuits," *proceedings of the Design Automation Conference*, 1994.

[9]   National Institute of Standards and Technology, Advanced Encryption Standard (AES), http://csrc.nist.gov/publications

[10] Ong, P., and Yan, R., "Power conscious software design - a framework for modeling software on hardware," *proceedings of the IEEE Symposium on Low Power Electronics*, 1994.

[11] Oppenheim, A., and Schafer, R., *Discrete-Time Signal Processing*. Prentice-Hall International, 1989.

[12] Peymandoust, A., et al., "Automatic instruction set extension and utilization for embedded processors," proceedings of the International Conference on Application-specific *Systems, Architectures and Processors,* 2003.

[13] Powell, S., et al., "Estimating power dissipation of VLSI signal processing chips," *in VLSI Signal Processing IV*, 1990.

[14] Puri, R., et al., "Pushing ASIC Performance in a Power Envelope," *proceedings of the Design Automation Conference*, 2003.

[15] Sequence Design, *PowerTheater User Guide*, http://www.sequencedesign.com

[16] Sun, F., et al., "A scalable application-specific processor synthesis methodology," *proceedings of the International Conference on Computer-Aided Design*, 2003.

[17] Svensson, C., and Liu, D., "A power estimation tool and prospects for power savings in CMOS VLSI chips," *proceedings of the International Workshop on Low Power Design*, 1994, pp. 171-176.

[18] Tiwari, V., Malik, S., and Wolfe, A., "Power analysis of embedded software: a first step towards software power minimization," *IEEE Transactions on Very Large Scale Integration (VLSI) Systems*, vol. 2, no. 4, 1994, pp. 437-445.

[19] Wang, A., et al., "Hardware/software instruction set configurability for system-on-chip processors," *proceedings of the Design Automation Conference*, 2001.

Chapter 6

LINEAR PROGRAMMING FOR GATE SIZING

David Chinnery, Kurt Keutzer
Department of Electrical Engineering and Computer Sciences
University of California at Berkeley
Berkeley, CA 94720, USA

For many ASIC designs, gate sizing is the main low level design technique used to reduce power. Gate sizing is a classical circuit optimization problem for which the same basic method has been used for the past 20 years. The standard approach is to compute a sensitivity metric, for example for the power versus delay tradeoff for upsizing, and then greedily resize the gate with highest sensitivity, iterating this process until there is no further improvement. Such methods are relatively fast, with quadratic runtime growth versus circuit size, but they are known to be suboptimal. The challenge has been to find a better approach that still has fast runtimes.

Our linear programming approach achieves 12% lower power even on the smallest ISCAS'85 benchmark c17, as shown in Figure 6.1. The linear program provides a fast and simultaneous analysis of how each gate affects gates it has a path to. Versus gate sizing using the commercial tool Design Compiler with a 0.13um library, we achieve on average 12% lower power at a delay constraint of 1.1 times the minimum delay (T_{min}), and on average 17% lower at $1.2T_{min}$ – in one case 31% lower. The runtime for posing and solving the linear program scales between linearly and quadratically with circuit size.

6.1 INTRODUCTION

We wish to find the minimum power for a circuit to satisfy given delay constraints. To limit the solution space, we consider a gate-level combinational circuit with fixed circuit topology. The circuit may be represented as a directed acyclic graph (DAG), where each node is a logic gate and edges are connections between gates. The logic gate at each node is fixed; we do not allow nodes to be inserted or removed, or graph edges to change – for example pin swapping is not allowed.

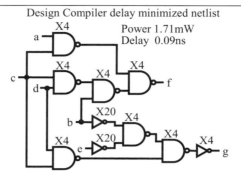

Design Compiler delay minimized netlist

Power 1.71mW
Delay 0.09ns

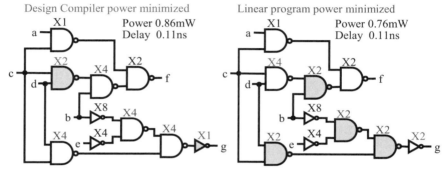

Design Compiler power minimized

Power 0.86mW
Delay 0.11ns

Linear program power minimized

Power 0.76mW
Delay 0.11ns

Figure 6.1 At a delay constraint of $1.2T_{min}$ for ISCAS'85 benchmark c17, we achieve lower power than Design Compiler. The two shaded gates on the lower left circuit are suboptimally downsized by Design Compiler. In contrast, the linear programming approach downsizes four of the NAND2 gates and achieves 12% lower power.

For each logic gate, there are a number of logic cell implementations that may be chosen from the available standard cell libraries. Each cell has different delay, dynamic power, and leakage power characteristics. These characteristics are determined by factors such as the gate oxide thickness, width, length and threshold voltage of transistors composing the logic cell; transistor topology – for example stack forcing [16] and alternate XOR implementations; and the supply voltage. We shall limit discussion in this chapter to gate sizing. However, the same delay and power tradeoffs need to be considered for all these factors, and the optimization problem does not fundamentally differ except in the case of gate supply voltage, where there can be topological constraints. Our objective is to minimize the power subject to a delay constraint, but the approach herein is equally applicable to minimize the area subject to a delay constraint. Using libraries with multiple supply and threshold voltages is a relatively new low power technique and will be discussed in chapters 7 and 8.

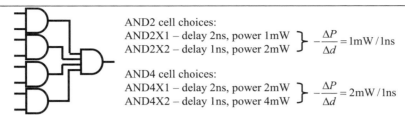

AND2 cell choices:
AND2X1 – delay 2ns, power 1mW
AND2X2 – delay 1ns, power 2mW $\left.\right\}$ $-\dfrac{\Delta P}{\Delta d} = 1\text{mW}/1\text{ns}$

AND4 cell choices:
AND4X1 – delay 2ns, power 2mW
AND4X2 – delay 1ns, power 4mW $\left.\right\}$ $-\dfrac{\Delta P}{\Delta d} = 2\text{mW}/1\text{ns}$

Figure 6.2 This simple example shows that greedily choosing the gate with the maximum sensitivity is suboptimal. If all the gates are initially size X2, the critical path is 2ns and power is 12mW. Consider a 3ns delay target. Picking the max power_reduction/delay_increase sensitivity results in sizing down the AND4 gate, giving total power of 10mW. If the four AND2 gates are sized down instead, the power is only 8mW.

6.1.1   Gate sizing approaches

Gate sizing algorithms have changed little in the past 20 years. In 1985, Fishburn and Dunlop proposed a fast method (TILOS) to minimize area and meet delay constraints, greedily picking the transistor with maximum delay_reduction/transistor_width_increase at each step [6]. Variants of this are still standard in commercial sizing tools.

Approaches similar to TILOS can be used to minimize power when gate sizing. Srivastava et al. used max delay_reduction/power_increase to meet delay constraints, after reducing power by assigning gates to low supply voltage (Vdd) [22]. Downsizing a gate gives a linear reduction in the power to charge and discharge its internal capacitances and input pin capacitances, as the capacitance is proportional to the transistor widths in the gate. Leakage power is also reduced linearly as gate width is reduced.

Greedy heuristics that pick the gate with the maximum sensitivity fail to consider the whole circuit and are suboptimal – for example see Figure 6.2. The challenge is to find a better approach with a global view of the circuit that has fast runtimes.

Several groups have used convex optimization to find a globally optimal solution. Convex optimization requires convex delay and power models, such as linear or posynomial models. In our experience, linear models are inaccurate. Least squares fits of linear models versus gate size and load capacitance of 0.13um library data had delay inaccuracy of 19% to 30%, and least squares fits of piecewise linear models also has sizable error of 10% to 30% [20]. The analysis in [20] assumed a fixed input slew of 0.07ns, so these errors will be even larger once variable slew is taken into account. Linear program (LP) solvers with linear models can scale to problems with millions of variables. Higher order convex models, such as posynomials [13], are at best accurate to within 5% to 10% [4][13][20][24]. The accuracy is limited because real data for delay and power is not a convex function of gate size – standard cell layouts change significantly as the gate width changes, due to transistor folding for layout of larger cells and other cell layout concerns. Sacrificing delay accuracy is unacceptable, when a 10% delay increase can give 20% power savings (e.g. compare power at $1.1T_{min}$ and $1.2T_{min}$ in Table 6.2). Optimization with posynomial models requires using a geometric program solver with runtimes that scale cubically [4]. Thus geometric programming optimization of circuits of tens of thousands of gates or more is computationally infeasible. In addition, convex models must assume at least a piecewise continuous range for optimization variables that are typically discrete, which introduces suboptimality when the resulting intermediate values must be rounded up or down in some manner to a discrete value – though this is less of an issue for a library with very fine-grained sizes.

It is possible to formulate a linear program to perform optimization at a global level with more accurate delay and power models. The basic approach is to use the linear program to specify the delay constraints, and the power and delay changes if the cell for a gate is changed. A heuristic is required to choose which cell change is the best to encode in the linear program, for example the cell that gives the best power_reduction/delay_increase. The solution to the linear program indicates which cells may be changed, or how much timing slack is available to change a cell to one that consumes less power. Cells with sufficient slack are then changed. This procedure of specifying the best alternate cells in the linear program, solving it, and assigning cell changes is iterated.

The linear program formulation requires some timing slack for the circuit to be downsized and upsized in an iterative manner to converge on a good solution. A 0.13um standard cell library was used. At a delay constraint of 1.1 times the minimum delay (T_{min}), we achieve on average 12% lower power by sizing than Design Compiler at the same delay constraint, and on average 17% lower at $1.2T_{min}$ – in one case 31% lower. Design Compiler is the commercial EDA synthesis tool which is most commonly used in industry, and it is generally considered to produce high quality results compared to other EDA tools [8]. The timing and power results for the optimized netlists have been verified in Design Compiler.

An overview of gate sizing approaches along the lines of TILOS is provided in Section 6.2. The linear programming formulation is detailed in Section 6.3. The optimization flow is detailed in Section 6.4. Section 6.5 compares our gate sizing results versus gate sizing in Design Compiler, and then Section 6.6 discusses computational runtime. Section 6.7 concludes with a short summary of this gate sizing work.

6.2   OVERVIEW OF TILOS GATE SIZING

Starting with a circuit that violates delay constraints, TILOS aims to meet the delay constraints with the minimum increase in area. Transistors on critical paths, that is paths that don't satisfy the delay constraint, were analyzed with the following sensitivity metric [6]

$$Sensitivity_{reduce\ delay} = -\frac{\Delta d}{\Delta w} \tag{6.1}$$

where Δd is the change in delay on the path and $\Delta w > 0$ is the increase in transistor width. Δd was determined from convex delay models for the distributed RC network representing the circuit. The total circuit area was measured as the sum of the transistor widths, so the aim was to get the best delay reduction with the minimum transistor width increase. The transistor with the maximum sensitivity was upsized to reduce the path delay. This greedy approach proceeded iteratively upsizing transistors with maximum sensitivity until delay targets are met, or there are no upsizing moves to further reduce delay [6].

Dharchoudhury et al. used a similar approach for transistor-level sizing of domino circuits in 1997. By this time, distributed RC networks had fallen out of favor in industry due to inaccuracy and it was essential for timing accuracy to model individual timing arcs. The sensitivity to upsizing a transistor was computed by [5]

$$Sensitivity_{reduce\ delay} = -\frac{1}{\Delta w} \sum_{l \in timing\_arcs} \frac{\Delta d_l}{Slack_l - Slack_{min} + k} \tag{6.2}$$

where $Slack_{min}$ is the worst slack of a timing arc seen in the circuit; $Slack_l$ is the slack on the timing arc; Δd_l is the change in delay on the timing arc if the transistor is upsized ($\Delta w > 0$); and k is a small positive number for numerical stability purposes. The weighting $1/(Slack_l - Slack_{min} + k)$ more heavily weights the timing arcs on critical paths.

Srivastava and Kulkarni in [14] and [22] used a delay reduction metric similar to Dharchoudhury et al., for gate-level sizing to minimize power and meet delay constraints in their TILOS-like optimizer. The sensitivity metric was [22]

$$Sensitivity_{reduce\ delay} = -\frac{1}{\Delta P} \sum_{l \in timing\_arcs} \frac{\Delta d_l}{Slack_l - Slack_{min} + k} \tag{6.3}$$

where $\Delta P > 0$ is the change in total power for upsizing a gate. The same analysis was used in [21] for the delay versus leakage power trade-off. The timing slack on a timing arc through a gate from input i to output j is computed as

$$Slack_{arc\ ij} = (t_{required\ at\ output\ j} - t_{arrival\ at\ input\ i}) - d_{arc\ ij} \tag{6.4}$$

where $t_{required\ at\ output\ j}$ is the time the transition must occur at gate output j for the delay constraint to be met on paths that include the timing arc; $t_{arrival\ at\ input\ i}$ is the arrival time at input i; and $d_{arc\ ij}$ is the delay on the timing arc. The slack is the maximum increase in delay of the timing arc that will satisfy the delay constraints, and it will be negative if a delay constraint is not met on a path through the gates with that timing arc. Note here that the impact on delay of slew changing is not included in Equation (6.3), and that for accuracy the delay change on the timing arc Δd_l should include the delay change of the gate that drives gate input i due to the change in input capacitance of pin i as the gate is upsized. This TILOS gate sizing approach does not include gate downsizing, but that is not important as the starting point for the TILOS gate sizer is with all gates at minimum size.

Similar metrics have been used for greedy power minimization approaches. For example, for leakage reduction [25]

$$Sensitivity_{reduce\ power} = -\frac{\Delta P_{leakage}}{\Delta d} \tag{6.5}$$

and for power reduction [22]

$$Sensitivity_{reduce\ power} = -\Delta P \sum_{l \in timing\_arcs} \frac{Slack_l}{\Delta d_l} \tag{6.6}$$

The worst case theoretical complexity of these TILOS-like sizers is $O(|V||E|)$, as iteratively the gate with the maximum delay reduction is picked, then static timing analysis must be updated over the timing arc edges from gate to gate, and the number of size increases for a gate is limited. For our benchmarks, $|E|$ ranged from 1.59× to 2.01×$|V|$. Consequently, the worst case runtime behavior is $O(|V|^2)$, as was observed in Section 6.6.2 for gate sizing with Design Compiler.

The greedy approach of optimizing the gate with the greatest sensitivity is a *peephole* optimization approach. The optimizer considers only changing one gate at a time and only looks at the impact on the immediate neighborhood. For example, reducing a gate's delay will provide some timing slack there, but the primary output delay on that path may not be reduced as there may be other convergent paths that are timing critical. An approach with a *global* view is needed to consider multiple gates simultaneously and determine the overall impact of them changing.

6.3   LINEAR PROGRAMMING FORMULATION

Linear programming has been proposed previously for gate sizing with linear delay models [1][12]. As described above, linear delay models are very inaccurate for today's technologies. Instead, we encode the delay constraints and the impact of cells changing in the linear program. The linear program formulation provides a global view of the circuit for optimization.

The linear program gives a fast and simultaneous analysis of how changing each gate affects the gates it has a path to. From the LP solution, all the gates may be sized simultaneously, which avoids greedily sizing individual gates.

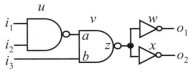

Figure 6.3 A combinational circuit for illustrating the delay constraints. Primary inputs are denoted i_1 to i_3, and the primary outputs are o_1 and o_2.

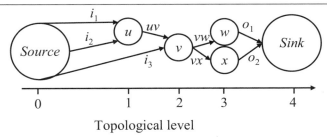

Topological level

Figure 6.4 Directed acyclic graph (DAG) representation of the circuit in Figure 6.3. Primary inputs connect to the source, and primary outputs connect to the sink. The topological level is determined in a breadth first manner from the source which is level 0. Vertex names are noted inside the circle representing the node, and edge names are next to the edge.

The idea for the linear programming approach came from the zero-slack algorithm, which determines a maximal safe assignment of additional delays to each node that avoids violating the delay constraints at the outputs, but if any further delay was added to a node then a delay constraint would be violated [15]. No timing slack remains in the resulting circuit, hence the name "zero-slack". The essential idea is formulating a set of delay constraints that determine how delays along a path add up and what additional delays can be added at each node without violating output constraints. In this constraint formulation, the delay constraints are linear. Our linear programming approach differs from [15], using the change in total power as the objective function, effectively a weighted sum over the additional delay at each node.

A combinational circuit (Figure 6.3) is represented as a directed acyclic graph $G(V,E)$ (Figure 6.4), where each gate is represented by a vertex in V, and edges in E between vertices represent wires. Assuming no gate drives more than one input on another gate, we can uniquely represent a directed edge from gates u to gate v, as uv; u is a fanin of v, and v is a fanout of u.

For each gate v, we can determine the best alternate library cell that implements the same logic, by examining the following sensitivity metric:

$$\text{Sensitivity metric for changing cell} = \frac{\Delta P_v}{\Delta d_v}, \text{ where } \Delta P_v < 0, \ \Delta d_v > 0 \quad (6.7)$$

where Δd_v and ΔP_v are respectively the change in delay and power if the cell is changed. The cell alternative with the *minimum* value for this metric is the best alternative, giving the largest power_reduction/delay_increase. If the only cell alternates have $\Delta P_v > 0$ then the gate will not be changed.

For each gate, the best cell alternative, if one exists that reduces power, is encoded in the linear program using a *cell choice* variable $\gamma_v \in [0,1]$, which determines if it is changed. $\gamma = 1$ if the alternate cell is used, $\gamma = 0$ if not. The LP solution may give a γ value in-between 0 and 1, in which case appropriate thresholds must be chosen.

The linear program's objective function to minimize the total power is

$$\text{minimize} \quad \sum_{v \in V} \gamma_v \Delta P_v \qquad\qquad (6.8)$$

and there are constraints on the cell choice variables,

$$0 \le \gamma_v \le 1, \text{ for all } v \in V \qquad\qquad (6.9)$$

These cell choice variables are the only "free variables"; they determine the delay constraint variables, t_{vw}, on each vw between gates. The delay constraints on edges between gates are

$$t_{vw} \ge t_{uv} + d_v + \gamma_v \Delta d_v, \text{ for all } v \in V, \; w \in fanout(v), \; u \in fanin(v) \qquad (6.10)$$

Namely, the arrival time at the output of gate v (t_{vw}, on edge vw) is equal to the arrival time at the input of gate v on edge uv (t_{uv}), plus the delay of gate v, plus the change in delay Δd_v of gate v if its cell is changed.

For simplicity, we assume that all circuit paths are subject to the same maximum delay constraint, T_{max}, noting that it is straightforward to encode different delay constraints in the linear program if so desired. A circuit sink node is added to V', such that all primary outputs of the combinational circuit connect to the sink and are subject to the constraint T_{max}. Delay constraints to the circuit sink are

$$t_{wSink} \le T_{max}, \text{ for all } w \in fanin(Sink) \qquad\qquad (6.11)$$

Similarly, we add a circuit source to V', such that all primary inputs connect to the source, $V' = V \cup \{Source, Sink\}$. We assume that arrival times from the circuit source are at $t = 0$, though it is trivial to specify individual arrival times by input if so desired, giving

$$t_{Source\ u} = 0, \text{ for all } u \in fanout(Source) \qquad\qquad (6.12)$$

where $t_{Source\ u}$ is the arrival time from the source to gate u. As there may be more than one connection from a gate to the circuit source or sink, to uniquely identify those edges we can use the primary input or output name.

The complete linear program to minimize power, subject to delay constraints, is minimize $\sum_{v \in V} \gamma_v \Delta P_v$

subject to $\quad t_{vw} \geq t_{uv} + d_v + \gamma_v \Delta d_v$, for all $v \in V$, for all $w \in \mathit{fanout}(v)$, $\quad$ (6.13)

$$\text{for all } u \in \mathit{fanin}(v)$$

$$t_{\mathit{Source}\ u} = 0, \text{ for all } u \in \mathit{fanout}(\mathit{Source})$$

$$t_{w\mathit{Sink}} \leq T_{\max}, \text{ for all } w \in \mathit{fanin}(\mathit{Sink})$$

$$0 \leq \gamma_v \leq 1, \text{ for all } v \in V$$

This was our initial formulation for the linear program [17]. The signal slew is not included, nor are rise and fall delays considered separately. This leads to significant delay inaccuracy. As the fanin delay impact due to changed input capacitance C_{in} is not modeled, gates can not be upsized to avoid increasing fanin delay. There is no method for reducing delay to fix violated delay constraints. With the help of Fujio Ishihara and Farhana Sheikh, we tested this approach on a 17,000 gate inverse discrete cosine transform block to implement dual supply voltages with a 0.13um library – the simplistic models resulted in a 24% increase in the clock period when measured in Synopsys Design Compiler, despite a tight delay constraint.

6.3.1 Improving power and delay accuracy

This subsection discusses how the linear program is accurately formulated using the data from incremental static timing and power analysis.

Static timing and power analysis are performed using the standard cell libraries to determine gate delay, slew and power values used in the linear program. Considering alternate cells for a gate is a core part of the inner optimization loop for setting up the linear program, so it is essential that it be fast. In particular, when analyzing changing the gate for a cell, only a very limited range of interactions must be considered to minimize computational overheads. It is also important to be able to roll back this change and consider alternatives for another gate, without any additional overheads to recompute the original timing and power values.

Software modules were written to perform incremental timing and power analysis, with the ability to store temporary alternative values when considering cell changes and to store the best alternatives found. Design Compiler was used as a reference to debug the software for timing and power analysis, and later validate the results for optimized netlists.

The changes in delay and power determined by incremental analysis are encoded in the linear program. The LP remains limited in accuracy, because only first order changes, one gate's cell changing at a time, can be encoded. Static timing and power analysis are performed after the linear program has been solved and cells have been changed, to more accurately determine the power of the new circuit and whether delay constraints have been met.

6.3.1.1   What variables must be modeled in the LP for accuracy?

As outlined earlier, a central optimization issue is the accuracy of the delay and power models. The linear program constraints must model the impact on delay and power due to changing gate size, Vdd, or Vth.

Consider changing a single gate's cell. The input capacitance C_{in} of the gate's input pins loads the fanin gates, affecting their delay and switching power. The gate's drive strength affects its delay and output slew, which may increase the delay of paths the gate is on, and the internal power of fanouts may be affected by the change in output slew. If the gate's voltage changes, that affects the switching power for the load it drives. The gate's subthreshold leakage increases exponentially with decreasing Vth. The gate size primarily determines Cin, which affects the load on fanins. Size (transistor width), Vdd, and Vth all affect the gate drive strength.

The impact of changing a gate on its power and delay and on that of neighboring gates was examined. Incremental timing analysis allowed exploring what neighborhood of affected gates needs to be considered for accurate analysis. The fanin level of logic must be considered, as there can be substantial delay and slew changes due to changing the load capacitance. Analysis was also performed with one or two fanout logic levels, from both the gate that is changed and its fanins.

More than 95% of the change in power occurs at the gate whose cell changes: switching power due to C_{in}; switching power of the load with Vdd; leakage power; and internal power. Slew changes affect the short circuit power of neighboring gates, but short circuit power is only a small part of the total power – typically less than 10% [3]. Usually 99% or more of the total power impact is accounted for at the gate, though in some cases it may be as low as 95%. Thus to determine with reasonable accuracy the change in power by changing a gate, we only need to consider the gate itself and can avoid computing the impact on other gates.

In contrast, changing C_{in} and output slew significantly impacts the delay of neighboring gates. The impact of C_{in} is limited to the immediate fanins – the fanins of fanin gates do not see the change in load capacitance, as complementary static CMOS logic decouples this. However, the delay and slew changes of fanins and of the gate itself propagate forwards topologically. It is computationally expensive to calculate this impact over more than the fanin level of logic. Instead, we conservatively determine the worst case impact on the transitive fanout from the gate being changed and its fanins. This was sufficient to produce good power minimization results with fast computation time – though the impact of slew propagation to the fanouts must be considered as described in Section 6.3.1.2. Delay propagation is handled in the usual way for static timing analysis, with output arrival time constraints in terms of the arrival time at the inputs and delay of the gate.

To determine the impact of a gate x's input capacitance changing on a fanin v, we assume the cell of the fanin gate has not changed and calculate the change in delay due to x changing, adding this to the delay constraint. For example, for the constraint on the arrival time of the rising output of gate v on edge vw is

$$t_{vw,rise} \geq t_{uv,fall} + d_{uv,rise} + \gamma_v \Delta d_{uv,rise,v} + \sum_{x \in fanout(v), x \neq w} \gamma_x \Delta d_{uv,rise,x} \qquad (6.14)$$

where $t_{uv,fall}$ is the falling arrival time at v from gate u; $d_{uv,rise}$ is the delay from the signal on uv to the output of v rising; and $\Delta d_{uv,rise,x}$ is the change in delay out of this timing arc if the cell of x changes. Here we have assumed gate v is of negative polarity, namely that a falling input may cause a rising output transition. Wire RC delays, for example on the edge from gate v to gate w, are also included in the delay calculations.

6.3.1.2   Modeling the impact of signal slew propagation

We will now address modeling the delay impact of signal slew in the linear program. For cells in our 0.13um libraries, $\Delta s_{out}/\Delta s_{in}$ ranges from -0.23 to 0.67, where Δs_{out} is the change in output slew due to change in input slew Δs_{in}. $\Delta d/\Delta s_{in}$ sensitivity ranges from -0.32 to 0.54, where Δd is the delay change due to Δs_{in}. Thus a slew change can significantly impact delay. Larger magnitude $\Delta d/\Delta s_{in}$ values occur with larger NOR and NAND drive strengths, and when the input voltage swing exceeds the gate supply voltage, for example -0.29 in the case of Vin=1.2V and Vdd=0.6V.

A simple approach was proposed in [19] to analyze the transitive fan-out delay impact of slew on a given path. Their delay models were linear versus slew, and they did not consider rise and fall delay separately. So it is straightforward just to sum the delay impact along the path, giving, in our terminology, the transitive fanout delay/slew sensitivity β as

$$\beta_{uv} = \frac{\Delta d_z}{\Delta s_a} + \beta_{vw} \frac{\Delta s_z}{\Delta s_a} \qquad (6.15)$$

where a and z are respectively the input and output pins of gate v shown in Figure 6.3, which has output delay d_z and output slew s_z due to a signal on a with input slew s_a; $\Delta d_z/\Delta s_a$ and $\Delta s_z/\Delta s_a$ are the gradients for gate delay and output slew from the linear models; and the path goes through the connected gates u, v and w in the order $u \rightarrow v \rightarrow w$. A slew change Δs at the output of gate u increases the path delay by $\beta_{uv}\Delta s$.

Our approach to calculate the transitive fanout delay impact of slew considers multiple paths simultaneously and is more accurate, providing upper and lower bounds on the slew impact. To consider simultaneous changes, we account for gates in the transitive fanout changing by determining the worst case slew impact over all alternate cells available for a gate. In reverse topological order, we then calculate the maximum and minimum transitive fanout delay/slew sensitivity β:

$$\beta_{uv,\max} = \max_{s_a, C_{load}} \left\{ \frac{\Delta d_z}{\Delta s_a} \right\} + \max_{w \in fanout(v)} \left\{ \beta_{vw,\max} \max_{s_a, C_{load}} \left\{ \frac{\Delta s_z}{\Delta s_a} \right\} \right\}$$

$$\beta_{uv,\min} = \min_{s_a, C_{load}} \left\{ \frac{\Delta d_z}{\Delta s_a} \right\} + \min_{w \in fanout(v)} \left\{ \beta_{vw,\min} \min_{s_a, C_{load}} \left\{ \frac{\Delta s_z}{\Delta s_a} \right\} \right\}$$

(6.16)

To conservatively bound the slew impact on delay, we multiply by $\beta_{uv,\min}$ if the output slew of u decreases, and by $\beta_{uv,\max}$ if it increases. We do model multiple outputs and separate rise/fall delays, but omit these in Equation (6.16) for clarity.

A change in slew propagating may reduce the delay if the lower bound $\beta_{uv,\min} > 0$ and the change in input slew is $\Delta s < 0$, as $\beta_{uv,\min}\Delta s < 0$. However, as the output slew is the maximum over the timing arcs, this decrease in delay may not propagate if the slew on this arc is not the maximum slew, even if this arc is on the critical path in terms of delay. Consequently, $\beta_{uv,\min} > 0$ may be optimistic. In practice, this does not appear to be a significant issue in most cases, but it might explain why delay reduction can perform poorly with more aggressive slew analysis, where $\beta_{uv,\min}$ is typically around 0.1.

Several approaches may be used when calculating β. Firstly, to be conservative and reduce computation time, β may be calculated over all the alternate cells for a gate, not just the current cell. Alternately, β may be calculated only for the current cell, which avoids re-computation after the best cell is chosen for each gate. β could also be calculated over the current cell and the best alternate cell. This last option was not tried, because it requires additional computation in the inner optimization loop that sets up the linear program.

Secondly, we can conservatively calculate β over all alternate possible input slew and load conditions, or about the current input slew and load conditions only – optimistically assuming that they will not change substantially.

In practice, it is not clear what the best approach to calculate β is. The results achieved with these different options typically vary within about 2% of the best solution, and no single approach is always best. Starting with a more conservative approach, calculating β over all alternate cells for a gate and over all possible load and input slew conditions, and then trying more aggressive settings, calculating β over only the cell for a gate and only the current load and input slew conditions, produces better results on average than starting with more aggressive settings. With the conservative settings, typical values are 0.0 for $\beta_{\min}$ and 0.3 for $\beta_{\max}$. With the aggressive settings, typical values are 0.1 for $\beta_{\min}$ and 0.2 for $\beta_{\max}$.

Adding the additional slew terms to Equation (6.14), we have

$$
\begin{aligned}
t_{vw,rise} \geq t_{uv,fall} + d_{uv,rise} &+ \gamma_v (\Delta d_{uv,rise,v} + \beta_{vw,rise} \Delta s_{uv,rise,v}) \\
&+ \sum_{x \in fanout(v), x \neq w} \gamma_x (\Delta d_{uv,rise,x} + \beta_{vw,rise} \Delta s_{uv,rise,x})
\end{aligned}
\qquad (6.17)
$$

where $\Delta s_{uv,rise,x}$ is the change in slew out of this timing arc if the cell of x changes. Multiplying by β_{vw} gives the worst case transitive slew impact on delay, where we use $\beta_{vw,min}$ if $\Delta s_{uv} < 0$, or $\beta_{vw,max}$ if $\Delta s_{uv} > 0$. The change in delay and slew of v due to the cell of w changing is included in the delay and slew changes for w, thus there is no $\gamma_w(\Delta d_{uv,rise,w} + \beta_{vw}\Delta s_{uv,rise,w})$ term for $t_{vw,rise}$.

Comparing Equation (6.17) to Equation (6.10), there are additional terms here that consider the impact of slew, the impact of the cell of output gates x changing, and rise and fall timing arcs are considered separately.

We examined the importance of including the transitive fanout delay impact of slew by setting $\beta = 0$ and performing optimization. The inaccuracy due to ignoring slew resulted in delay constraints violations and less power savings due to the timing inaccuracy and reduced timing slack available for power minimization [4]. This shows how important it is to account for slew both in static timing analysis and in the optimization formulation.

6.3.2   Formulating cell changes

Now that we have identified what is necessary for delay and power accuracy, we can again address identifying the best alternative cells for a gate. Instead of computing our sensitivity metric with a single value for the delay change in Equation (6.7), $\Delta P_v/\Delta d_v$, we must consider multiple timing arcs. To allow the linear programming approach to be used for delay reduction, we must allow $\Delta d_v < 0$ to be encoded in the LP. As discussed in Section 6.3.1.1, ΔP_v is determined by summing over the change in leakage power, change in internal power, change in switching power of the gate's inputs, and change in switching power of the gate's outputs if Vdd changes.

When considering multiple timing arcs and Δd_v, there are two options that were considered. We could just use the worst change in delay and additionally the transitive delay impact of slew, or we could combine multiple timing arcs into the metric. The latter approach was performed by averaging the delay and transitive delay impact of slew over the timing arcs. Averaging the timing arcs is equivalent to the worst case if pull-up and pull-down drive strengths are balanced and individual timing arcs have similar delays. However, in practice this is not the case (e.g. see Figure 7.4 and Figure 7.5). Generally, the more conservative approach using the worst case delay change on a timing arc to determine Δd_v produces slightly better results, because this is less likely to result in a delay change that violates the delay constraint. A third possible approach, which has not been tried, would be to weight by slack on each timing arc. Note that when Δd_v is calculated, it includes the impact of the cell changing on the delay and the slew of the fanins of v on timing arcs that propagate to v.

The best alternative cell for a gate is chosen as follows. If a cell change reduces power and delay ($\Delta P < 0$, $\Delta d < 0$), pick the cell which best reduces the objective, delay or power. Otherwise: to reduce power pick the cell with maximum power_reduction/delay_increase (min $\Delta P / \Delta d$, $\Delta P < 0$, $\Delta d > 0$); to reduce delay pick the cell with the max delay_decrease/power_increase (max $\Delta P / \Delta d$, $\Delta P > 0$, $\Delta d < 0$). Here, Δd is the maximum delay change over its timing arcs, including the slew impact, or the average change over the timing arcs as discussed above. Note that for different gates we may encode cells that reduce delay or reduce power in the same linear program. For example when minimizing power, if a gate is already minimum size, then there may be no cell change that reduces power further, but if there is a size increase that reduces delay at the expense of power, we encode that in the LP to allow for the situation where other gates can better use the slack that would be created by upsizing the gate – the LP determines whether this is a worthwhile trade-off or not.

The best delay and power cell alternatives for a gate are cached by its input slews, input arrivals and load capacitance values, and that of its fanins along with their supply voltages. Caching provides substantial speed ups for setting up the LP on later iterations.

The LP solution gives values for the cell choice variables γ_v between zero and one. A threshold is used to determine when to change a cell: if a cell reduced delay and $\gamma_v > 0.01$, the alternate cell was used; if a cell reduced power and $\gamma_v > 0.99$, the alternate cell was used. A number of different thresholds were tried, these produced the best results. The philosophy behind using a threshold of 0.01 for delay is that if that gate's delay needs to be reduced to meet delay constraints, then the alternate cell must be used. Conversely, a high threshold was set for power reduction, because if γ_v wasn't close to 1, we couldn't guarantee that delay constraints would be met if the cell was changed. In practice, this threshold approach produces very good power minimization results as described in Section 6.5.2.

6.3.3   Input drivers

An additional accuracy improvement is modeling the impact of gates loading the primary inputs, by including input drivers in the circuit representation. The cell for the input drivers is user specified, and might be set to a drive strength X1 inverter for example. The switching power of the driver input pins is not included, nor is the internal power of the input drivers. The set of drivers is denoted D. As the cell for a driver v cannot change, $\gamma_v = 0$, the delay constraint on an input driver is

$$t_{vw,rise} \geq d_{Source\,v,rise} + \sum_{x \in fanout(v), x \neq w} \gamma_x (\Delta d_{uv,rise,x} + \beta_{vw} \Delta s_{uv,rise,x}) \qquad (6.18)$$

where $d_{Source\,v,rise}$ is the rise delay of the input driver.

6.3.4 The linear program

To minimize the total power, the complete formulation for the linear program formulation is

minimize $\sum_{v \in V} \gamma_v \Delta P_v$

subject to $t_{Source\,u,fall} = 0$, for all $u \in D$

$t_{Source\,u,rise} = 0$, for all $u \in D$

$t_{wSink,rise} \leq T_{max}$, for all $w \in fanin(Sink)$
$t_{wSink,fall} \leq T_{max}$, for all $w \in fanin(Sink)$

$0 \leq \gamma_v \leq 1$, for all $v \in V$

$\gamma_v = 0$, for all $v \in D$

For all $v \in V \cup D$, $w \in fanout(v)$, $u \in fanin(v)$,

timing arc constraints:

$$t_{vw,rise} \geq t_{uv,fall} + d_{uv,rise} + \gamma_v (\Delta d_{uv,rise,v} + \beta_{vw,rise} \Delta s_{uv,rise,v}) \qquad (6.19)$$
$$+ \sum_{x \in fanout(v), x \neq w} \gamma_x (\Delta d_{uv,rise,x} + \beta_{vw,rise} \Delta s_{uv,rise,x})$$
$$t_{vw,fall} \geq t_{uv,rise} + d_{uv,fall} + \gamma_v (\Delta d_{uv,fall,v} + \beta_{vw,fall} \Delta s_{uv,fall,v})$$
$$+ \sum_{x \in fanout(v), x \neq w} \gamma_x (\Delta d_{uv,fall,x} + \beta_{vw,fall} \Delta s_{uv,fall,x})$$

where $t_{uv,fall}$ is the falling arrival time at v from gate u; $d_{uv,rise}$ is the delay from the signal on uv to the output of v rising; and $\Delta d_{uv,rise,x}$ and $\Delta s_{uv,rise,x}$ are the changes in delay and slew out of this timing arc if the cell of x changes. Multiplying by β_{vw} gives the worst case transitive slew impact on delay, where we use $\beta_{vw,min}$ if $\Delta s_{uv} < 0$, or $\beta_{vw,max}$ if $\Delta s_{uv} > 0$. The change in delay and slew of v due to the cell of w changing is included in the delay and slew changes for w. This is why we don't have a $\gamma_w (\Delta d_{uv,rise,w} + \beta_{vw,rise} \Delta s_{uv,rise,w})$ term for $t_{vw,rise}$ for example. The LP with linear approximations cannot model the higher order delay impact of multiple cells changing ($\gamma_v \gamma_x$ terms). Solving such higher order problems would be much slower.

The delay constraints specified in Equation (6.19) assume that the gates have negative polarity, that is a rising input causes a falling output if there is a logical transition, or a falling input causes a rising output. Constraints for positive polarity and nonunate transitions can be handled similarly. Positive polarity means that a rising (falling) output is caused by a rising (falling) input transition. A nonunate transition can be caused by both a rising input or a falling input, depending on the value of other inputs – for example for an XOR gate. Our software handles all timing arcs, including positive polarity and nonunate gates, and gates with multiple outputs. Wire delays are included in the timing analysis. The wire load model is specified in the library, or extracted post-layout wiring parasitics could be specified for each wire.

We can also use the same LP formulation approach to reduce delay when $T > T_{\max}$. When reducing delay, the objective is

$$\text{minimize} \quad \max\{\tau T_{\max}, T\} + k \sum_{v \in V} \gamma_v \Delta P_v \tag{6.20}$$

where k is a weight to limit the power increase when reducing delay, and τ limits the delay reduction. If the ratio of total power to the critical path delay is large, then k should be small to allow delay reduction. For our benchmarks, the best values were k of 0.01 and τ of 0.99, so that after delay reduction there is timing slack for further power minimization. In several cases to meet T_{max}, k of 0.001 and τ of 0.98 were used.

The complete formulation of the linear program for delay reduction with a weighting on power is

$$\text{minimize} \quad \max\{\tau T_{\max}, T\} + k \sum_{v \in V} \gamma_v \Delta P_v$$

$$\text{subject to} \quad t_{Source\ u, fall} = 0, \text{ for all } u \in D$$

$$t_{Source\ u, rise} = 0, \text{ for all } u \in D$$

$$t_{wSink, rise} \leq T_{\max}, \text{ for all } w \in fanin(Sink)$$
$$t_{wSink, fall} \leq T_{\max}, \text{ for all } w \in fanin(Sink)$$

$$0 \leq \gamma_v \leq 1, \text{ for all } v \in V$$

$$\gamma_v = 0, \text{ for all } v \in D$$

For all $v \in V \cup D$, $w \in fanout(v)$, $u \in fanin(v)$,

timing arc constraints:

$$t_{vw, rise} \geq t_{uv, fall} + d_{uv, rise} + \gamma_v (\Delta d_{uv, rise, v} + \beta_{vw, rise} \Delta s_{uv, rise, v}) \tag{6.21}$$
$$+ \sum_{x \in fanout(v), x \neq w} \gamma_x (\Delta d_{uv, rise, x} + \beta_{vw, rise} \Delta s_{uv, rise, x})$$

$$t_{vw, fall} \geq t_{uv, rise} + d_{uv, fall} + \gamma_v (\Delta d_{uv, fall, v} + \beta_{vw, fall} \Delta s_{uv, fall, v})$$
$$+ \sum_{x \in fanout(v), x \neq w} \gamma_x (\Delta d_{uv, fall, x} + \beta_{vw, fall} \Delta s_{uv, fall, x})$$

The next section describes how these delay reduction and power minimization linear programs are used iteratively to reduce the circuit power consumption while meeting delay constraints.

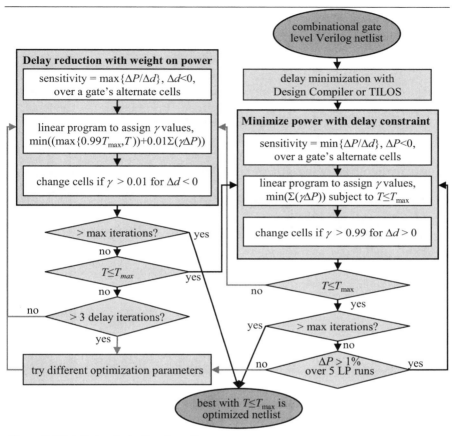

Figure 6.5 Detailed optimization flow diagram.

6.4 OPTIMIZATION FLOW

The optimization flow is shown in Figure 6.5. We start with a combinational gate-level netlist (in a Verilog file), accompanying switching activity and leakage state probabilities (in a SAIF file), and standard cell libraries (Liberty .lib format). Input drivers or input slew and output port load capacitance are user specified. Power minimization is performed subject to a delay constraint. If delay constraints are violated after optimization, delay reduction with a weighting on power is performed. The optimization is iterated until the maximum number of iterations is reached.

The starting point for optimization matters little providing that if a delay constraint is violated, for example after power minimization, we can reduce the delay to satisfy delay constraints. For example, multi-Vth experiments starting with all gates at low Vth rather than high Vth gave only marginally better results after optimization. However, at a fairly tight delay constraint, such as 1.1× the minimum critical path delay, the delay reduction phase may have trouble reducing delay, in which case it is essential to start with a delay minimized netlist.

Alternate cells for a logic gate are chosen by the best power/delay sensitivity as described in Section 6.3.2; we set up the linear program constraints; and then the open source COIN-OR LP solver [7] is used to choose γ values to minimize the power subject to delay constraints. Gates with γ close to 1 in the LP solution are then changed – a threshold of $\gamma > 0.99$ generally worked best. If $\gamma < 0.99$ then there is insufficient slack in the circuit for the cell to be changed without violating a delay constraint, assuming that changing the cell results in a delay increase.

Gates are changed simultaneously without fully considering the impact of other gate changes. If a gate is upsized increasing C_{in} and its fanin is downsized, then the fanin delay is larger than modeled in the LP which may lead to violating T_{max}. If this occurs, we perform delay reduction to satisfy T_{max}. For delay reduction, a threshold of $\gamma > 0.01$ worked well – if a gate has to be upsized to satisfy the delay constraint, we do so.

Thus the solution converges iteratively to reduce power and satisfy T_{max}. At a tight delay constraint, the delay reduction may fail to satisfy T_{max}. Design Compiler with greedy delay_reduction/power_increase reduces delay better than our tool. The output of our optimization is the optimized Verilog netlist, for which the static timing and power analysis can then be verified in Design Compiler.

As the solution converges, the power reduction that is achieved per iteration decreases. To ensure that the solution is close to the optimal that can be achieved by the linear programming approach, several different parameter settings can be tried to see if any additional savings may be achieved.

6.4.1   The importance of the delay reduction phase

The importance of the delay reduction phase is illustrated by Figure 6.6, which shows the typical progress of the optimization flow. After a couple of iterations of power minimization, the timing slack in the initial delay-minimized circuit has been used to perform gate downsizing. If cell changes were not allowed to cause the delay constraint to be violated, the optimization would stop at a solution with total power of about 5.9mW. Alternatively, we can allow the power minimization flow to make cell changes that may end up violating delay constraints, changing the cells of gates that have $\gamma > 0.99$ from the LP solution, without additional computational overheads to double-check that the delay constraints are not violated.

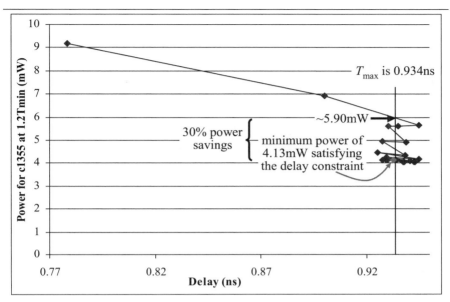

Figure 6.6 This graph shows the power and delay after each iteration of the optimization flow. This is for the c1355 benchmark at a delay constraint of $1.2T_{min}$ for the Design Compiler delay-minimized netlist which is the starting point. The Vdd=1.2V/Vth=0.23V PowerArc characterized 0.13um library was used for this gate sizing optimization.

After cells have been changed and static timing analysis reports a circuit delay that violates the delay constraint T_{max}, we then perform delay reduction, but we add a weight on the power in the objective to ensure that the power does not increase too much. This helps ensure that power minimization phase steps have steeper $\Delta P/\Delta T$ than the delay reduction steps, thus multiple iterations of power minimization with delay reduction can achieve further power savings. In the example shown in Figure 6.6, a substantial 30% power savings are achieved beyond the 5.9mW point, with a minimum power of 4.13mW at a point that meets the delay constraint.

The delay reduction phase allows hill climbing, by allowing the T_{max} constraint to be violated by power minimization to see if additional power savings are possible. The cell changes from delay reduction may be different to those performed in delay minimization of the initial netlist, thus providing some slack back into the circuit for power minimization, without reversing all the power minimization steps that resulted in violating T_{max}.

Unfortunately, the delay reduction phase can sometimes perform poorly at a tight delay constraint. For a delay constraint of $1.1T_{min}$, starting with TILOS-optimized netlists sized for $1.1T_{min}$ results in 4.6% worse results on average than starting with delay minimized netlists [4]. Delay reduction can perform poorly for several reasons.

Table 6.1 The functions performed by the ISCAS'85 benchmark circuits. The smallest benchmark, c17, wasn't in [10].

Circuit	Function
c17	not detailed in their paper
c432	27-channel interrupt controller
c499	32-bit single-error-correcting circuit
c880	8-bit ALU
c1355	32-bit single-error-correcting circuit
c1908	16-bit single-error-correcting/double-error-detecting circuit
c2670	12-bit ALU and controller
c3540	8-bit ALU
c5315	9-bit ALU
c6288	16x16 multiplier
c7552	32-bit adder/comparator

Firstly, the weight on power in the delay reduction objective may prevent certain cell changes that are essential to reduce delay below T_{max}, but cause too large an increase in power. This can be solved by reducing the weighting on power, and allowing additional delay reduction. Setting $\tau = 0.98$ and $k = 0.001$ in Equation (6.20) is a parameter change that is tried to do this.

Secondly, the aggressive slew analysis setting may underestimate the delay increase due to a slew increase and overestimate the delay reduction due to a slew decrease. In contrast, the conservative slew analysis setting is pessimistic and often provides better delay reduction results.

Thirdly, in both phases of the optimization, all cells are changed simultaneously after the optimization. However, the linear program is a linear approximation, and there are no second order terms of the form $\gamma_u \gamma_v$ to directly account for the delay impact of say a cell being downsized while its fanout is upsized. This case can result in delay actually getting worse after delay reduction is attempted. This is more difficult to solve without additional computation overheads, as cells need to be sized individually. For example, a cell being downsized but its fanout upsized could be disallowed, though either one of these on its own would be acceptable, and analysis would be required to decide which of the two cells is better to resize.

6.5   COMPARISON OF GATE SIZING RESULTS

We shall compare our results versus the commercial synthesis tool Design Compiler [23], which is most commonly used in industry today and has a gate sizing approach that is based on TILOS. Design Compiler is generally considered to produce high quality results compared to other commercially available EDA tools [8]. Section 6.5.1 discusses the combinational benchmarks on which we compare results. Section 6.5.2 compares our results versus Design Compiler, which performs gate sizing based on a TILOS sizing approach. Section 6.5.3 then discusses how optimal the linear program sizing results are, and whether any additional improvements can be made.

6.5.1   Benchmarks

Two sets of circuit benchmarks were used in this chapter. The first set of benchmarks was the combinational ISCAS'85 benchmark set [2]. Besides these netlists being small, one of the criticisms of them has been that they are not realistic circuit benchmarks. In particular, what circuits they represent and how to stimulate them properly with input vectors is not detailed in the benchmark set. The ISCAS'85 benchmarks were reverse-engineered by Hansen, Yalcin and Hayes [10], and the functions that they determined for these circuits are listed in Table 6.1. The behavioral Verilog netlists for these reverse-engineered netlists are available [9], and they were synthesized and delay minimized using Design Compiler with the PowerArc characterized Vdd=1.2V/Vth=0.23V 0.13um library. Assignment statements and redundant outputs were removed manually in the synthesized netlists. These gate-level synthesized netlists were simulated with VCS using independent random inputs with equal probabilities of 0 or 1 to produce SAIF (Switching Activity Interchange Format) files with switching activity and gate input state probabilities for power analysis. For comparison to our LP power minimization approach, the delay minimized netlists were then power minimized in Design Compiler restricted to sizing only changes.

The second set of three benchmark circuits was provided by Professor Nikolić's group at the Berkeley Wireless Research Center. The SOVA EPR4 circuit is an enhanced partial response class-4 (EPR4) decoder [26]. There is also a Huffman decoder [18]. These are typical datapath circuits that appear on chips for communication systems. These circuits were mapped to the PowerArc characterized Vdd=1.2V/Vth=0.12V 0.13um library by Sarvesh Kulkarni and Ashish Srivastava, who provided the combinational gate-level netlists. The SOVA EPR4 and R4 SOVA benchmarks are substantially larger than the ISCAS'85 benchmarks.

6.5.2   Comparison versus Design Compiler

For results in this section, the PowerArc characterized 0.13um library at 25°C with Vdd of 1.2V and Vth of 0.23V was used. The channel length was 0.13um. This was characterized for STMicroelectronics 0.13um HCMOS9D process. The library consisted of nine inverter sizes; and four sizes of NAND2, NAND3, NOR2 and NOR3 logic gates. The output port load capacitance was set to 3fF, which is reasonable if the combinational outputs drive flip-flops, and in addition there is a wire load to the port. The wire load model used was 3+2×num_fanout fF. The input slew was 0.1ns for the 1.2V input drive ramps – typical slews within the circuits ranged from 0.05ns to 0.15ns.

The same wire load, input slew, and load conditions were used in [14] and [22]. The switching activities used were directly from the SAIF files. Leakage was about 0.1% of total power, due to characterization at 25°C and high Vth. This avoids the problem with versions of Design Compiler before 2004.12 where either dynamic power or leakage power had to be prioritized, rather than total power – as leakage is so small, prioritizing dynamic power is equivalent to total power, which is minimized by the linear programming approach.

The starting point for LP optimization was the netlists that were synthesized and delay minimized using Design Compiler, and that was also the starting point for sizing_only power minimization in Design Compiler. Results were verified in Design Compiler.

The linear programming approach does better than Design Compiler in all cases, except for c880 at a delay constraint of $1.1 \times T_{min}$ where the LP power is 2.4% higher (shown in bold in Table 6.2). This is not surprising as the LP approach is heuristic and may still get stuck in a local minimum. The LP approach performs better in most cases because it has a global view, rather than the greedy peephole approach of TILOS.

Table 6.2 Here we compare our sizing results (LP) with sizing only power minimization results from Design Compiler (DC), at delay constraints of $1.1T_{min}$ and $1.2T_{min}$, where T_{min} (shown in column 7) is the critical path delay after delay minimization by Design Compiler. Circuit statistics such as the number of logic levels, the numbers of inputs and outputs, the number of gates, and the number of edges between gates in the circuit are also listed. The "LP then DC" results are discussed in Section 6.5.3.

Netlist	# logic levels	# inputs	# outputs	# gates	# edges	Min Delay (ns)	Power (mW) $1.1T_{min}$ DC	$1.1T_{min}$ LP	$1.1T_{min}$ LP then DC	$1.2T_{min}$ DC	$1.2T_{min}$ LP	$1.2T_{min}$ LP then DC
c17	4	5	2	10	17	0.094	1.11	0.96	0.95	0.86	0.76	0.76
c432	24	36	7	259	485	0.733	2.78	2.21	2.18	2.22	1.74	1.70
c499	25	41	32	644	1,067	0.701	5.83	4.59	4.48	4.98	3.73	3.64
c880	23	60	26	484	894	0.700	**3.37**	**3.45**	**3.13**	2.83	2.60	2.54
c1355	27	41	32	764	1,322	0.778	6.88	5.42	5.26	5.97	4.12	4.04
c1908	33	33	25	635	1,114	0.999	3.26	3.08	3.01	2.67	2.40	2.36
c2670	23	234	139	1,164	1,863	0.649	9.23	8.42	8.28	8.08	6.87	6.79
c3540	36	50	22	1,283	2,461	1.054	6.69	5.79	5.70	5.60	4.64	4.53
c5315	34	178	123	1,956	3,520	0.946	10.39	9.48	9.15	8.82	7.81	7.66
c6288	113	32	32	3,544	6,486	3.305	6.91	6.07	5.89	6.08	4.69	4.61
c7552	31	207	86	2,779	4,759	0.847	18.02	16.65	16.34	15.60	13.44	13.23
Huffman	29	79	42	774	1,286	0.845	6.02	4.81	4.62	5.07	3.72	3.61
SOVA EPR4	110	791	730	15,686	27,347	3.039	17.07	15.82	15.61	15.28	13.89	13.73
R4 SOVA	144	1,177	815	33,344	59,178	4.811	24.26	21.81	21.22	20.82	19.16	18.69
Minimum power savings vs. Design Compiler:								-2.4%	7.0%		8.0%	10.1%
Average power savings vs. Design Compiler:								12.0%	14.5%		16.6%	18.1%
Maximum power savings vs. Design Compiler:								21.4%	23.5%		30.9%	32.3%

We achieved 12.0% and 16.6% average power savings versus Design Compiler at delay constraints of $1.1 \times T_{min}$ and $1.2 \times T_{min}$ respectively – see Table 6.2. We achieved lower power even on the smallest ISCAS'85 benchmark, c17, as was illustrated in Figure 6.1. This illustrates the sub-optimal choices made by greedy optimization approaches that only consider individual gates, rather than the whole circuit. These results were versus the 2003.03 version of Design Compiler – version 2004.12 was also tried, but it produced worse power results on average.

6.5.3 Post-pass cleanup with Design Compiler

After our linear programming optimization returns the lowest power solution that satisfies the delay constraint, there is a little timing slack left, up to about 0.6% of the delay target, which can be used to further downsize individual gates. Each LP optimization pass sizes multiple gates, whereas to fully utilize the remaining slack, individual gates should be sized. A power_reduction/delay_increase sensitivity approach such as provided by Design Compiler's sizer is appropriate for this.

On average, a post-pass with Design Compiler on the LP power minimized netlists achieved another 2% to 3% power savings versus the LP results, as listed in the "LP then DC" columns in Table 6.2. Interestingly, for c880 where the LP results were worse than Design Compiler at $1.1T_{min}$ by 2.4%, the post-pass by Design Compiler improves the result by 9.3%, giving 7% overall power reduction with "LP then DC" versus the Design Compiler result. After running Design Compiler, there is typically at most 0.001ns slack, i.e. less than 0.1% of the delay constraint. The average power savings of the "LP then DC" results versus Design Compiler were 14.5% and 18.1% at delay constraints of $1.1T_{min}$ and $1.2T_{min}$ respectively.

It should be noted that multiple passes of power minimization sizing by Design Compiler on its own does not provide any significant benefit (<1%) over a single incremental power minimizing sizing compilation in Design Compiler. Design Compiler gets stuck in a local minimum where it has greedily downsized the wrong gates.

6.6   COMPUTATIONAL RUNTIME

This section examines the runtime for the linear program, and then compares it to the TILOS-like sizing runtimes in Design Compiler.

6.6.1   Theoretical runtime complexity and actual runtimes

It is not straightforward to determine the theoretical worst case runtime complexity. The open source COIN-OR LP solver [7] uses the simplex method to solve the linear program, which has exponential runtime growth with problem size in the worst case. There are linear programming methods with guaranteed polynomial runtime; however, typically the simplex method is a fast method for solving linear programs.

Each vertex in the directed acyclic graph representation has one "free" cell choice variable, and edges between vertices (i.e. wires between gates) determine the constraints. As a result, our linear program constraint matrix is sparse, because the number of edges is of similar order to the number of vertices ($O(|E|)$ is $O(|V|)$ for our benchmarks). Thus we might expect that runtime growth with circuit size will be reasonable.

To measure the actual runtimes, we need to consider when optimization should be terminated. The LP optimization flow consists of two approaches: power minimization subject to a delay constraint, and delay reduction with a weighting on power. In both of these, a linear program is posed and solved to determine which cells to change. These alternating optimization phases shift back and forth in the power-delay space about the delay constraint, as was illustrated in Figure 6.6. In addition, the more sophisticated optimization approach changes parameter settings if optimization progress is slow. Consequently, there is no clearly defined optimization endpoint. However, the point at which any further power savings are minimal can be measured. To do this, we can run a large number of iterations, where each iteration refers to a run of setting up the LP, solving it, and changing cells, whether this is for power minimization or delay reduction.

Table 6.3 Number of iterations for gate sizing with the LP optimization flow to find a solution that satisfies the delay constraint and is within 1% of the minimum power found in 40 LP iterations. Runtimes for 20 iterations at a delay constraint of $1.2 \times T_{min}$ with the 0.13um Vdd=1.2V, Vth=0.23V library are listed. The number of iterations to get within 1% does not depend on the circuit size, whereas the LP solver runtime grows roughly quadratically.

Benchmark	# gates	# iterations to get within 1%		Runtime for 20 iterations (s)		
		At $1.1T_{min}$	At $1.2T_{min}$	Total	LP Solver	Total - LP Solver
c17	10	14	4	0.6	0.2	0.4
c432	259	15	21	28.2	8.8	19.4
c880	484	16	14	49.8	16.4	33.4
c1908	635	5	12	70.5	26.6	44.0
c499	644	21	14	63.7	23.7	40.0
c1355	764	13	15	79.4	30.5	49.0
Huffman	774	16	20	67.7	27.3	40.4
c2670	1,164	12	17	105.4	42.4	63.1
c3540	1,283	12	9	223.5	114.6	108.9
c5315	1,956	9	9	231.6	96.7	134.9
c7552	2,779	13	12	383.8	165.2	218.6
c6288	3,544	21	15	1,811.2	1,450.5	360.7
SOVA EPR4	15,686	8	7	3,427.1	2,106.1	1,321.0
R4 SOVA	33,344	5	5	20,078.2	17,679.1	2,399.0

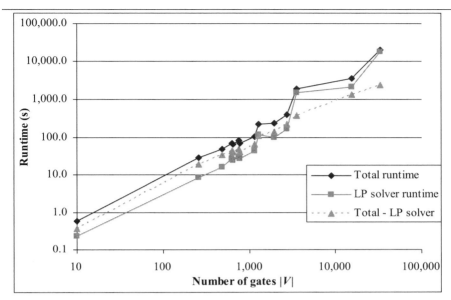

Figure 6.7 Runtimes for the LP approach in Table 6.3 shown on a log-log scale.

We examined the number of iterations required to get within 1% of the best solution found in 40 iterations. It appears that the number of iterations to get a good solution with gate sizing is not dependent on the circuit size or circuit depth in terms of logic levels. This is because gates are sized simultaneously on each iteration. From the data in Table 6.3, about 20 iterations is sufficient to get good results. Fewer iterations were required for the two largest benchmarks (SOVA EPR4, R4 SOVA). If fewer iterations are required for larger benchmarks, the growth of runtime with circuit size will be less.

The runtime for twenty iterations for sizing the benchmarks on a 2GHz Athlon XP with 512KB of L2 cache is shown in Figure 6.7 on a log-log scale, as there is a wide range of circuit sizes, over three orders of magnitude. The runtime for static timing and power analysis, posing the LP, and changing cells using the LP solution grows linearly with the circuit size and dominates the total runtime for smaller netlists, below about 1,000 gates. The runtime for the linear program solver can grow quadratically with circuit size, and is the dominant portion of the total runtime for the larger circuits.

The LP solver runtime is substantially larger at three points, benchmarks c3540, c6288 and R4 SOVA. The runtime for the LP solver grows between $O(|V|)$ and $O(|V|^2)$, where $|V|$ is the number of gates [4].

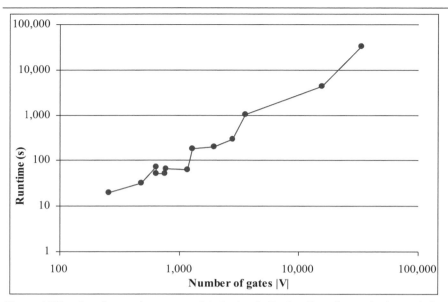

Figure 6.8 Runtime for running power minimization in Design Compiler on the best solution found by linear programming, that is for the "LP then DC" Design Compiler run. These runtimes were on a 300MHz Sun Ultra II. Design Compiler's runtimes grow faster than $O(|V|)$, but slower than $O(|V|^2)$ [4].

6.6.2   Runtime comparison versus Design Compiler

It is interesting to compare runtimes versus Design Compiler's runtimes, run on a 300MHz Sun Ultra II, shown in Figure 6.8. This was for Design Compiler performing power minimization on the netlists after linear programming optimization, where the analysis is similar to what occurs in a single iteration of linear programming power minimization. Accounting for the much slower computer used to run the benchmarks, power minimization in Design Compiler is about an order of magnitude faster than our runtimes. Our runtimes in Figure 6.7 and Design Compiler's runtimes in Figure 6.8 have quite similar shapes – performing faster on certain benchmarks and slower on others.

There are several ways that the runtimes for the LP approach could be reduced. Firstly, the standard settings for the linear program solver have been used, which performs analysis with the simplex algorithm to converge to a precise local minimum. However, we don't need the same degree of precision, and relaxing the accuracy requirement would reduce the number of simplex iterations and speed up the LP solver. Secondly, the first few LP iterations provide the biggest power savings. Later iterations tend to bounce around the optimal solution as too many gates are being changed simultaneously. It may be possible to run fewer LP iterations and then use a TILOS sizing post-pass to clean up the result. Thirdly, additional analysis could be performed with the solution from the LP solver to avoid changing gates that would cause a delay constraint violation, or to use an alternate cell that reduces power but avoids the delay constraint violation. Lastly, a better delay minimization approach, getting closer to the delay constraint, may help speed up convergence. These latter suggestions have computational overheads too, so experiments are needed to see what benefit they offer.

6.7   SUMMARY

The gate sizing results in this chapter demonstrated that commonly used greedy TILOS-like circuit sizing approaches are suboptimal. It was known that this traditional approach to gate sizing could be suboptimal for small circuit examples, but it was not clear how to address the problem, nor whether there was significant suboptimality on typical circuits.

Our linear programming optimization flow simultaneously optimizes all gates in the circuit. Comparing the LP approach to the commercial implementation of a TILOS-like sizer in Design Compiler, the power savings on average were 12.0% and 14.5% at $1.1 \times T_{min}$ and $1.2 \times T_{min}$ respectively. We achieved a power reduction of 31% on one circuit.

Iterating cycles of reducing power then reducing delay to meet the delay constraint provides more power savings than stopping power minimization when the delay constraint is reached: further cycles of delay reduction then power reduction get out of this local minimum. Results also demonstrated the importance of having accurate delay and power analysis within the optimization formulation. In particular, it is important to consider slew and separate timing arcs, which much academic optimization research tends to avoid.

The runtime for posing the linear program constraints and changing cells using the linear programming solution scales linearly with circuit size. The LP solver runtimes scale between linearly and quadratically with circuit size, so this approach is applicable for larger circuits. Some approaches that may be useful for reducing the runtime of the linear programming solver have been outlined.

There are two improvements that may be made to our approach. (1) A traditional sizing tool, like Design Compiler, is better for pure delay minimization. This greedy, one gate at a time, optimization approach is also useful for a slight further improvement after our optimization. It was observed that a post-pass sizing individual gates with Design Compiler improved on the linear programming results by a further 2% to 3%. (2) If γ < threshold to change a gate's cell, other cells which require less slack could be considered. In particular, the current linear programming formulation is not applicable for a tight delay constraint, as the delay reduction phase is incapable of meeting a very tight delay constraint. In practice, this will not generally be a major issue when power is a significant constraint, as the delay constraint is usually relaxed a little to allow a more energy optimal solution, rather than many gates being upsized at a tight delay constraint.

Given the computational complexity for this non-convex gate sizing optimization problem, it is not possible to compare the results found to the global minimum except for very small circuits. In comparison to other heuristic approaches, there was only one case where the linear programming approach was worse than Design Compiler. This was for c880 at $1.1 \times T_{min}$ in Table 6.2, where the power was 2.4% higher than Design Compiler, and 11.4% worse than the result found by running the LP approach then Design Compiler. The best results were found by using the linear programming approach with a post-pass by Design Compiler, which averages 2% lower power than just using the LP approach.

We did compare our LP results versus the equivalent integer linear programming (ILP) formulation with the cell choice variables γ_v restricted to 0 or 1. The integer problem is too computationally expensive to solve completely, except for the smallest benchmark c17. However, we compared ILP results from CPLEX's solver with a 1,000s time limit per iteration to the LP results. The ILP results were not better on average than the LP results, and were worse than the LP results after a post-pass by Design Compiler [4]. Given the prohibitive computation times for ILP and negligible benefit, the LP relaxation of the ILP problem should be used.

Our LP approach can be extended to a second order conic program (SOCP) to include the impact of process variation in the manner described in Chapter 12, taking advantage of the timing analysis accuracy improvements detailed in this chapter.

This chapter focused on power minimization subject to a delay constraint, but our approach is equally applicable to area minimization subject to a delay constraint.

Chapter 7 analyzes the power savings that can be achieved with use of multiple threshold voltages and multiple supply voltages versus the strong gate-sizing approach provided in this chapter.

6.8   REFERENCES

[1] Berkelaar, M., and Jess, J., "Gate sizing in MOS digital circuits with linear programming," in *Proceedings of the European Design Automation Conference*, 1990, pp. 217-221.

[2] Brglez, F., and Fujiwara, H., "A neutral netlist of 10 combinational benchmark circuits and a target translator in Fortran," in *Proceedings of the International Symposium Circuits and Systems*, 1985, pp. 695-698.

[3] Chandrakasan, A., and Brodersen, R., "Minimizing Power Consumption in Digital CMOS Circuits," in *Proceedings of the IEEE*, vol. 83, no. 4, April 1995, pp. 498-523.

[4] Chinnery, D., *Low Power Design Automation*, Ph.D. dissertation, Department of Electrical Engineering and Computer Sciences, University of California, Berkeley, 2006.

[5] Dharchoudhury, A., Blaauw, D., Norton, J., Pullela, S., and Dunning, J., "Transistor-level sizing and timing verification of domino circuits in the Power PC microprocessor," in *Proceedings of the International Conference on Computer Design*, 1997, pp. 143-148.

[6] Fishburn, J., and Dunlop, A., "TILOS: A Posynomial Programming Approach to Transistor Sizing," in *Proceedings of the International Conference on Computer-Aided Design*, 1985, pp. 326-328.

[7] Forrest, J., Nuez, D., and Lougee-Heimer, R., CLP User Guide, http://www.coin-or.org/Clp/userguide/

[8] Goering, R., "The battle for logic synthesis," EE Times, September 9, 2004. http://www.eetimes.com/news/design/columns/tool_talk/showArticle.jhtml?articleID=46200731

[9] Hansen, M., Yalcin, H., Hayes, J., ISCAS High-Level Models. http://www.eecs.umich.edu/~jhayes/iscas.restore/benchmark.html

[10] Hansen, M., Yalcin, H., Hayes, J., "Unveiling the ISCAS-85 Benchmarks: A Case Study in Reverse Engineering," *IEEE Design & Test of Computers*, vol. 16, no. 3, 1999, pp. 72-80.

[11] ILOG, *ILOG CPLEX 10.1 User's Manual*, July 2006, 476 pp.

[12] Jacobs, E., "Speed-Accuracy Trade-off in Gate Sizing," in *Proceedings of the workshop on Circuits, Systems and Signal Processing*, 1997, pp. 231-238.

[13] Kasamsetty, K., Ketkar, M. and Sapatnekar, S., "A New Class of Convex Functions for Delay Modeling and their Application to the Transistor Sizing Problem," *IEEE Transactions on Computer-Aided Design of Integrated Circuits and Systems*, vol. 19, no. 7, 2000, pp. 779-788.

[14] Kulkarni, S., Srivastava, A., and Sylvester, D., "A New Algorithm for Improved VDD Assignment in Low Power Dual VDD Systems," *International Symposium on Low-Power Electronics Design*, 2004, pp. 200-205.

[15] Nair, R., et al., "Generation of Performance Constraints for Layout," *IEEE Transactions on Computer-Aided Design*, vol. 8, no. 8, 1989, pp. 860-874.

[16] Narendra, S., et al., "Comparative Performance, Leakage Power and Switching Power of Circuits in 150 nm PD-SOI and Bulk Technologies Including Impact of SOI History Effect," *Symposium on VLSI Circuits*, 2001, pp. 217-8.

[17] Nguyen, D., et al., "Minimization of Dynamic and Static Power Through Joint Assignment of Threshold Voltages and Sizing Optimization," *International Symposium on Low Power Electronics and Design*, 2003, pp. 158-163.

[18] Nikolić, B., et al., "Layout Decompression Chip for Maskless Lithography," in *Emerging Lithographic Technologies VIII, Proceedings of SPIE*, vol. 5374, 2004, 8 pp.

[19] Pant, P., Roy, R., and Chatterjee, A., "Dual-Threshold Voltage Assignment with Transistor Sizing for Low Power CMOS Circuits," *IEEE Transactions on VLSI Systems*, vol. 9, no. 2, 2001, pp. 390-394.

[20] Satish, N., et al., "Evaluating the Effectiveness of Statistical Gate Sizing for Power Optimization," Department of Electrical Engineering and Computer Science, University of California, Berkeley, California, ERL Memorandum M05/28, August 2005.

[21] Sirichotiyakul, S., et al., "Stand-by Power Minimization through Simultaneous Threshold Voltage Selection and Circuit Sizing," in *Proceedings of the Design Automation Conference*, 1999, pp. 436-41.

[22] Srivastava, A., Sylvester, D., and Blaauw, D., "Power Minimization using Simultaneous Gate Sizing Dual-Vdd and Dual-Vth Assignment," in *Proceedings of the Design Automation Conference*, 2004, pp. 783-787.

[23] Synopsys, *Design Compiler User Guide*, version U-2003.06, June 2003, 427 pp.

[24] Tennakoon, H., and Sechen, C., "Gate Sizing Using Lagrangian Relaxation Combined with a Fast Gradient-Based Pre-Processing Step," in *Proceedings of the International Conference on Computer-Aided Design*, 2002, pp. 395-402.

[25] Wei, L., et al., "Mixed-V_{th} (MVT) CMOS Circuit Design Methodology for Low Power Applications," in *Proceedings of the Design Automation Conference*, 1999, pp. 430-435.

[26] Yeo, E., et al., "A 500-Mb/s Soft-Output Viterbi Decoder," *IEEE Journal of Solid-State Circuits*, vol. 38, no. 7, 2003, pp. 1234-1241.

Chapter 7

LINEAR PROGRAMMING FOR MULTI-VTH AND MULTI-VDD ASSIGNMENT

David Chinnery, Kurt Keutzer
Department of Electrical Engineering and Computer Sciences
University of California at Berkeley
Berkeley, CA 94720, USA

Having provided a strong gate sizing benchmark using only a single transistor threshold voltage (Vth) and single supply voltage (Vdd) in Chapter 6, we now examine the impact of additionally using multiple-Vth and dual Vdd to minimize power. Comparing cells with different Vth values is no different to comparing cells with different sizes, providing that the leakage is included in the total circuit power. Multiple supply voltages can also be handled similarly, with level converter overheads for restoring to high Vdd.

Our dual-Vdd/dual-Vth/sizing results achieve on average 5% to 13% power savings versus the two alternate dual-Vdd/dual-Vth/sizing optimization approaches suggested in [6] and [10]. Importantly, the linear programming approach has runtimes that scale between linearly and quadratically with circuit size, whereas other algorithms that have been proposed for multi-Vdd, multi-Vth and gate size assignment have cubic runtime growth. This chapter examines in detail optimization with multiple supply voltages and multiple threshold voltages.

7.1 INTRODUCTION

A high supply voltage and a low threshold voltage may be necessary to meet circuit delay constraints. However, using a lower Vdd can quadratically reduce the dynamic power, and using a higher Vth can exponentially reduce the leakage power. Thus it is possible to substantially reduce power while meeting delay constraints by using high Vdd with low Vth on delay critical paths, and low Vdd with high Vth where there is sufficient timing slack. There are significant design costs for using multiple supply voltages and multiple threshold voltages, so circuit designers are concerned about how much power saving multi-Vdd and multi-Vth can truly provide.

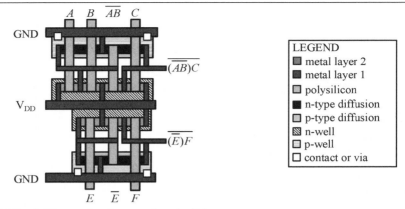

(a) Single V_{DD} requires less area than dual V_{DD}

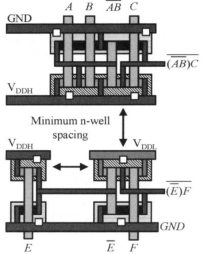

(b) Horizontal and vertical well isolation
issues when PMOS n-wells connect to
different Vdd

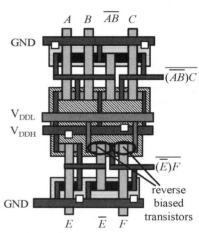

(c) Connecting the n-well of the PMOS
transistors to VDDH reverse biases the
substrate of the VDDL NAND gate (by
$V_{DDH} - V_{DDL}$), increasing the PMOS V_{th}

Figure 7.1 This diagram illustrates some of the differences between single Vdd and dual Vdd layout. Single Vdd layout is more compact as shown in (a). If the PMOS n-wells are connected to different supply voltages, then there are minimum spacing requirements as shown in (b). An alternative is to connect the n-wells of PMOS transistors in both VDDH and VDDL gates to VDDH, but this reverse biases the PMOS transistors in the VDDL gate as shown in (c). Note that the PMOS n-wells in (c) are all connected to V_{DDH}.

Each additional PMOS and NMOS threshold voltage requires another mask to implant a different density of dopants, which substantially increases processing costs. A set of masks costs on the order of a million dollars today and an additional Vth level increases the fabrication cost by 3% [8]. Each additional mask also increases the difficulty of tightly controlling process yield, which strongly motivates manufacturers to limit designs to a single NMOS and single PMOS threshold voltage. From a design standpoint, an advantage of multiple threshold voltages is that changing the threshold voltage allows the delay and power of a logic gate to be changed without changing the cell footprint, and thus not perturbing the layout.

Each additional supply voltage requires an additional voltage regulator and power supply rails for that voltage. The logic needs to be partitioned in some manner into voltage regions where a single supply is used. The regions of each supply voltage are not usually fully utilized and some spacing is required between them, increasing chip area. Wire lengths also increase between cells in different Vdd regions. The area overhead for gate-level dual Vdd assignment in modules of a media processor was 15% [13].

An alternative is to route the two supply rails along every standard cell row, which increases the cell height and has a similar area overhead. In bulk CMOS there are also minimum spacing issues between the PMOS n-wells at different biases to prevent latchup, as shown in Figure 7.1(b). The PMOS n-wells in high Vdd (VDDH) gates cannot be connected to low Vdd (VDDL) as this forward biases the transistors, increasing leakage substantially, and can cause other problems. VDDL gates can have the PMOS n-well connected to VDDH as shown in Figure 7.1(c), but this reverse biases the transistors, making the VDDL gate even slower – though this can be compensated for by using a lower PMOS Vth for VDDL gates.

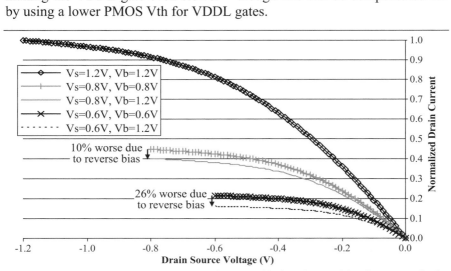

Figure 7.2 This graph shows the impact of reverse biasing the PMOS substrate. V_s is the source voltage (the supply voltage for an inverter), V_b is the body bias, and the input voltage is fixed at 0V. The body is reverse biased when $V_b > V_s$, which increases the threshold voltage and reduces the current. The thicker lines show drain current without reverse biased PMOS substrate, and the thin lines show the drain current with substrate reverse biased at 1.2V.

If the substrate is reverse biased at 1.2V for PMOS transistors with 0.8V and 0.6V drain-source voltage, the drain current is 10% lower at 0.8V and 26% lower at 0.6V, as shown in Figure 7.2. Even without being reverse biased, the PMOS transistor drain current is 55% and 78% lower respectively than a PMOS transistor connected to a high supply voltage of 1.2V. In silicon-on-insulator (SOI) technology, the transistors are isolated and spacing between wells at different biases is not an issue.

If the alternate supply voltage rails are routed on each standard cell row, it is much easier to change the supply voltage of cells without substantially perturbing the layout, simplifying post-layout optimization. It also makes it easy to connect to both power supply rails, which is needed for most voltage level restoration "level converters" [8]. When different supply voltages are routed next to each other, the second metal layer may be needed for internal wiring in logic cells, which creates blockages on metal layer two, reducing routing resources for wiring between cells.

Whether using separate voltage regions or routing the supply voltages next to each other, using multi-Vdd increases the chip area, which lowers the yield per wafer and increases fabrication costs.

Our optimization approach does not consider the increased area for multi-Vdd and the impact on yield of both multi-Vdd and multi-Vth approaches, but clearly there must be significant power savings to justify the cost. Large power savings have been suggested by a number of researchers, but in some cases the delay and the power models were inaccurate, or the initial comparison point was poor, causing power savings to be overstated. To justify the use of multi-Vdd and multi-Vth, we must show substantial power savings versus a good choice of single supply voltage and single threshold voltage with power minimization by gate sizing.

It is essential that gate sizing is considered with multi-Vdd and multi-Vth, as gate sizing can achieve greater power savings at a tight delay constraint. We found that the power savings achieved with multi-Vdd and multi-Vth are in most cases less than the power savings achieved by gate sizing with the linear programming approach versus the TILOS-like optimizers.

While multiple threshold voltages do not complicate the optimization problem, using multiple supply voltages requires insertion of voltage level converters to restore the voltage swing to higher Vdd gates as described in Section 7.2. As there is yet no standard approach for multiple supply voltage and multiple threshold voltage optimization, Section 7.3 summarizes previous research in the area, discussing the limitations and advantages of the various optimization approaches. Few papers have considered trying to perform simultaneous optimization with assignment of multiple supply voltages, multiple threshold voltages, and gate sizes. How the voltage level converter power and delay overheads are handled with the linear programming approach is detailed in Section 7.4.

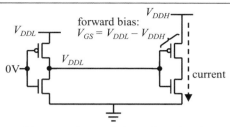

Figure 7.3 This diagram illustrates the need for voltage level restoration. The schematic shows an inverter with supply voltage of V_{DDH}, being driven by an inverter of supply voltage V_{DDL}. When the V_{DDL} inverter output is high, the driving voltage is only V_{DDL}, which results in a forward bias of $V_{DDL} - V_{DDH}$ across the PMOS transistor of the V_{DDH} inverter. The forward-biased PMOS transistor is not completely off resulting in large static currents.

Perhaps the best of the multi-Vdd/multi-Vth/sizing optimization methods proposed by other researchers are the two approaches proposed by Sarvesh Kulkarni, Ashish Srivastava, Dennis Sylvester, and David Blaauw in Chapter 8 [6][10]. Under the same conditions, the linear programming approach for multi-Vdd/multi-Vth/sizing is compared versus their results in Section 7.5. On average, the linear programming approach reduces power 5% to 13% versus their results across a range of delay constraints.

Having established that the linear programming approach performs well for supply voltage assignment and threshold voltage assignment as well as gate sizing, Section 7.6 examines how much power can be saved with multi-Vth and multi-Vdd versus using a single Vdd and single Vth. Section 7.7 discusses the impact of multi-Vdd and multi-Vth assignment in addition to sizing on the runtimes. Section 7.8 gives a summary of our results.

7.2 VOLTAGE LEVEL RESTORATION FOR MULTI-VDD

If a low Vdd (VDDL) input drives a high Vdd (VDDH) gate, the PMOS transistors are forward biased by $V_{DDL} - V_{DDH}$ which results in static current. This is illustrated with two inverters in Figure 7.3. To avoid this, a voltage level converter is needed to restore the signal to full voltage swing, restoring the signal from 0V↔VDDL to 0V↔VDDH. A VDDH gate may drive a VDDL gate.

Algorithms for gate level supply voltage assignment can be broadly sepa-rately into two methodologies depending on where voltage level restoration may occur. Clustered voltage scaling (CVS) [12] refers to when voltage level restoration only occurs at the registers, to reduce the level converter power and delay overhead. All VDDL gates must either drive VDDL gates or drive a level converter latch. Hence there are distinct clusters of VDDL combina-tional gates that have only VDDL combinational gates in their transitive fanout. A gate may only be changed from VDDH to VDDL if the fanouts are all VDDL, or changed from VDDL to VDDH if the fanins are all VDDH. Extended clustered voltage scaling (ECVS) [14] additionally allows "asynchronous" level converters to be placed between combinational logic gates, removing the restrictions on when a gate's Vdd may be changed. In ECVS, VDDL gates are still clustered in order to amortize the power and delay overheads for the level converters. CVS and ECVS algorithms are detailed in Section 8.2

Combining a level converter with a flip-flop minimizes the power overhead for voltage level restoration. As typical level converter and flip-flop designs essentially include a couple of inverters acting as a buffer, the level converter can replace these in the flip-flop. The power consumed by an LCFF can be less than that of a VDDH flip-flop, particularly if a low-voltage clock signal is used [2]. An LCFF with high-speed pulsed flip-flop design is comparable in delay to a regular D-type flip-flop [4], thus avoiding the level converter delay overhead. The delay and power overheads for voltage level restoration are minimal in CVS.

There has been concern about the noise immunity of asynchronous level converters [4]. The asynchronous level converters that we use for ECVS were shown to be robust and have good noise immunity [5].

We now look at the algorithms that have been proposed previously for multi-Vth and multi-Vdd optimization.

7.3   PREVIOUS MULTI-VDD AND MULTI-VTH OPTIMIZATION RESEARCH

A number of researchers have explored use of multiple threshold voltages, and/or multiple supply voltages. Some papers report large power savings versus an initial configuration that is substantially sub-optimal. For example when starting with a circuit with high leakage power with all transistors at low threshold voltage, introducing a second higher threshold voltage will provide significant power savings, but the real question is how much power would be saved with dual Vth versus choosing a more optimal single threshold voltage and sizing gates optimally. To address this, multi-Vth and multi-Vdd results in Section 7.6 are compared versus the optimal sizing results with the best threshold voltage available from a choice of three Vth values – though if a finer granularity of Vth were available, that would no doubt provide some additional power savings. In particular, we generally look at dual Vth/sizing power savings versus sizing with a higher Vth library, as it is much more difficult to reduce power versus an initial configuration that is already low on leakage power that has gate sizes optimized to reduce dynamic power.

Another major shortcoming of many academic papers is using simplified delay and power models that are inaccurate, such as ignoring slew and failing to use separate rise and fall timing arcs. Few algorithmic papers state the accuracy of their models. In some cases, it is not clear how optimization approaches with such models can be extended to real circuits using full static timing analysis with standard cell libraries, as their algorithmic approach or computation speed depend on the underlying simplified models. Merging rise and fall delays halves the number of delay constraints in optimization. Simplified power and delay analysis that ignores slew and doesn't use lookup tables for analysis, for example using linear interpolation versus load capacitance, can speed up analysis runtimes by an order of magnitude. This has a major impact on the total computational runtime as analyzing trade-offs is an essential portion of the inner loop of any optimizer. For example, runtimes reported for our initial linear programming approach [7] are about 10× faster with 0.18um simplified logical effort delay models and no internal power analysis versus interpolating 0.13um library data.

Given the range of computers used to run benchmarks in different papers, it is difficult to directly compare runtimes across tools. For the purposes of sizing large circuits in industry, it is more interesting to compare the runtime complexity. It is essential that runtime complexity be less than $O(|V|2)$, where $|V|$ is the number of gates in a circuit, to run on circuits of any appreciable size [11].

We summarize some of the better multi-Vth and multi-Vdd optimization approaches below.

7.3.1   Summary of papers on optimization with multi-Vth

TILOS-like optimizers have been used for multi-Vth assignment by a number of researchers, including Wei et al. [15]; Sirichotiyakul et al. [9]; and Wei, Roy, and Koh [16]. These optimizers proceed in a greedy manner, picking the gate with the best power or area versus delay tradeoff to change, and iterating. A number of other optimization heuristics have also been tried.

Most threshold voltage assignment approaches concentrate on reducing leakage power, though Wei, Roy, and Koh minimized total power [16]. Power dissipation is a major constraint in today's technologies, so minimizing total power is the more appropriate optimization objective. If so desired, total power could be minimized until a given constraint is reached, and then the objective could be set to reducing leakage power with constraints on both total power and delay. The only way to encode multiple objectives in an optimization is to weight the objectives according to their priority and include appropriate constraints, but it is generally best to find feasible solutions, for example satisfying the delay constraint, before focusing on a secondary objective.

Sirichotiyakul et al. began with a high Vth circuit and iteratively assigned transistors to low Vth, with transistors prioritized in a TILOS-like manner for the delay reduction versus the increase in leakage power. After a transistor was assigned to low Vth, transistors in neighboring gates over three levels of logic were resized. In the 0.25um process with supply voltage of 0.9V, their dual Vth approach reduced leakage by 3.1× to 6.2× with at most 1.3% delay penalty [9] compared to gate sizing with all gates at low Vth,. The circuit sizer was a TILOS-like sizer that provides a good baseline, but, as noted earlier, large savings can be achieved versus an all low Vth configuration. They did not consider the impact on dynamic power and the total circuit power consumption. Analysis of their algorithm indicates a theoretical complexity of $O(|V|^2)$.

Wei, Roy and Koh began with all gates at minimum size and high Vth. The sensitivity metric for gate upsizing or reducing a gate's threshold voltage was $-\Delta d/\Delta P$. In a 0.25um process with Vdd of 1V and threshold voltages of 0.2V and 0.3V with 0.1 switching activity, total power was reduced by 14% using dual Vth and gate sizing versus gate sizing with low Vth [16]. The theoretical worst case runtime complexity of this approach is the same as TILOS, $O(|V|^2)$.

Wei et al. compared gate-level assignment versus stack-level, and versus assignment at the level of series connected transistors, and found that series-level assignment provided 25% better power reduction than gate-level assignment [15]. Our work does not directly consider transistor-level Vth assignment, but if standard cell libraries are available with characterized cells of mixed-Vth, it is straightforward to use them in the optimization.

These TILOS-like algorithms appear to be the best of the Vth assignment algorithms, as other multi-Vth research has not shown better results than TILOS. We do not compare the LP approach versus TILOS for gate-level threshold voltage assignment, as Chapter 6 has already shown that our linear programming approach produces better sizing results than TILOS.

7.3.2   Summary of papers on optimization with multi-Vdd

For multiple supply voltage assignment, the approach in common to many of the algorithms is starting with a VDDH netlist, and prioritizing assignment to VDDL, typically in reverse topological order. The approaches by Srivastava and Kulkarni (see Chapter 8) take this one step further by forcing additional VDDL assignment using slack gained from starting with all gates at low Vth and gate upsizing. Having changed as many gates as possible to VDDL in either a CVS or ECVS methodology, they pick the best multi-Vdd/low Vth configuration, then look at assigning gates to high Vth.

Clustered voltage scaling (CVS) with voltage level restoration by level converters combined with latches was proposed in 1995 by Usami and Horowitz [12]. Their CVS algorithm proceeded in depth-first search manner from the combinational outputs, assigning VDDH gates to VDDL if they have sufficient slack and only VDDL or level converter latch fanouts. Assigning gates to VDDL was prioritized by the gate load capacitance or the slack in descending order. For two benchmarks in 0.8um process technology, dual supply voltages of 5V and 4V gave power savings of 9% and 18% when using a library with fine grained gate sizes [12]. Their delay models did not include slew nor separate rise/fall delays, and short circuit power was not included. The complexity of CVS is $O(|V|^2)$ (see Section 8.6.1).

In 1997, Usami et al. proposed an ECVS algorithm with the delay constraint relaxed to allow all flip-flops to be set to LCFFs or VDDL [13]. The combinational gates were examined in reverse topological order. If a VDDH gate had all VDDL fanouts and there was sufficient timing slack, it was set to VDDL. If the gate had some VDDH fanouts, an asynchronous level converter must be inserted. However, one gate may be insufficient to amortize the power overhead for the level converter. Thus in addition to the reduction in power for the gate changing from VDDH to VDDL, the potential power reduction for changing the gate's fanins to VDDL was estimated. They fabricated a dual supply voltage media processor chip in 0.3um technology with VDDH of 3.3V. In the initial circuit, more than 60% of the paths had slack of half the cycle time, suggesting that the initial circuit was not sizing power minimized, thus giving larger power savings with dual Vdd. They achieved on average 28% power savings for the combinational logic with VDDL of 1.9V. The area overhead in the dual Vdd modules was 15% due to level converters, additional VDDL power lines, and reduced cell density due to constrained placement on a VDDH row or VDDL row [13]. The theoretical runtime complexity for this heuristic is also $O(|V|^2)$.

More recent optimization approaches have included gate sizing with CVS and ECVS multiple supply voltage assignment. The theoretical runtime complexity of these algorithms is $O(|V|^3)$ or worse, which is too slow given the typical size of circuits of interest to designers today.

Chapter 8 details multi-Vdd assignment algorithms proposed by Kulkarni et al. that include gate sizing and threshold voltage assignment with $O(|V|^3)$ runtime complexity. Their GVS algorithm for ECVS-multi-Vdd/multi-Vth/ sizing gives 21.6% average power saving versus their TILOS-like gate sizer, but the average power saving is only 15.0% versus our better LP sizing results (see Table 7.1).

Previous multi-Vdd algorithms have handled level converter power and delay overheads by iteratively changing gates to VDDL and choosing the best configuration found along the way, or by estimating the power savings of changing multiple gates to VDDL. Both of these approaches can be computationally expensive.

We now examine how to account for the voltage level converter delay and power overheads with the linear programming approach that was detailed in Chapter 6.

7.4   OPTIMIZING WITH MULTIPLE SUPPLY AND THRESHOLD VOLTAGES

Using multiple threshold voltages does not complicate our optimization approach. The different delay and power – particularly leakage power – must be accounted for, but this just changes the values in the corresponding lookup tables for the standard cell library. Static timing and power analysis do not otherwise change. The power and delay trade-offs for different cell sizes and different threshold voltages for a cell are considered when determining the best alternate cell for a gate to encode in the linear program. Then optimization proceeds normally.

In contrast, using multiple supply voltages not only complicates optimization, but can hinder getting out of local minima. A voltage level converter must be inserted between low supply voltage and high supply voltage gates. The additional power and delay for level converters must be encoded in the linear program. The level converter overheads create a very "bumpy" optimization surface with many local minima, hindering gates changing from low Vdd to high Vdd or vice versa.

We assume that level converters are placed by the driven input of the VDDH gate to avoid any additional wiring overheads. The same assumption is made in Chapter 8, which we shall compare our results to.

With multi-Vth and multi-Vdd, the linear program and optimization parameters are the same as used for gate sizing, detailed in Section 6.3.4. The change for adding or removing a level converter as necessary for multi-Vdd is included in the ΔP, Δd and Δs values encoded with the alternate cell choices in the linear program.

7.4.1   Voltage level converter power and delay overheads

If a gate is changed from VDDH to VDDL, it needs level converters to VDDH fanouts, and it no longer needs any fanin level converters. If a gate is changed from VDDL to VDDH, it needs level converters on any VDDL fanins, and no longer needs any fanout level converters. The level converters overheads are not represented directly in the linear program, to avoid adding unnecessary variables and unnecessary constraints. Instead the change in power for adding or removing level converters is added to the change in power for changing the gate's cell, and the delay change is added to the gate's delay change on the appropriate delay constraint edge.

The overheads for level converter flip-flops and "asynchronous" level converters are shown in Figure 7.4 and Figure 7.5. This data is from examination of the cell alternatives for a particular gate before encoding the best alternative for each gate in the linear program.

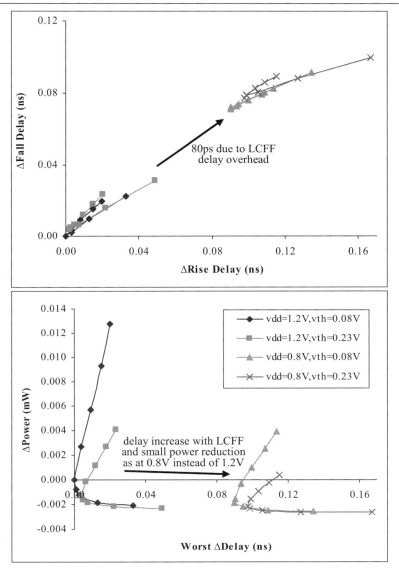

Figure 7.4 These graphs show delay and power trade-offs for alternate cells for a gate, including the level converter flip-flop (LCFF) overhead. We assumed that LCFFs do not consume more power than a normal flip-flop, but that they do impose an 80ps delay penalty. The effect of the 80ps delay penalty is shown here on alternate cell choices for a size X4 Vdd=1.2V/Vth=0.08V inverter that drives an output port in ISCAS'85 benchmark c17. Some power is saved by changing to VDDL and inserting the level converter flip-flop. Each point on a curve represents a different gate size – the largest gates consume the most power and have higher delay than the current cell with size X4 that is at the origin.

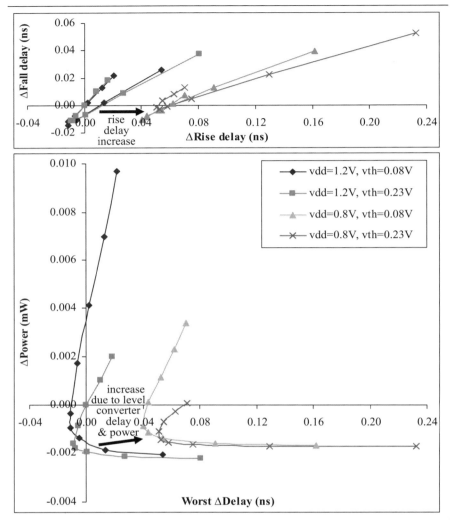

Figure 7.5 These graphs show delay and power trade-offs for alternate cells for a gate, including the asynchronous level converter overhead. These graphs show the alternate cell choices for a Vdd=1.2V/Vth=0.23V drive strength X12 inverter in ISCAS'85 benchmark c5315. The rise delay increases, because of the delay for a falling VDDH input to reach VDDL. In contrast, there is only a small increase in the fall delay. The power overhead for an asynchronous level converter is large, shown by the shift in the curves on the power versus delay graph, and cannot be amortized across a single gate. Each point on a curve represents a different gate size – the largest gates consume the most power.

For level converter flip-flops at the VDDL outputs of combinational logic, we assumed that there is no power overhead. There is some delay overhead – typically about 2 FO4 delays [2], which corresponds to 80ps in our 0.13um process. The impact of the additional 80ps LCFF delay on VDDH and VDDL cell alternatives is shown in Figure 7.4. The 80ps delay penalty is considerable compared to the inverter delay. The power savings by changing to VDDL are often less than by downsizing a gate, which argues against prioritizing assignment to low Vdd over gate downsizing.

To analyze the ECVS approach, we used the strength 5 asynchronous level converter shown in Figure 7.6 [5]. This design is a higher speed and more energy efficient modification of a pass gate level converter. Transistor M1 is an NMOS pass gate that isolates the input from the level converter's VDDH supply. Feedback from the inverter, composed of M2 and M3 transistors, to transistor M4 pulls up a VDDL input to VDDH. The logic connected to the transistor gate of M1 serves to raise the transistor gate voltage to VDDL + 0.11V, ensuring that transistor M1 is still off when the input voltage is VDDL to isolate the input from VDDH, but improving the performance of M1 and reducing contention with the inverter's feedback [5]. The two characterized drive strengths of the level converter have similar input capacitance and delay to a Vdd=1.2V/Vth=0.23V X2 drive strength inverter, but their leakage is about 30× more due to use of low Vth transistors and more leakage paths from Vdd to ground.

A VDDH input to a VDDL gate reduces the fall delay, but increases the rise delay. For example, an inverter's input falling from Vdd=1.2V, with slew of 0.15ns, only reaches 0.8V after 0.05ns, which delays when the inverter's output starts to rise by at least 0.05ns. Compared to 0.8V input drivers, using input drivers of 1.2V to 0.8V gates generally reduces the fall delay more than the increase in rise delay for our 0.13um libraries, giving about 1% to 3% net reduction in circuit delay for most of the ISCAS'85 benchmarks. The increased rise delay is apparent for Vdd of 0.8V in Figure 7.5, which shows the VDDH and VDDL cell alternatives for an inverter with the overhead for inserting an asynchronous level converter at the output.

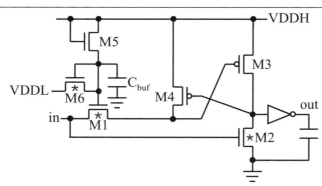

Figure 7.6 The strength 5 asynchronous voltage level converter [5]. The NMOS transistors labeled with ★ have Vth of 0.11V. The other NMOS and PMOS transistors have Vth of 0.23V and –0.21V respectively. VDDH is 1.2V and VDDL is 0.8V or 0.6V, with transistors and capacitance C_{buf} sized appropriately.

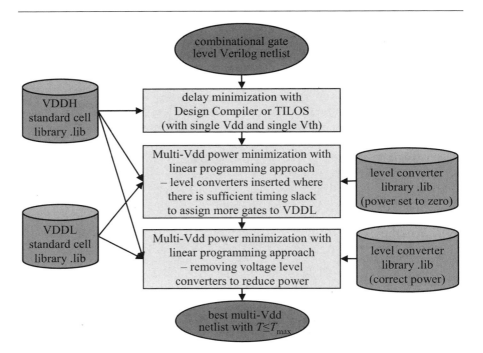

Figure 7.7 An overview of the optimization flow for multi-Vdd with asynchronous level converters. The linear programming flow for each multi-Vdd power minimization run is the same as for sizing in Figure 6.5, with removal and insertion of level converters included when considering alternate cells for a gate.

Changing more than one gate to VDDL is required to amortize the asynchronous level converter power overhead, as can be seen from the power-delay trade-offs in Figure 7.5. This poses a significant barrier to our optimization formulation, as we pick the best cell alternative to encode in the linear program by considering only a single gate.

7.4.2   Climbing the optimization barrier posed by level converter power overheads

Linear programming CVS results showed some power savings versus only using gate sizing. LP results for ECVS, where asynchronous level converters were allowed, provided minimal (1%) or no additional power savings versus CVS [3]. The optimized circuits did not have asynchronous level converters, because the power overhead for a level converter was too high to amortize across a single gate.

The LP optimization had no method of climbing the "hills" posed on the optimization surface by the power overhead for a level converter. In contrast, iteratively forcing gates to VDDL enables hill climbing in an ECVS multi-Vdd methodology, which is the approach taken with multi-Vdd in Chapter 8.

We examined the power savings possible when the level converter power and delay overheads were reduced. With reduced level converter overheads, asynchronous level converters were used in the LP optimized circuits. From this came the idea of setting the level converter power overheads to zero, enabling use of the level converters, without violating the delay constraints.

It was essential to not change the level converter delays for two reasons. Firstly, the linear programming optimization approach is not as good at delay reduction, which suggests that trading delay for power reduction, then trying to correct it later would be a mistake. Secondly, increased path delay with the delay overhead of a level converter has a major impact on optimization choices and the power, for example causing gates to be upsized.

Setting level converter power consumption to zero, running the LP approach, correcting the level converter power, then running the LP approach again provided good results and level converters were used [3]. The run with level converter power set to zero power results in using a larger number of level converters. The LP run with the correct level converter power then substantially reduces the number of level converters, but more gates remain at VDDL than in a CVS approach – that is VDDL regions are clustered. The ECVS multi-Vdd optimization flow is shown in Figure 7.7. A simplified example to illustrate what happens is shown in Figure 7.8.

7.4.3   Climbing the optimization barrier posed by level converter delay overheads

The delay overhead for level converters and larger rise delays when Vin > Vdd can still pose a significant barrier to achieving better results with multi-Vdd. To illustrate the severity of this problem, consider the circuit in Figure 7.8(e), but suppose that the delay overhead for an asynchronous level converter is 2 units of delay. The larger level converter delay causes the path delay from gate 3→2→5 to be 5 units, violating the delay constraint. We cannot change gate 5 to VDDL to remove the level converter, as this would introduce a level converter from gate 6 to gate 5 which is a critical path. Changing gate 2 to VDDH shifts the level converter to its input, which does not reduce the delay on the path. The solution is to change both gate 2 and gate 3 to VDDH, but as we determine the best alternatives to encode in the LP for a single gate at a time, we cannot find this solution.

The problem with the multi-Vdd delay overheads can occur in situations where the best solution would be to change multiple gates to VDDH, or in situations where the best solution would be to change multiple gates to VDDL. For the example in Figure 7.8, if the combinational outputs can be driven at VDDL, setting all gates to VDDL would be the lowest power solution. Indeed, setting all gates to VDDL is a better solution in the case described in Section 7.6.5.2, but it is not a solution that we can find without forcing all gates to VDDL in the first place.

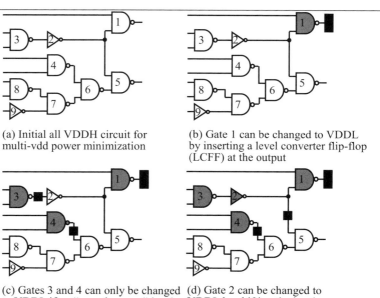

(a) Initial all VDDH circuit for multi-vdd power minimization

(b) Gate 1 can be changed to VDDL by inserting a level converter flip-flop (LCFF) at the output

(c) Gates 3 and 4 can only be changed to VDDL if an "asynchronous" level converter is inserted between gates 2 and 5, and between gates 4 and 6.

(d) Gate 2 can be changed to VDDL by shifting the level converter from its input to its output to gate 5.

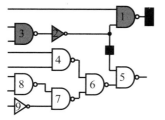

(e) The power overhead for an asynchronous level converter is too large to amortize by a single gate changing to VDDL. Thus the optimal configuration has gate 4 at VDDH.

Symbol	Represents	Delay	Power
	VDDH NAND2	1	5
	VDDH inverter	1	3
	VDDL NAND2	1	2
	VDDL inverter	1	1
	asynchronous level converter	1	4
	level converter flip-flop	1	0

Figure 7.8 This illustrates where level converters may be needed on a simple circuit. The legend at the bottom right lists the delay and power of gates for this example. Suppose we start in (a) with a circuit where all gates are at VDDH, and that the circuit delay constraint is 4 units. We require that the outputs are driven at VDDH. Gate 1 may be changed to VDDL, if we use a level converter flip-flop at its output, shown in (b). The power overhead for an asynchronous level converter is too large to amortize over only a single gate. Thus we temporarily set the level converter power to zero to try and change more gates to VDDL. Changing gates 3 and 4 to VDDL give power savings of 3 each, and we insert level converters at their outputs in (c). Then in (d), we can propagate the level converter from the input of gate 2 to its output, to get additional savings. Now we restore the power overhead for the level converters, and find that it is best to change gate 4 back to VDDH, as shown in (e).

It might be possible to surmount the optimization barrier posed by the level converter delay overheads by temporarily setting the delay overhead to zero. Actually setting level converter delays to zero may be unacceptable, because the level converters then act as very effective buffers that can reduce delay by reducing the load on a gate. A better approach would be to set the net delay impact to zero, that is no change in delay of the fanin gate or on the timing arc where the level converter would be inserted. However, this will usually result in too many gates being at VDDL and thus slower, causing the delay constraint to be violated once correct delay overheads are used. Performing delay reduction to meet the delay constraints may then give a suboptimal power result where too many gates have to be upsized to reduce the delay. In particular, it would be helpful to have a better delay minimizer than the linear programming approach currently provides.

Given a good delay reduction approach to fix violated delay constraints, relaxing the delay constraint would allow more gates to be assigned to VDDL, and then delay reduction could be performed to satisfy the tightened delay constraint. Thus the level converter delay penalty is not an optimization barrier to reducing power by assigning more gates to VDDL at a relaxed delay constraint.

Another approach would be to find groups of gates to assign to VDDL or VDDH amortizing the delay penalty across them. However, this will be very expensive computationally.

More gates could be forced to VDDL in the manner proposed in Chapter 8, but prioritizing assignment to lower supply voltage over gate downsizing or increasing threshold voltage will be suboptimal, except in situations where there is greater power sensitivity to Vdd, which is often not the case. These approaches are also too computationally expensive as they are $O(|V|^3)$.

It is not clear how to resolve this problem. Further optimization experiments with the multi-Vdd delay overhead barrier require a better delay minimizer than the linear programming approach. For now, we will also examine solutions where all gates are at the lower supply voltage, noting that there may be other intermediate multi-Vdd solutions that would be better that we cannot find due to the delay barrier.

7.5   COMPARISON OF MULTI-VDD AND MULTI-VTH RESULTS

We shall now compare the linear programming results versus the multi-Vdd/multi-Vth/sizing CVS and ECVS results provided by Sarvesh Kulkarni and Ashish Srivastava in Chapter 8 for Vth values of 0.23V and 0.12V, and Vdd values of 1.2V, 0.8V and 0.6V. The gate delays with Vdd of 0.6V were estimated by scaling from 0.8V to 0.6V using Equation (4.4); the other libraries were characterized in PowerArc for STMicroelectronics' 0.13um HCMOS9D process. To provide a range of Vth values for the characterization with PowerArc, the zero bias threshold voltage parameter vth0 [1] in the SPICE technology files was adjusted.

We used the same libraries and conditions as they did. The port loads were 3fF, excluding the additional wire load. The wire loads were 3+2×num_fanout fF, and slews were 0.1ns for the 1.2V input drive ramps. Switching activities were multiplied by a fraction such that leakage was about 20% of total power at Vdd=1.2V/Vth=0.12V. We used an 80ps delay overhead for level converter flip-flops, and used the two characterized sizes of the strength 5 asynchronous level converter described in [5]. The ISCAS'85 and Huffman benchmarks for comparison were discussed in Section 6.5.1.

There were twelve inverter cell sizes, and seven sizes for NAND2, NAND3, NOR2 and NOR3 logic gates[1]. For multi-Vdd with a low supply voltage of 0.8V or 0.6V, we encountered slews that were outside the cell input slew characterization range. To avoid input slews exceeding 1.02ns, the maximum cell capacitance was set to prevent a gate having an output slew of more than 1.02ns.

The multi-Vth/multi-Vdd/sizing approaches in Chapter 8 start from a circuit where gates have been upsized from minimum size by their TILOS-like sizer to meet the delay constraint with high Vdd and low Vth. They do not perform any gate downsizing, but may perform further gate upsizing to allow more gates to be changed to low Vdd and high Vth. As the linear programming approach outperforms their TILOS-like sizer, some power savings will be simply due to better gate sizing. Their multi-Vdd approaches climb the voltage level converter power and delay hills in the optimization surface by forcing as many gates as possible to VDDL. Despite lacking a global circuit view when assigning gates to VDDL, they may achieve lower power by assigning more gates to VDDL, as we do not have any method for climbing the level converter delay hills in the multi-Vdd optimization surface.

The linear programming approach performs better if optimization starts with a delay minimized netlist. Thus, the netlists sized by the TILOS-like sizer for minimum delay ($1.0 \times T_{min}$) with high Vdd and low Vth were the starting point for the LP runs. The logic gates in the netlists are the same, but the sizes differ from the starting point used for the University of Michigan optimization, which starts with the circuit sized to meet the particular delay constraint.

The TILOS sizing power results were on average 7.6% worse than the LP sizing results, thus we use the LP sizing results to provide the sizing baseline with single Vdd and single Vth, as shown in Table 7.1.

---

[1] The inverter gate sizes were X1, X2, X3, X4, X6, X8, X10, X12, X14, X16, X18 and X20. The other gate sizes were X1, X2, X4, X5, X6, X7 and X8.

Dual Vth with sizing in the LP approach is on average 13.7% lower power than the LP sizing baseline, which is not surprising given that leakage is 20% of the total power with single Vth of 0.12V at the delay constraint of $1.1 \times T_{min}$. The LP CVS and ECVS Vdd=1.2V&0.8V/Vth=0.23V&0.12V results are respectively on average 17.4% and 21.7% lower power than the LP sizing results. The largest power saving is 37.6% for benchmark c7552. For the lower power ECVS results, using VDDL of 0.6V provides 2% or less power savings versus VDDL of 0.8V. As a lower supply voltage is less robust to noise and will have slower LCFFs, though we assume 80ps LCFF for both VDDL values in these results, VDDL of 0.6V is probably not worthwhile. The LP ECVS results are on average about 5% lower power than the LP CVS results, as shown on the right in Table 7.2.

The University of Michigan sizing/multi-Vth/multi-Vdd results for netlists c432 and c1355 are worse than the LP sizing results, because multi-Vdd is not particularly helpful for these netlists and their TILOS-like sizing results are more than 20% worse than the LP sizing results for these two benchmarks. The linear programming results are on average about 6% lower power than the University of Michigan CVS and ECVS results, as shown in Table 7.2.

Table 7.1 This table shows the percentage power savings versus the sizing baseline provided with the linear programming (LP) approach with Vdd=1.2V/Vth=0.12V at $1.1 \times T_{min}$. University of Michigan (UM) results from Chapter 8 are reported in the last two columns for their CVS and ECVS approaches that include dual Vth and gate sizing (they refer to these respectively as VVS and GVS in Chapter 8). They found that using 0.6V for VDDL gave worse results for their multi-Vdd approaches, so those results are not included.

				LP				UM	
	TILOS			**CVS**		**ECVS**		**CVS**	**ECVS**
Vth (V)			0.23	0.23	0.23	0.23	0.23	0.23	0.23
	0.12	0.12	0.12	0.12	0.12	0.12	0.12	0.12	0.12
Vdd (V)	1.2	1.2	1.2	1.2	1.2	1.2	1.2	1.2	1.2
				0.8		0.8		0.8	0.8
					0.6		0.6		
Netlist				**Power savings versus LP sizing**					
c432	-23.0%	0.0%	6.6%	4.5%	9.2%	6.6%	8.7%	-7.4%	-11.2%
c880	-1.4%	0.0%	15.9%	22.4%	21.9%	23.7%	22.6%	22.9%	26.3%
c1355	-21.4%	0.0%	5.7%	6.0%	6.7%	8.2%	8.0%	-10.8%	-9.9%
c1908	-8.9%	0.0%	9.8%	11.1%	11.2%	14.4%	13.8%	11.1%	7.4%
c2670	-1.8%	0.0%	17.7%	29.0%	31.8%	32.3%	33.1%	26.5%	27.9%
c3540	-3.9%	0.0%	15.1%	16.6%	17.0%	21.8%	20.9%	13.1%	11.7%
c5315	-1.7%	0.0%	15.2%	18.0%	19.5%	26.7%	28.1%	16.9%	25.3%
c7552	-2.1%	0.0%	21.8%	30.3%	28.0%	37.6%	37.1%	22.0%	33.1%
Huffman	-4.4%	0.0%	15.8%	18.5%	19.3%	23.7%	23.9%	16.2%	24.1%
Average	**-7.6%**	**0.0%**	**13.7%**	**17.4%**	**18.3%**	**21.7%**	**21.8%**	**12.3%**	**15.0%**

Table 7.2 This table compares the linear programming CVS and ECVS results versus the University of Michigan CVS and ECVS results respectively, and compares our LP results with gate sizing and dual Vth for ECVS versus CVS.

	LP savings versus UM				LP ECVS savings versus LP CVS	
	LP CVS		LP ECVS			
Vth (V)	0.23	0.23	0.23	0.23	0.23	0.23
	0.12	0.12	0.12	0.12	0.12	0.12
Vdd (V)	1.2	1.2	1.2	1.2	1.2	1.2
	0.8		0.8		0.8	
		0.6		0.6		0.6
c432	11.1%	15.4%	16.0%	17.8%	2.1%	-0.5%
c880	-0.6%	-1.2%	-3.5%	-5.1%	1.7%	0.9%
c1355	15.2%	15.8%	16.5%	16.3%	2.3%	1.5%
c1908	-0.1%	0.1%	7.6%	7.0%	3.8%	2.9%
c2670	3.3%	7.1%	6.1%	7.1%	4.7%	1.9%
c3540	4.1%	4.5%	11.4%	10.3%	6.2%	4.6%
c5315	1.4%	3.1%	1.9%	3.8%	10.6%	10.7%
c7552	10.6%	7.7%	6.8%	6.0%	10.5%	12.6%
Huffman	2.8%	3.7%	-0.4%	-0.2%	6.4%	5.7%
Average	**5.3%**	**6.3%**	**6.9%**	**7.0%**	**5.4%**	**4.5%**

In most cases the LP results are lower power, but they are up to 5.1% higher power for c880. In the few cases where the University of Michigan results are better, they were able to assign more gates to VDDL to achieve lower total power. This emphasizes the importance of level converter "delay hill" climbing in the optimization space, which our linear programming approach lacks. Several possible approaches to climbing these delay barriers between local minima with the LP approach were discussed in Section 7.4.3. It is a difficult non-convex optimization problem.

We would expect that an optimization approach which uses only a subset of the optimization space (e.g. CVS with level converter flip-flops only) would provide suboptimal or at best equivalent results to approaches with additional options (e.g. ECVS which also has asynchronous level converters). This is not always the case as the optimization approaches are heuristic, with no guarantee of finding the global minimum, and they can get stuck in local minima. The benefits of multi-Vdd may be underestimated due to the difficulty of assigning more gates to VDDL with the level converter delay penalty creating a nonconvex "bumpy" optimization space where it is difficult to get out of local minima.

Our multi-Vdd/multi-Vth power savings of 28% or more versus sizing are sufficient to justify use of multi-Vdd and multi-Vth. We compare multi-Vdd and multi-Vth results versus single Vdd and single Vth in more detail in the next section. In particular, we start with high Vth netlists for which leakage is 1% of the total power, from which it is more difficult to achieve substantial power savings with multi-Vth.

7.6 ANALYSIS OF POWER SAVINGS WITH MULTI-VTH AND MULTI-VDD

Comparisons versus Design Compiler in Section 6.5.2 showed that the linear programming approach provided very good gate sizing results. In Section 7.5, we saw that the LP approach also provides good multi-Vdd/multi-Vth/sizing results, averaging 5% to 7% lower power for both CVS and ECVS multi-Vdd methodologies. By comparison to the good gate sizing baseline with single Vdd and single Vth, we can now carefully analyze the power savings that may be possible with multi-Vdd and multi-Vth in addition to gate sizing.

With the large amount of data presented in this section, we italicize the more significant results.

7.6.1 General experimental conditions

The starting gate sizes for multi-Vth and multi-Vdd optimization are the Design Compiler netlists that were sized to minimize delay with the Vdd=1.2V/Vth=0.23V 0.13um PowerArc characterized library. Using high Vdd and low Vth cells provides the minimum delay starting point for optimization (see Table 4.1 and accompanying discussion).

We used fewer gate sizes for results in this section. There were nine inverter sizes, and four sizes for NAND2, NAND3, NOR2 and NOR3 logic gates[1]. Using more gate sizes did not significantly change the results from Design Compiler or our LP approach, as only the larger gate sizes were not included. Inclusion of the larger gate sizes does not change the results significantly as the larger gates with substantial power consumption are seldom used in the power minimized netlists.

For multi-Vdd with a low supply voltage of 0.8V or 0.6V, we encountered slews that were outside the cell input slew characterization range. These cell sizes were characterized with input slew of up to 1.8ns. To avoid input slews exceeding 1.8ns, the maximum cell capacitance was set to prevent a gate having an output slew of more than 1.8ns.

As in the earlier analysis, the port loads were 3fF, not including the wire load, and wire loads were 3+2×num_fanout fF. Switching activities were multiplied by a fraction such that leakage was about 1% of total power at Vdd=1.2V/Vth=0.23V. Input slew was set to 0.1ns. The input drive ramps have voltage swing from 0V to 1.2V, except in Section 7.6.5 for the single 0.8V Vdd results where we look at the impact of using 0.8V drivers. With multi-Vdd, the optimization is not allowed to set the input drivers to VDDL to further reduce power.

---

[1] The inverter gate sizes were XL, X1, X2, X3, X4, X8, X12, X16 and X20. The other gate sizes were XL, X1, X2, X4.

7.6.2   Experimental conditions for multi-Vth comparison

Power savings are compared at a tight delay constraint for high Vth to avoid exaggerating savings versus low Vth, where leakage and thus total power are substantially higher. Starting with all gates at low Vth, it is easy to reduce leakage by going to high Vth, as many gates are not on timing critical paths. The power savings are less when starting with all gates at high Vth, because using low Vth causes a substantial increase in leakage power that can only be justified by gate downsizing on timing critical paths, or using the resulting timing slack to reduce Vdd.

We shall examine multi-Vth with three Vth values: 0.23V, 0.14V, and 0.08V. The leakage at Vth of 0.14V and 0.08V is respectively about 10× and 50× than at 0.23V, but they also provide a substantial delay reduction versus Vth=0.23V, ranging from 12% to 43% less depending on Vdd. There is minimal benefit for choosing Vth higher than a value that results in leakage being 1% of total power – the power savings are at best 1%, and in practice less due to the reduced slack for gate downsizing. Thus we consider a scenario with 1% of total power being leakage at high Vth.

The delay constraints were $1.0 \times T_{min}$ and $1.2 \times T_{min}$ for multi-Vth results, where T_{min} is the minimum delay for the Design Compiler delay minimized netlists at Vdd=1.2V/Vth=0.23V. Using a lower threshold voltage provides sufficient timing slack to get good multi-Vth results with the linear programming approach at a delay constraint of $1.0 \times T_{min}$ for Vdd=1.2V/Vth=0.23V. We expect that using a lower Vth may provide most benefit at the tight $1.0 \times T_{min}$ delay constraint, as the additional slack allows gates to be downsized, backing away from the sharp rise in dynamic power on the gate sizing power versus delay "banana" curves. Analysis is also performed at $1.2 \times T_{min}$, as this is where we found the greatest power savings with geometric programming optimization of multi-Vdd and multi-Vth for benchmarks c499 and c880, and where other researchers have performed multi-Vdd analysis [6][10].

7.6.3   Experimental conditions for multi-Vdd comparison

We shall examine three possible supply voltages: 1.2V, 0.8V and 0.6V. Earlier multi-Vdd research suggested as a rule of thumb to use a low supply voltage of about 70% of VDDH, while some more recent research has suggested that VDDL should be 50% of VDDH [6]. Thus if VDDH is 1.2V, VDDL should be 0.8V or 0.6V. The gate delays with Vdd of 0.6V were estimated by scaling from 0.8V to 0.6V using Equation (4.3) with $\alpha = 1.66$; the other libraries were characterized in PowerArc for STMicroelectronics' 0.13um HCMOS9D process. To provide a range of Vth values for the characterization with PowerArc, the zero bias threshold voltage parameter vth0 [1] in the SPICE technology files was adjusted.

There is a 50% delay increase when using Vdd=0.6V even with Vth=0.08V, though Vdd=0.6V/Vth=0.08V does reduce power by about 65% versus Vdd=1.2V/Vth=0.23V from analysis without gate sizing in Chapter 4. In comparison, Vdd=0.8V/Vth=0.08V has only a 10% delay increase and reduces power by 28%. For multi-Vdd, the smaller delay penalty at VDDL=0.8V will allow it to be used for more gates in the circuit than VDDL=0.6V. A smaller VDDL delay penalty leaves more slack for power minimization by gate down sizing or increasing Vth.

We used a delay constraint of $1.2 \times T_{min}$ to look at the benefits of multi-Vdd, where T_{min} is the minimum delay for the Design Compiler delay minimized netlists at Vdd=1.2V/Vth=0.23V. There is sufficient slack for the linear programming approach to work well at Vth=0.23V. 10% to 20% relaxed delay constraints from the TILOS sized netlists were used in [6] and [10] to allow sufficient slack for good power savings with multi-Vdd. The Design Compiler delay minimized netlists average 22% faster than the TILOS delay minimized netlists that were used as a starting point in Section 7.5. Thus, there may be significantly less timing slack at $1.2 \times T_{min}$ for the Design Compiler netlists than for the TILOS netlists. However, it is difficult to compare the netlists as they differ because delay minimization in Design Compiler used both technology mapping and gate sizing, whereas TILOS was limited to gate sizing. Reducing Vth below 0.23V provides multi-Vdd scenarios with more timing slack.

We must account for the delay and power overheads for restoring a low voltage swing signal. Compared to high speed flip-flops, a level converter flip-flop has a delay overhead of about 2 FO4 delays [2], which corresponds to 80ps in our 0.13um process. Making the same assumption for results in Section 7.6.4 as in Chapter 8, we assume an 80ps delay overhead for voltage level restoration with an LCFF and no power overhead.

An LCFF delay overhead of 0ps is appropriate if comparing to the typical D-type flip-flops in an ASIC standard cell library [4], rather than high speed pulsed flip-flops. Results with 0ps LCFF delay overhead are discussed in Section 7.6.5. The additional timing slack permits lower power results with Vdd of 0.8V. With 80ps LCFFs, VDDH of 1.2V is required to meet the delay constraints with sufficient timing slack to reduce power.

For ECVS, we used two characterized drive strengths of the strength 5 asynchronous level converters [5]. This higher speed and energy efficient modification of a pass gate level converter was described in Section 7.4.1.

We make the same assumption as made by Kulkarni and Srivastava et al. in [6], [10] and Chapter 8 that level converters are placed next to the input pin of the gate that they drive, and that there are no additional wiring overheads. This assumption is optimistic, unless there is a standard cell library with level converters incorporated into the logic cells. However, there are level converter designs incorporating additional logic, (e.g. see Figure 13.8), so this assumption may be reasonable.

The optimization approach is only a heuristic, so in some cases the results found given a larger possible state space, for example dual Vdd versus single Vdd, can be worse. Section 7.6.5 will dwell on multi-Vdd results with 0ps LCFF delay overhead that are substantially worse than using a single Vdd of 0.8V. As we are interested in looking at the benefits of multi-Vdd and multi-Vth, a suboptimal result obscures the benefits. Instead, if there was a better solution found with a subset of the Vdd or Vth values, that solution has been tabulated, except for the multi-Vdd results in Section 7.6.5 where their suboptimality is discussed.

7.6.4 Results with 80ps level converter flip-flop delay overhead

We begin analysis assuming an 80ps LCFF delay overhead for voltage level restoration at the outputs. This applies to the multi-Vdd results and to the single Vdd results where the supply voltage has been scaled to 0.8V. We have made the same assumptions as in Chapter 8 for comparison of our results to theirs. In Section 7.6.5, we analyze results with a 0ps LCFF delay overhead, which substantially improves the single Vdd=0.8V results.

7.6.4.1 Impact of multi-Vth with single Vdd at 1.0×Tmin

To examine the benefits of using multiple supply and threshold voltages, we must first provide a sizing only baseline with single Vdd and single Vth. At a delay constraint of $1.0 \times T_{min}$, the only possible choice of supply voltage if only a single Vdd is used is 1.2V, as the delay is too large otherwise. For Vth of 0.23V, there is no timing slack, and the linear programming approach cannot minimize power without violating the delay constraint. Thus the Design Compiler power minimization results are reported for Vth=0.23V at the $1.0 \times T_{min}$ delay constraint.

The best single Vth gate sizing results at $1.0 \times T_{min}$ are with Vth of 0.14V, reducing the power on average by 12.0% from the Vth=0.23V sizing results. The lower threshold voltage gives sufficient slack for gate downsizing to reduce the dynamic power without resulting in excessive leakage, unlike using Vth of 0.08V which is 19.7% higher power on average as listed in Table 7.3.

The largest power saving with dual Vth versus the single Vth=0.14V baseline is 7.0% with Vth of 0.23V and 0.14V, and on average they provide 5.2% power savings. The additional leakage with low Vth of 0.08V is too great to justify using it with dual Vth, though sparing use of it on the critical path for *triple Vth provides up to 5.1% power savings versus dual Vth.*

Table 7.3 This table compares dual Vth and triple Vth sizing power minimization results versus the best sizing only results with single Vth of 0.14V. The delay constraint was $1.0 \times T_{min}$ and Vdd was 1.2V for all these results. Results for single Vth of 0.23V are from Design Compiler (DC); the other results are from the linear programming approach (LP).

		Single			Dual			Triple
Vth (V)		0.23	0.14	0.08	0.23 0.14	0.23 0.08	0.14 0.08	0.23 0.14 0.08
		Power savings vs. Vdd=1.2V/Vth=0.14V						
Netlist	**DC**		LP					
c17	-13.7%	0.0%	-20.1%	7.0%	-8.0%	0.0%	7.0%	
c432	-19.6%	0.0%	-17.9%	2.5%	-5.6%	0.8%	3.3%	
c499	-20.9%	0.0%	-17.1%	7.0%	-0.2%	2.8%	8.1%	
c880	-4.2%	0.0%	-20.8%	5.1%	4.6%	2.8%	10.0%	
c1355	-26.1%	0.0%	-17.2%	4.6%	-1.8%	2.6%	5.9%	
c1908	-11.1%	0.0%	-19.9%	6.1%	-1.6%	1.0%	6.1%	
c2670	-13.5%	0.0%	-23.5%	5.2%	0.5%	1.4%	7.1%	
c3540	-17.4%	0.0%	-20.9%	4.5%	-1.7%	0.8%	5.9%	
c5315	-6.4%	0.0%	-24.1%	6.2%	0.0%	1.4%	6.8%	
c6288	-12.7%	0.0%	-16.9%	2.4%	-3.5%	1.6%	3.8%	
c7552	-8.0%	0.0%	-18.3%	6.2%	0.8%	3.2%	7.6%	
Average	**-14.0%**	**0.0%**	**-19.7%**	**5.2%**	**-1.5%**	**1.7%**	**6.5%**	

7.6.4.2   Impact of multi-Vth with single Vdd at 1.2×Tmin

With a relaxed delay constraint of $1.2 \times T_{min}$, in some cases single Vth of 0.23V produced the best results for gate sizing and in other cases Vth of 0.14V was the best choice, as shown in Table 7.4. The best single Vth/single Vdd/gate sizing results were with Vdd of 1.2V, though the relaxed delay constraint does allow Vdd of 0.8V with Vth reduced to 0.08V in some cases. The best of these gate sizing only results are used for a baseline to compare multi-Vth against.

Dual threshold voltages of 0.23V and 0.14V with Vdd of 1.2V provide the best dual Vth power savings, except for benchmark c6288, averaging 5.0% lower power than the single Vth baseline. The leakage with Vth of 0.08V is too high to justify its use with Vdd of 1.2V. Triple Vth provides at most 1.3% power savings versus dual Vth.

Interestingly, *optimization of c6288 with single Vdd of 0.8V and Vth values of 0.14V and 0.08V achieves the largest dual Vth power savings of 7.8% versus the single Vth gate sizing baseline,* and this result is 4.1% lower power than the multi-Vth results with Vdd of 1.2V. This is the only case where the multi-Vth results with Vdd of 0.8V achieve lower power. With Vdd of 0.8V, low Vth of 0.08V is essential to try and meet the delay constraint. For the single Vth results, using only Vth of 0.08V results in too much leakage and worse total power. For those gates with sufficient slack to change to high Vth, the higher Vth of 0.14V provides about a 5× reduction in leakage power. The 80ps LCFF delay overhead is only 2% of the delay constraint for c6288, but for the other netlists it is 6% or more of the delay constraint, leaving less slack for gates to be downsized or changed to high Vth. This is why the results for c6288 with Vdd of 0.8V are different.

For the Vdd=0.8V/multi-Vth results where the delay constraints were not violated, the delay reduction phase of the LP approach did manage to meet the delay constraint after iterations of the power reduction phase, which was not the case for the Vdd=0.8V/single Vth results. To reduce delay with Vth, a gate can be changed back to low Vth, which only slightly increases the capacitive load on fanin gates and doesn't reduce their speed substantially. Whereas to reduce delay with sizing, a gate must be upsized, which substantially increases the capacitive load on the fanins and increases their delay. Thus the delay reduction phase can reduce delay better with multi-Vth than with gate sizing alone.

Table 7.4 This table compares dual Vth and triple Vth sizing power minimization results versus the best sizing only results with single Vth. The delay constraint was $1.2 \times T_{min}$ and input drivers were ramps with voltage swing from 0V to 1.2V. There was an 80ps LCFF delay overhead at the outputs for the Vdd=0.8V results. Vdd=0.8V results were not included if there was no power savings versus the baseline. All these results are for the LP approach.

	Single			Dual			Triple	Dual	Triple
Vth (V)	0.23	0.14	0.08	0.23 0.14	0.23 0.08	0.14 0.08	0.23 0.14 0.08	0.14 0.08	0.23 0.14 0.08
				Single					
Vdd (V)	1.2	1.2	1.2	1.2	1.2	1.2	1.2	0.8	0.8
Netlist	Power savings vs. single Vdd/single Vth baseline								
c17	-8.8%	0.0%	-20.4%	5.0%	-2.3%	0.0%	5.0%	failed delay constraint	
c432	-3.5%	0.0%	-24.3%	6.7%	1.0%	0.0%	6.7%	failed delay constraint	
c499	-7.2%	0.0%	-30.0%	1.8%	-5.7%	-3.7%	1.8%	failed delay constraint	
c880	0.0%	-3.1%	-31.5%	5.0%	1.9%	-2.3%	5.0%	failed delay constraint	
c1355	-1.8%	0.0%	-26.2%	4.5%	-0.8%	0.0%	4.7%	failed delay constraint	
c1908	0.0%	-0.2%	-29.3%	7.1%	3.3%	-0.8%	7.7%	-18.1%	-8.3%
c2670	0.0%	-2.8%	-31.1%	4.3%	2.5%	-2.8%	4.6%	failed delay constraint	
c3540	-0.4%	0.0%	-27.9%	6.9%	2.3%	0.8%	7.5%	-8.8%	-8.8%
c5315	0.0%	-4.9%	-36.0%	3.7%	1.4%	-4.7%	3.9%	-53.3%	-53.3%
c6288	-1.0%	0.0%	-24.9%	3.9%	1.5%	0.7%	5.2%	7.8%	7.8%
c7552	0.0%	-1.3%	-29.0%	5.6%	2.3%	-0.9%	5.6%	failed delay constraint	
Average	**-2.1%**	**-1.1%**	**-28.2%**	**5.0%**	**0.7%**	**-1.2%**	**5.2%**	**-18.1%**	**-15.7%**

7.6.4.3 Summary of multi-Vth results with 80ps LCFF delay overhead

Using dual threshold voltages provides only a power saving of up to 11.4%, which does not justify the additional processing costs and yield impact. Using three different threshold voltages provides no significant additional benefits. However, we will revisit this with Vdd of 0.8V and 0ps LCFF delay overheads in Section 7.6.5, where dual Vth is found to provide larger power savings.

In the next section we examine the power savings with multiple supply voltages at $1.2 \times T_{min}$. Then we look at using multiple supply voltages in conjunction with multiple threshold voltages in Section 7.6.4.5. We might anticipate that multi-Vth is more beneficial with multi-Vdd, as the lower Vth can help provide sufficient slack to change more gates to low Vdd.

7.6.4.4 Impact of multi-Vdd with single Vth at 1.2×Tmin

The largest power saving with ECVS dual Vdd versus the single Vdd baseline was 13.9% for c2670 with Vdd values of 1.2V and 0.6V and Vth of 0.14V, as shown in Table 7.5. *The largest power saving with CVS dual Vdd was 12.2%* was for the same benchmark and Vdd values, but Vth was 0.23V. *On average ECVS provides only 1.2% power savings versus CVS, though the maximum power saving is 7.3%.* Dual Vdd with single Vth provided no power savings for c17 and c499.

Comparing the best CVS dual Vdd/single Vth/sizing results against the baseline, *CVS dual Vdd gives on average 2.5% power savings.* The results with VDDL of 0.8V and 0.6V were quite similar, indicating a somewhat flat optimization space in terms of the choice for VDDL. In most cases, the best choice for a single Vth with dual Vdd was 0.14V, as it provides more slack for gates to change to VDDL than a Vth of 0.23V. *The best ECVS dual Vdd with single Vth results versus the baseline give an average 4.1% power saving.*

The dual Vdd/single Vth results range from 6.6% better to 11.2% worse than the single Vdd/dual Vth results at $1.2 \times T_{min}$, and are 1% better on average. The power savings depend on the particular design and the optimizer, for example the LP approach works well with multi-Vth but has problems in some cases with multi-Vdd. Depending on the power savings and the design cost, it may be preferable to use only dual Vth, only dual Vdd, or both. However, the maximum power savings of 7.8% with single Vdd/dual Vth and 13.9% with dual Vdd/single Vth are probably insufficient to justify the additional design cost.

We now look at the benefits of multi-Vdd with multiple threshold voltages. Using a low Vth provides slack for greater use of low Vdd, and voltage level converter designs may utilize more than one threshold voltage. For example, the strength 5 level converters [5] use Vth of 0.23V and 0.11V.

Table 7.5 This table compares CVS and ECVS dual Vdd/single Vth/sizing power minimization results versus the sizing only baseline results in Table 7.4. At the bottom are shown the ECVS power savings versus CVS.

	Single							
Vth (V)	0.23 / 0.14	0.23 / 0.14	0.23 / 0.14	0.23 / 0.14	0.23 / 0.14	0.23 / 0.14	0.23 / 0.14	0.23 / 0.14
	CVS dual Vdd				**ECVS dual Vdd**			
Vdd (V)	1.2 / 0.8	1.2 / 0.8	1.2 / 0.6	1.2 / 0.6	1.2 / 0.8	1.2 / 0.8	1.2 / 0.6	1.2 / 0.6
Netlist	**Power savings vs. single Vdd/single Vth baseline**							
c17	-8.8%	0.0%	-8.8%	0.0%	-8.8%	0.0%	-8.8%	0.0%
c432	-2.4%	0.6%	-2.9%	0.0%	-2.4%	0.6%	-2.9%	0.0%
c499	-7.1%	0.0%	-7.1%	0.0%	-7.1%	0.0%	-7.1%	0.0%
c880	2.3%	0.0%	3.0%	1.4%	2.3%	6.2%	3.0%	4.2%
c1355	-1.8%	0.0%	-1.8%	0.0%	-1.8%	0.8%	-1.5%	0.0%
c1908	1.4%	2.2%	0.4%	2.6%	1.6%	3.4%	1.9%	3.6%
c2670	9.8%	7.8%	12.2%	11.3%	10.0%	10.4%	12.2%	13.9%
c3540	1.8%	2.1%	2.2%	1.9%	3.0%	7.6%	2.2%	5.5%
c5315	3.3%	-0.1%	3.2%	0.4%	6.6%	7.1%	3.9%	7.7%
c6288	0.4%	1.3%	-0.1%	1.2%	0.4%	1.3%	-0.1%	1.2%
c7552	2.3%	0.6%	1.7%	0.7%	2.3%	3.7%	1.7%	3.2%
Average	**0.1%**	**1.3%**	**0.2%**	**1.8%**	**0.5%**	**3.7%**	**0.4%**	**3.6%**

Netlist	**ECVS power saved vs. CVS**			
c17	0.0%	0.0%	0.0%	0.0%
c432	0.0%	0.0%	0.0%	0.0%
c499	0.0%	0.0%	0.0%	0.0%
c880	0.0%	6.2%	0.0%	2.8%
c1355	0.0%	0.8%	0.3%	0.0%
c1908	0.1%	1.2%	1.4%	1.0%
c2670	0.2%	2.8%	0.0%	3.0%
c3540	1.1%	5.6%	0.0%	3.7%
c5315	3.4%	7.2%	0.7%	7.3%
c6288	0.0%	0.0%	0.0%	0.0%
c7552	0.0%	3.1%	0.0%	2.5%
Average	**0.4%**	**2.5%**	**0.2%**	**1.8%**

7.6.4.5   Impact of multi-Vdd with multi-Vth at 1.2×Tmin

In most cases, the best Vdd values were 1.2V and 0.6V for dual Vdd/ dual Vth, and the best Vth values were 0.23V and 0.14V. The largest power saving seen with ECVS dual Vdd/dual Vth versus the single Vdd/single Vth baseline was 18.6%, as shown in Table 7.6, and the average power saving with the best dual Vdd/dual Vth results for each benchmark is 8.5%. *The best dual Vdd/dual Vth power saving versus single Vdd/dual Vth is 6.9%, and the results are 4.6% better on average. The best dual Vdd/dual Vth power saving versus dual Vdd/single Vth is 15.0%, and the results are 3.4% better on average.* The largest power saving for dual Vdd/triple Vth versus dual Vdd/dual Vth was 2.3% for c5315.

Table 7.6 This table compares CVS and ECVS dual Vdd/multi-Vth/sizing power minimization results versus the sizing only baseline results in Table 7.4. At the bottom are shown the ECVS power savings versus CVS. Suboptimal dual Vdd/dual Vth results with low Vth of 0.08V are omitted here.

	Dual	Triple	Dual	Triple	Dual	Triple	Dual	Triple
Vth (V)	0.23 0.14	0.23 0.14 0.08	0.23 0.14	0.23 0.14 0.08	0.23 0.14	0.23 0.14 0.08	0.23 0.14	0.23 0.14 0.08
	CVS dual Vdd				**ECVS dual Vdd**			
Vdd (V)	1.2 0.8	1.2 0.8	1.2 0.8 0.6	1.2 0.8 0.6	1.2 0.8	1.2 0.8	1.2 0.8 0.6	1.2 0.8 0.6

Netlist	Power savings vs. single Vdd/single Vth baseline							
c17	5.0%	5.0%	5.0%	5.0%	5.0%	5.0%	5.0%	5.0%
c432	6.9%	7.7%	7.5%	7.5%	7.0%	7.7%	7.5%	7.5%
c499	1.8%	1.9%	1.8%	1.8%	1.8%	1.9%	1.8%	1.8%
c880	6.9%	6.9%	7.0%	7.0%	10.4%	10.4%	7.3%	9.3%
c1355	4.7%	4.7%	4.8%	4.9%	4.7%	4.9%	4.8%	4.9%
c1908	8.4%	8.6%	10.2%	10.2%	9.4%	9.4%	10.2%	10.2%
c2670	13.8%	14.0%	16.7%	16.8%	15.1%	16.4%	18.6%	18.6%
c3540	8.1%	8.8%	8.5%	9.0%	10.6%	10.8%	10.4%	10.9%
c5315	6.6%	6.8%	6.7%	6.7%	11.0%	11.5%	11.9%	13.9%
c6288	4.7%	5.2%	5.1%	5.2%	4.7%	5.2%	5.1%	5.1%
c7552	6.2%	6.6%	6.8%	7.0%	6.9%	7.7%	8.1%	8.1%
Average	**6.6%**	**6.9%**	**7.3%**	**7.4%**	**7.9%**	**8.3%**	**8.2%**	**8.7%**

Netlist	ECVS power savings vs. CVS			
c17	0.0%	0.0%	0.0%	0.0%
c432	0.1%	0.0%	0.0%	0.0%
c499	0.0%	0.0%	0.0%	0.0%
c880	3.8%	3.8%	0.3%	2.5%
c1355	0.0%	0.2%	0.0%	0.0%
c1908	1.0%	0.8%	0.0%	0.0%
c2670	1.5%	2.8%	2.3%	2.2%
c3540	2.7%	2.1%	2.1%	2.1%
c5315	4.8%	5.0%	5.5%	7.7%
c6288	0.0%	0.0%	0.0%	0.0%
c7552	0.8%	1.2%	1.4%	1.2%
Average	**1.3%**	**1.5%**	**1.1%**	**1.4%**

Results from these small benchmarks suggest that dual Vth with dual Vdd doesn't provide substantial benefits over using single Vth with dual Vdd, and that the processing costs for additional mask layers would not be justified. An additional Vth value doesn't provide much power saving: either by using a second higher Vth to reduce leakage, or a second lower Vth to provide more timing slack to reduce Vdd.

Comparing ECVS versus CVS for the best dual Vth values of 0.23V and 0.14V, ECVS provided up to 5.5% power savings. This may be insufficient improvement to justify use of asynchronous level converters, given that they are not as robust to noise as level converter flip-flops, and thus require tighter design constraints on voltage IR drop and more careful noise analysis.

The gate sizing results for the LP approach in Chapter 6 had average power savings of 16.6% versus Design Compiler for the delay constraint of $1.2T_{min}$. Thus LP gate sizing provides about twice the average improvement of 8.5% seen with dual Vdd/dual Vth/sizing, without any additional processing costs for multi-Vth or area overhead for multi-Vdd. Given the additional design costs, use of dual Vdd and dual Vth appear dubious.

In the next section with 0ps LCFF delay overheads, we will examine how optimal these multi-Vdd results are, and see situations where multi-Vth can provide larger power savings.

7.6.5   Results with 0ps level converter flip-flop delay overhead

Thus far, an 80ps LCFF delay penalty has been assumed for voltage level restoration to 1.2V at the primary outputs if the driving gate has a 0.8V or 0.6V supply. However, the delay of a level converter flip-flop is comparable to that of a typical D-type flip-flop in an ASIC standard cell library [4], though slower than fast pulsed flip-flops. Thus a 0ps delay penalty is appropriate if comparing to D-type flip-flops.

The output signals may also not require voltage level restoration. For example, the whole circuit may use a 0.8V supply voltage. In this case, there will be some additional delay due to using registers with 0.8V supply. A typical D-type flip-flop used for a register in an ASIC has delay of 2 to 4 FO4 delays, corresponding to 80ps to 160ps in this 0.13um process. Voltage scaling from Vdd=1.2V/Vth=0.23V to Vdd=0.8V/Vth=0.14V increases the delay by about 15%. So the flip-flops may be 12ps to 24ps slower, which is substantially less than an 80ps delay penalty.

In this section, we look at LP results with single Vdd=0.8V and 0ps LCFF delay overhead. Then we will examine why optimization with multi-Vdd has problems finding result as good as these.

Table 7.7 This table compares single Vdd=0.8V results versus the baseline of the best sizing only results with 1.2V input drivers from Table 7.4 and the second column here. At the bottom left, the gate sizing results with Vdd of 0.8V are compared against the best gate sizing results from Table 7.4. The input drivers had voltage swing of 0.8V or 1.2V. At the bottom right are shown the power savings with 0.8V drivers versus 1.2V drivers.

	with 1.2V drivers				with 0.8V drivers			
	Single	**Dual**		**Triple**	**Single**	**Dual**		**Triple**
Vth (V)		0.23		0.23		0.23		0.23
			0.14	0.14			0.14	0.14
	0.08	0.08	0.08	0.08	0.08	0.08	0.08	0.08
Netlist	**Power savings vs. single Vdd/single Vth baseline with 1.2V drivers**							
c17	failed delay constraint				3.3%	3.3%	24.4%	24.4%
c432	-8.6%	-8.7%	2.3%	3.4%	5.5%	6.0%	14.0%	16.0%
c499	-2.7%	0.7%	9.2%	10.1%	7.8%	10.0%	19.7%	20.7%
c880	-5.7%	8.9%	16.7%	16.7%	8.8%	19.5%	27.7%	27.7%
c1355	-15.5%	-1.1%	6.7%	6.7%	7.5%	9.4%	18.2%	18.3%
c1908	-11.8%	7.8%	15.5%	16.5%	9.0%	15.1%	24.1%	24.6%
c2670	0.0%	9.4%	16.3%	16.6%	3.2%	22.2%	28.6%	29.5%
c3540	0.0%	9.2%	16.9%	17.3%	8.7%	17.8%	24.3%	25.4%
c5315	0.0%	14.0%	18.1%	18.6%	13.2%	21.4%	27.7%	28.9%
c6288	0.0%	2.6%	9.5%	9.5%	5.9%	5.9%	14.3%	15.0%
c7552	0.0%	10.5%	18.0%	18.5%	11.9%	18.1%	25.1%	26.0%
Average	**-4.4%**	**5.3%**	**12.9%**	**13.4%**	**7.7%**	**13.5%**	**22.5%**	**23.3%**

Netlist	**Saved vs. Vdd=1.2V single Vth baseline**
c432	-8.6%
c499	-2.7%
c880	-5.7%
c1355	-15.5%
c1908	-11.8%
c2670	4.1%
c3540	0.3%
c5315	3.9%
c6288	3.3%
c7552	3.5%
Average	**-2.9%**

Netlist	**Saved vs. 1.2V drivers**			
c432	13.0%	13.4%	12.0%	13.0%
c499	10.3%	9.4%	11.6%	11.8%
c880	13.8%	11.6%	13.2%	13.2%
c1355	19.9%	10.4%	12.2%	12.4%
c1908	18.6%	8.0%	10.1%	9.6%
c2670	3.2%	14.1%	14.7%	15.4%
c3540	8.7%	9.5%	8.9%	9.8%
c5315	13.2%	8.7%	11.7%	12.7%
c6288	5.9%	3.4%	5.3%	6.0%
c7552	11.9%	8.4%	8.6%	9.2%
Average	**11.8%**	**9.7%**	**10.8%**	**11.3%**

7.6.5.1   Impact of multi-Vth with Vdd of 0.8V at 1.2×Tmin

Vth of 0.08V is necessary with Vdd=0.8V to meet the delay constraint. With 1.2V drivers, the single Vdd=0.8V/Vth=0.08V results are up to 4.1% better than the Vdd=1.2V gate sizing baseline, but on average are 2.9% worse as shown in Table 7.7. The increased rise delay with 1.2V input swing to the 0.8V gates prevents the delay constraint being satisfied for c17, where the 0.033ns for the input to fall from 1.2V to 0.8V (calculated from the 0.1ns input slew of the drivers) is 29% of the 0.112ns delay constraint.

When voltage is scaled from Vdd=1.2V/Vth=0.23V to Vdd=0.8V/ Vth=0.08V, there is about a 26× increase in leakage, but the dynamic power is reduced substantially by about a factor of 2.2× from the analysis that excluded gate sizing in Chapter 4. As the dynamic power was 99% of the total power at Vdd=1.2V/Vth=0.23V, this trade-off can be worthwhile. In the absence of gate sizing, Vdd=0.8V/Vth=0.08V reduced power on average by 28% versus Vdd=1.2V/Vth=0.23V, but with gate sizing included the average power saving is only 10.4% with 0.8V input drivers. The extra timing slack at Vdd=1.2V/Vth=0.23V allows more gate downsizing, thus reducing Vdd provides less power savings.

With 1.2V drivers, the Vdd=0.8V/dual Vth results with Vth values of 0.14V and 0.08V are up to 18.1% lower power than the baseline, and average 12.9% less. In comparison for single Vdd of 1.2V at $1.2 \times T_{min}$, dual Vth was only 5.0% better than the baseline on average, and at best 7.1% better. Thus the dual Vth results at Vdd of 0.8V are much better. The low Vth of 0.08V provides sufficient slack for Vdd to be reduced to 0.8V, and the higher Vth of 0.14V is used where possible to reduce leakage. Comparing the Vdd=0.8V results, adding the high Vth of 0.14V allows power to be reduced by 16.5% on average versus the single Vth=0.08V results. This may be sufficient power savings to justify the additional process costs for dual Vth. At Vdd=0.8V, gates with Vth of 0.14V were only about 15% slower than at Vth of 0.08V, compared to Vth of 0.23V which had 50% larger delay. Consequently, Vth of 0.14V is a better choice for high Vth than 0.23V. Comparing triple Vth and dual Vth results at Vdd=0.8V in Table 7.7, there is at most 2.3% power savings by using the third threshold voltage.

Until now, we have assumed that the input drivers had voltage swing of 1.2V. Using 0.8V input drivers reduces the dynamic power to switch capacitances driven by the primary inputs. The results with 0.8V input drivers average 10.9% lower power than the results with 1.2V drivers in Table 7.7. With 0.8V gates and 0.8V input drivers, the single Vth of 0.08V results average 7.7% lower power than the other gate sizing results. *In comparison to these better gate sizing results, results for 0.8V gates and 0.8V drivers with dual Vth of 0.14V and 0.08V average 16.0% lower power, and the maximum power saving is 26.2%.* This shows the full extent of power savings that might be achieved by scaling Vdd from 1.2V to 0.8V. These savings are comparable to what we achieved with the gate sizing approach versus Design Compiler.

The multi-Vdd results were restricted to having 1.2V input drivers, so it is not fair to compare them to Vdd=0.8V results with 0.8V drivers. With 0ps LCFF delay overhead and 1.2V drivers, we will compare the multi-Vdd results with single Vdd=0.8V results in the next subsection.

7.6.5.2 Multi-Vdd results are suboptimal versus using single 0.8V Vdd

The ECVS multi-Vdd/multi-Vth results with 0ps LCFF delay overhead averaged 8.9% worse than Vdd=0.8V/Vth=0.14V&0.08V results with 1.2V drivers and 0ps LCFF delay overhead [3], excluding c17 where the delay constraint could not be met with Vdd=0.8V and 1.2V drivers. The multi-Vdd solutions with VDDL of 0.8V were clearly suboptimal as the single Vdd=0.8V solution is in a subspace of the multi-Vdd solution space.

Several different approaches were tried to improve the suboptimal multi-Vdd results [3]. The closest results started optimization with all gates at Vdd=0.8V and Vth=0.08V, for which the dual Vdd=1.2V&0.8V results were only 2.8% worse on average than the single Vdd=0.8V results.

The multi-Vdd results are suboptimal because the delay overhead of the level converters creates barriers in the optimization space to iteratively changing all the supply voltages to 0.8V, as level converters have to be inserted at intermediate steps. The problem posed by the delay overhead of level converters was discussed in Section 7.4.3, along with possible ways of overcoming the barrier. Forcing more gates to VDDL and keeping track of the best solution found, as proposed in Chapter 8, may allow a better solution to be found.

This concludes the analysis of using multi-Vdd and multi-Vth in comparison to using only a single Vdd and single Vth. We saw some power savings, but they were not comparable to the power savings from sizing alone with the linear programming approach versus TILOS, except for Vdd=0.08V/Vth=0.14V&0.08V with 0ps LCFF delay overheads. We now look at what additional computation time is required for multi-Vdd and multi-Vth optimization.

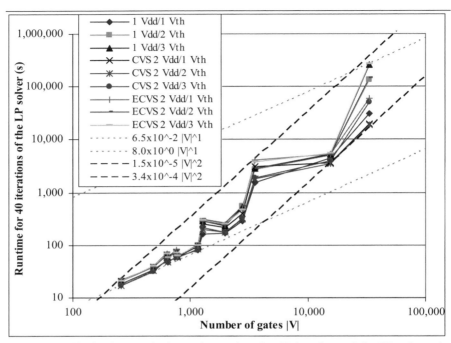

Figure 7.9 This log-log graph shows the runtime for 40 iterations of the LP solver. As illustrated with the lines showing O($|V|$) and O($|V|^2$) runtime growth, the linear program solver runtimes grow between linearly and quadratically with circuit size.

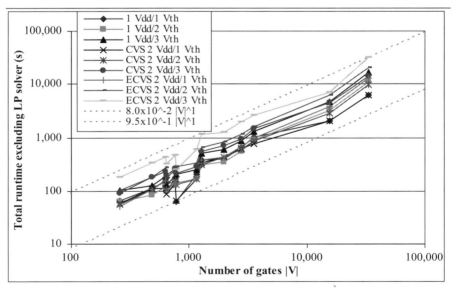

Figure 7.10 This log-log graph shows the runtime for 40 iterations, excluding the LP solver runtime. The runtimes excluding the linear program solver grow linearly with circuit size.

7.7   COMPUTATIONAL RUNTIMES WITH MULTI-VDD AND MULTI-VTH

The LP solver runtimes in Figure 7.9 are not substantially affected by the number of Vdd and Vth values, as that does not affect the number of variables or constraints in the linear program. Rather, the net runtime for multiple LP solver iterations depends on the timing slack and Vdd and Vth values available, as this determines the number of iterations for power minimization versus delay reduction and power minimization with the LP solver typically takes twice as long as delay reduction.

Excluding the LP solver, the CVS runtimes are not much worse than the single Vdd runtimes, as Vdd changes are considered only for gates on the VDDH to VDDL wavefront, that is gates with all fanouts at VDDL or primary outputs and all fanins at VDDH. The ECVS runtimes are up to 2.2 times the CVS runtimes, as changing Vdd is considered for all gates, doubling the setup runtime to consider the different Vdd alternatives, and there are additional computation overheads for insertion and removal of level converters. Considering alternate cells for a gate for multi-Vth can double setup runtimes for dual Vth and triple setup runtimes for triple Vth. The computation runtimes excluding the LP solver are shown in Figure 7.10.

With a CVS multi-Vdd methodology, each iteration allows at most one additional level of logic to change from high Vdd to low Vdd, or vice versa. Thus for a very deep circuit, for example c6288 with 113 logic levels, more iterations can be required for CVS multi-Vdd. The number of iterations required to get within 1% of the best solution varies substantially depending on the Vdd and Vth values available and the corresponding timing slack. Up to about 40 iterations is necessary in a few cases to get within 1% of the best solution when there is substantial timing slack, for example with Vth of 0.08V. For many cases 20 or fewer iterations are required, as for gate sizing.

As the delay overhead for level converters is substantial, an ECVS multi-Vdd methodology can also take more iterations for Vdd changes to propagate. The number of iterations required to get within 1% of the best solution are in the same range for ECVS and CVS. However, for ECVS we start with zero power for the level converters, and then run further optimization iterations with the correct level converter power. Usually less than 20 additional iterations are required to get within 1% of the best solution, as primarily gates on the boundary of the VDDH and VDDL regions change Vdd to reduce the number of level converters.

Runtimes for the Huffman, SOVA EPR4 and R4 SOVA benchmarks are included in these figures. The larger benchmarks have longer runtimes, so a less exhaustive range of results were collated, and the limited multi-Vdd and multi-Vth results for them have not been included in this chapter. Benchmark c17 was omitted due to its small size and small runtimes.

In summary, the runtimes for setting up the linear program grow linearly with the number of alternate cells available. For sizing only runs in Chapter 6, we saw that only about 20 iterations were necessary to find a good solution. However, with multi-Vth and CVS multi-Vdd there are cases where up to 40 iterations were necessary to converge within 1% of the best solution, and ECVS may take up to 60 iterations in total.

On smaller benchmarks, the Chapter 8 approaches for CVS and ECVS multi-Vdd with multi-Vth and gate sizing are substantially faster. However, as their runtime growth is $O(|V|^3)$ and our worst case runtime growth is $O(|V|^2)$, the LP approach was faster for c6288 and larger benchmarks.

7.8   SUMMARY

This chapter examined the power savings that the linear programming approach can achieve with single and multiple supply and threshold voltages. In comparison to the best optimization approaches without major simplifying assumptions that we know of for multi-Vdd/multi-Vth/sizing, our LP approach reduces power on average by 5% to 7%. The LP approach has runtime growth of $O(|V|)$ to $O(|V|^2)$, rather than $O(|V|^3)$, so the LP approach is also more applicable to larger benchmarks.

Scaling a single Vdd and single Vth optimally can provide significant power savings, reducing the power savings that may be found with multi-Vdd and multi-Vth. Versus the nominal Vdd=1.2V/Vth=0.23V, using Vdd=1.2V/Vth=0.14V reduces power on average by 12.0% at $1.0 \times T_{\min}$, and Vdd=0.8V/Vth=0.08V reduces power by 10.8% on average at $1.2 \times T_{\min}$ assuming 0ps level converter flip-flop delay overhead and that input drivers are also scaled down to 0.8V.

The optimal value of Vth depends greatly on Vdd. Vth of 0.08V provides a 22% speed increase at Vdd=1.2V versus Vth of 0.23V, but provides a 50% speed increase at Vdd=0.8V. In addition at Vdd=0.8V, the leakage is only about half the leakage at Vdd=1.2V. So the absolute increase in leakage power is less as Vth is reduced at Vdd=0.8V, reducing the penalty for using low Vth to reduce delay. Thus using a single Vth of 0.08V is acceptable at Vdd of 0.8V in the $1.2 \times T_{\min}$ scenario, but a poor choice with Vdd of 1.2V.

The multi-Vdd/multi-Vth results were compared against the best single Vdd/single Vth results at a given delay constraint. Our gate sizing results provided a strong baseline to compare results against.

In our detailed analysis of multi-Vdd and multi-Vth, the largest power savings were with Vdd=0.8V/Vth=0.14V&0.08V versus the optimal choice of Vdd=0.8V/Vth=0.08V at $1.2 \times T_{\min}$ assuming 0ps LCFF delay overhead and 0.8V input drivers. In this scenario, dual Vth reduced power on average by 16.0% and the maximum power saving was 26.2%. Triple Vth provided at most 5.1% power savings versus using dual Vth in the scenarios that we have considered, which is not sufficient to justify use of a third Vth.

Despite achieving better multi-Vdd/multi-Vth results than other known approaches, the LP multi-Vdd results with VDDL of 0.8V were suboptimal by up to 23.6% versus using a single Vdd of 0.8V with Vth values of 0.14V and 0.08V assuming 0ps LCFF delay overhead at $1.2 \times T_{min}$ [3].

The level converter delay and power overheads and larger rise delay when Vin > Vdd pose a significant barrier to optimization. Running optimization with the level converter power set to zero then running further iterations with the correct power did improve ECVS results substantially [3]. Experiments with reducing the level converter delay overheads would require a better delay reduction approach to ensure the final netlist meets the delay constraints, as the intermediate netlist with zero level converter delays will violate the delay constraints. The linear programming approach is not as good as Design Compiler at delay minimization, so some combined approach with a TILOS-like optimizer would be helpful.

The largest power saving with ECVS multi-Vdd versus single Vdd/single Vth at $1.2 \times T_{min}$ assuming 80ps LCFF delay overhead and 1.2V input drivers was 13.9%, but the average power saving was only 4.1%. In that scenario, ECVS with asynchronous voltage level converters averages only 1.2% power saving versus CVS, where only level converter flip-flops are allowed, though the maximum power saving with ECVS versus CVS is 7.3%.

We saw larger power savings with ECVS versus CVS and multi-Vdd/multi-Vth versus single Vdd/single Vth in the comparisons to the University of Michigan results, but Vdd=1.2V/Vth=0.12V was not the optimal choice for single Vdd/single Vth in that scenario.

There is no significant advantage for using VDDL of 0.6V versus 0.8V. The greatest saving for VDDL of 0.6V versus 0.8V was 4.2%, and the average saving was only 0.3%. We have not accounted for any additional LCFF delay for conversion from 0.6V to 1.2V compared to converting from 0.8V to 1.2V, assuming 80ps delay for both. In correcting the Vdd=0.6V delays, the $\alpha=1.66$ delay scaling with Equation (4.3) may still have been optimistic by 13% at Vth=0.23V to 6% at Vth=0.08V compared to the delay fit in Equation (4.4) that fit the Vdd=0.5V data as well. Lower Vdd cells also have other problems such as smaller voltage noise margins. These weak 0.6V VDDL results argue against the conclusion in [6] and other papers that VDDL should be 50% of VDDH. The rule of thumb to use a value of about 70% of VDDH for VDDL, that is 0.8V, provides sufficiently good results.

The delay reduction phase of the LP approach performs better with multi-Vth, as changing a gate from high Vth to low Vth to speed it up only slightly increases the load on fanin gates compared to upsizing a gate.

The multi-Vth power savings may be enough to justify the additional process costs for a second Vth. The weak multi-Vdd results do not justify use of multi-Vdd on these small benchmarks. More power savings may be available with multi-Vdd for an optimization approach that can overcome the optimization barrier posed by the level converter delay overheads.

For a larger sequential circuit with different delay constraints on portions of the circuit, different supply voltages may be justified. For example, Vdd of 1.2V at $1.0 \times T_{min}$ and Vdd of 0.8V at $1.2 \times T_{min}$. In the event that multiple supply voltages are justified by use at a module level in this manner, our results suggest that another 5% to 10% power saving may be available via a gate-level CVS or ECVS multi-Vdd assignment methodology.

7.9   REFERENCES

[1]   Avant!, *Star-Hspice Manual*, 1998, 1714 pp.
[2]   Bai, M., and Sylvester, D., "Analysis and Design of Level-Converting Flip-Flops for Dual-Vdd/Vth Integrated Circuits," *IEEE International Symposium on System-on-Chip*, 2003, pp. 151-154.
[3]   Chinnery, D., and Keutzer, K., "Linear Programming for Sizing, Vth and Vdd Assignment," in *Proceedings of the International Symposium on Low Power Electronics and Design*, 2005, pp. 149-154.
[4]   Ishihara, F., Sheikh, F., and Nikolić, B., "Level Conversion for Dual-Supply Systems," *IEEE Transactions on VLSI Systems*, vol. 12, no. 2, 2004, pp. 185-195.
[5]   Kulkarni, S., and Sylvester, D., "Fast and Energy-Efficient Asynchronous Level Converters for Multi-VDD Design," *IEEE Transactions on VLSI Systems*, September 2004, pp. 926-936.
[6]   Kulkarni, S., Srivastava, A., and Sylvester, D., "A New Algorithm for Improved VDD Assignment in Low Power Dual VDD Systems," *International Symposium on Low-Power Electronics Design*, 2004, pp. 200-205.
[7]   Nguyen, D., et al., "Minimization of Dynamic and Static Power Through Joint Assignment of Threshold Voltages and Sizing Optimization," *International Symposium on Low Power Electronics and Design*, 2003, pp. 158-163.
[8]   Puri, R., et al., "Pushing ASIC Performance in a Power Envelope," in *Proceedings of the Design Automation Conference*, 2003, pp. 788-793.
[9]   Sirichotiyakul, S., et al., "Stand-by Power Minimization through Simultaneous Threshold Voltage Selection and Circuit Sizing," in *Proceedings of the Design Automation Conference*, 1999, pp. 436-41.
[10] Srivastava, A., Sylvester, D., and Blaauw, D., "Power Minimization using Simultaneous Gate Sizing Dual-Vdd and Dual-Vth Assignment," in *Proceedings of the Design Automation Conference*, 2004, pp. 783-787.
[11] Stok, L., et al., "Design Flows," chapter in the *CRC Handbook of EDA for IC Design*, CRC Press, 2006.
[12] Usami, K., and Horowitz, M., "Clustered voltage scaling technique for low power design," in *Proceedings of the International Symposium on Low Power Design*, 1995, pp. 3–8.
[13] Usami, K., et al., "Automated Low-Power Technique Exploiting Multiple Supply Voltages Applied to a Media Processor," *IEEE Journal of Solid-State Circuits*, vol. 33, no. 3, 1998, pp. 463-472.
[14] Usami, K., et al., "Automated Low-power Technique Exploiting Multiple Supply Voltages Applied to a Media Processor,", in *Proceedings of the Custom Integrated Circuits Conference*, 1997, pp.131-134.
[15] Wei, L., et al., "Mixed-V_{th} (MVT) CMOS Circuit Design Methodology for Low Power Applications," in *Proceedings of the Design Automation Conference*, 1999, pp. 430-435.
[16] Wei, L., Roy, K., and Koh, C., "Power Minimization by Simultaneous Dual-V_{th} Assignment and Gate-Sizing," in *Proceedings of the IEEE Custom Integrated Circuits Conference*, 2000, pp. 413-416.

Chapter 8

POWER OPTIMIZATION USING MULTIPLE SUPPLY VOLTAGES

Sarvesh Kulkarni, Ashish Srivastava, Dennis Sylvester, David Blaauw
Department of Electrical Engineering and Computer Science,
University of Michigan,Ann Arbor, MI
shkulkar,ansrivas,dmcs@umich.edu

Multiple supply voltage design is an effective technique for power minimization in CMOS circuits. Clustered Voltage Scaling (CVS) and Extended Clustered Voltage Scaling (ECVS) are the two major methodologies used for assigning the voltage supply to gates in circuits having dual power supplies. This chapter presents current state of the art approaches that combine CVS and ECVS with threshold voltage assignment and gate sizing to enable the maximum reduction in power dissipation. Later we also present a comparison of achievable power savings using CVS and ECVS and point out that ECVS provides appreciably larger power improvements compared to CVS. However, ECVS rests on the availability of well designed asynchronous level converters. We also quantify the impact of the efficiency of level conversion on power savings.

8.1   INTRODUCTION

Dynamic power dissipation in CMOS circuits is proportional to the square of the supply voltage (VDD). A reduction in VDD thus considerably lowers the power dissipation of the circuit. Dual- (or more generally multi-) VDD design is an important scheme that exploits this concept to reduce power consumption in integrated circuits (ICs) [5][30]. Since a reduction in VDD degrades circuit performance, in order to maintain performance in dual-VDD designs, cells along critical paths are assigned to the higher VDD (VDDH) while cells along non-critical paths are assigned to a lower VDD (VDDL). Thus the timing slack available on non-critical paths is efficiently converted to energy savings by use of a second supply voltage. However, level conversion (from VDDL to VDDH) becomes essential at boundaries where a VDDL driven cell feeds into a VDDH driven cell to eliminate the undesirable static current that otherwise flows. This current flows since the logic "high" signal of the VDDL driven cell cannot completely turn off the PMOS pull-up network of the subsequent VDDH cell.

The use of level converters is largely determined by the algorithm used in assigning VDD to gates. The two major existing algorithms used for VDD assignment are (1) Clustered Voltage Scaling (CVS) [30], and (2) Extended Clustered Voltage Scaling (ECVS) [10]. In CVS, the cells driven by each power supply are grouped ("clustered") together and level conversion is needed only at sequential element outputs (referred to as *"synchronous level conversion"*). In ECVS, the cell assignment is more flexible, allowing level conversion anywhere (i.e., not just at the sequential element outputs) in the circuit. This is referred to as *"asynchronous level conversion"*. Since ECVS allows more freedom in VDD assignment, it has been suggested that it potentially provides greater power reductions than CVS [33].

Both CVS and ECVS assign the appropriate power supply to the gates by traversing the circuit from the primary outputs to the primary inputs in reverse topological level order. CVS is based on a topological constraint that only allows a single transition from a VDDH driven cell to a VDDL driven cell along any path from input to output (i.e., a VDDL driven cell may not feed into a VDDH driven cell). Depending on the design, this may greatly reduce the fraction of VDDL assigned gates and degrade the achievable power savings. Alternatively, ECVS relaxes this topological constraint by allowing a VDDL driven cell to feed a VDDH driven cell along with the insertion of a dedicated asynchronous level converter (ALC). However, since ECVS performs this assignment simply by visiting gates one at a time in reverse topological level order, it still assigns supply voltages in a funda- mentally constrained manner. Noting these drawbacks, an algorithm that removes the "levelization" approach will be discussed in Section 8.3. Since level converters consume power and timing slack, it is important to consider their effect on the power savings.

Techniques such as gate sizing and dual threshold voltage (Vth) assign- ment can be combined with dual-VDD assignment in order to realize a more optimized design. Many different approaches have been proposed that use the variables of VDD/Vth/sizing for power optimization. In [34] the authors address the problem of power optimization using simultaneous VDD and Vth assignment. They propose two different approaches depending on whether a system is dynamic or leakage power dominated. The approach for dynamic power dominated systems fails to consider that assigning a gate to high Vth negatively impacts the extent to which other gates in the circuit can be assigned to VDDL and thus fails to consider the optimization of total power. The approach for leakage dominated systems assigns gates to high Vth in the order of their level from the outputs. Since Vth assignment does not impose any topological constraints as in the case of VDD assignment, this approach unnecessarily limits the achievable power savings. Recently, [21] proposed a new method for slack redistribution to solve the leakage power optimization problem with dual-Vth and sizing by iteratively formulating and solving a linear program. However, the extension to dual-VDD assignment is formulated as an integer linear program resulting in unreasonable runtime complexity. Reference [7] uses a Lagrangian multiplier based optimization followed by heuristic clustering for dual-VDD and dual-Vth assignment. This is a general technique for solving optimization problems involving discrete variables, where the problem is initially solved while assuming the variables involved are continuous. This allows the problem to be solved in a computationally efficient manner (using well-known non-linear optimization techniques), and then heuristically clustering the obtained solution to the discrete domain [25][27][28]. The Lagrangian multiplier based approach is used to perform module level power optimization using path enumeration. The approach requires a power-aware partitioning of a circuit, which is a very difficult problem as acknowledged by the authors. The approach cannot be extended to perform gate-level power optimization due to its computational complexity and also does not consider other circuit issues such as level conversion. Reference [9] uses a genetic algorithm based approach to solve the problem of simultaneous VDD and Vth assignment with gate sizing, which is both computationally inefficient and performs poorly.

References [4] and [36] address the power minimization problem using dual-VDD assignment and sizing. In particular, [4] uses maximum weighted independent sets to identify gates for downsizing or assignment to VDDL by identifying sets of gates which have independent timing slacks. This technique is severely limited by the amount of slack available in the original circuit as there are no means to create additional slack by sizing gates, only to consume it. In [36], the authors use a sensitivity-based technique to optimize power dissipation using dual-VDD assignment. Another work employs a delay balancing approach to solve the problem of VDD assignment [28]. There has been a large amount of work recently in power optimization using dual-Vth and sizing [12][14][21][22][24][35]. References [22], [24] and [35] use sensitivity-based approaches to direct the optimization, whereas [12] solves the problem using a Lagrangian relaxation based tool. Reference [14] employs a state-space enumeration based approach along with efficient pruning methods and demonstrates better power savings for tight delay constraints and better run-time compared to [24]. However, all these approaches fail to integrate all three design variables that are crucial to low-power design, namely sizing, threshold and supply voltages, in a computationally efficient manner.

In Section 8.2 we describe the basic implementation of CVS and ECVS. In Section 8.3 we describe a recently developed algorithm (referred to as Greedy ECVS or GECVS) that avoids some of the pitfalls of the original ECVS. Section 8.4 presents power savings obtained using these algorithms.

In Section 8.5 we present techniques that extend CVS and GECVS by including Vth assignment and gate sizing to perform power optimization using all three (VDD, Vth and sizing) levers available to a designer. Finally, in Section 8.6 we present more results and conclude in Section 8.7.

8.2   OVERVIEW OF CVS AND ECVS

This section provides a comprehensive summary of the existing CVS and ECVS algorithms. For all the algorithms discussed, the starting point is a design having all cells assigned to VDDH and then VDDL is utilized according to the algorithm being applied. Both CVS and ECVS aim at using the available timing slack in a circuit by applying a lower supply voltage on gates that are off the critical paths. This results in reduced dynamic power dissipation and hence lowers system level power dissipation. However, they differ in the policies they follow in making this power supply assignment. As a result of this, the final structure of the resulting netlists after applying these algorithms differs. As stated in Section 8.1, voltage level conversion is required whenever a VDDL driven cell feeds a VDDH driven cell. An example of this is shown in Figure 8.1, where a VDDL driven inverter directly feeds into a VDDH driven inverter. The resulting DC current will result in extremely high static power dissipation without the use of level converters.

CVS and ECVS differ in the way they address the issue of level conversion. Since CVS does not allow VDDL driven cells to directly feed VDDH driven cells, level conversion is therefore implemented only at flip-flop (or sequential) boundaries. The level conversion functionality can be embedded into the flip-flop circuit [1][11] and such a flip-flop is referred to as a level converting flip-flop (LCFF).

ECVS relaxes this topological constraint and allows a VDDL driven cell to feed a VDDH driven cell after its output has undergone level conversion. ECVS thus has more freedom in finding portions of the circuit that can be operated at the lower supply and can potentially lead to higher power savings. However, the asynchronous level converters impose penalties in terms of their delay, power and area. Fast and low power ALCs are thus important in mitigating these penalties. Figure 8.2 depicts the nature of the final topologies attained by CVS and ECVS when applied to a given circuit. From this figure, it is seen that CVS partitions a circuit into two clusters that can be ordered topologically – one having only VDDH cells and the other having only VDDL cells. The scenario in which a VDDL driven cell directly feeds a VDDH driven cell is clearly precluded in this partitioning. On the other hand, ECVS allows interspersing of VDDL and VDDH cells with insertion of any required ALCs. We now discuss the implementation of these algorithms in greater detail.

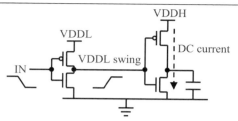

Figure 8.1 Demonstrating the need for level conversion.

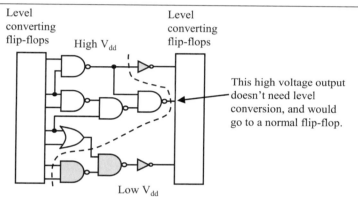

(a) Clustered Voltage Scaling (CVS) – combinational logic can be partitioned into separate VDDH and VDDL portions in topological order.

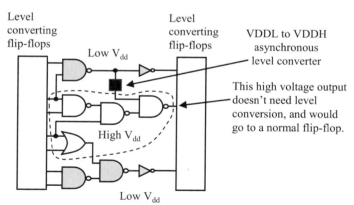

(b) Extended Clustered voltage scaling (ECVS) – a VDDL signal may go to a VDDH gate if an asynchronous level converter (ALC) is inserted. Here the logic cannot be partitioned into separate VDDH and VDDL portions that are topologically ordered.

Figure 8.2 This illustrates the difference between the circuit structures after the application of ECVS and CVS.

```
CVS( ) {
    minimum power found = power of initial VDDH circuit.
    Best configuration = all VDDH assignment.

    L = gates that only drive circuit primary outputs.
    While L is non-empty {
        STEP: "SET VDDL" –
        Select candidate A from L.
        Remove A from L.
        Set the supply voltage of A to VDDL.

        If A drives a primary output, insert an LCFF.

        Check timing.
        If circuit still meets timing constraints {
            STEP: "CONSTRAINED TOPOLOGY" –
            Add to L gates that fan into A but not into any VDDH gate.

            Check power consumption.
            If power < minimum power found {
                minimum power found = power.
                Best configuration = current VDDL assignment.
            }
        } else {
            Remove any added LCFFs.
            Set the supply voltage of A back to VDDH.
        }
    }
}
```

Figure 8.3 Pseudo-code for the CVS algorithm.

CVS maintains a list (referred to as *L*) of candidate cells that can be assigned to VDDL. New cells continue to be added to this list as the algorithm proceeds. The elements of *L* are ranked according to a heuristic and the first element is chosen to be assigned to VDDL at each step of the algorithm. The initial implementation of CVS [30] used a heuristic which ordered the cells in *L* on the basis of their slack. *L* is initialized to the set of gates that drive the circuit primary outputs. Pseudo-code for CVS is shown in Figure 8.3.

The step *"CONSTRAINED TOPOLOGY"*, guarantees that there will be no VDDL driven gate that feeds directly into a VDDH driven gate. However, this constraint acts to curtail many potential VDDL cell assignments as later results will demonstrate.

ECVS (as earlier implemented in [10][31][32][33]) begins by levelizing the circuit from the primary outputs to the primary inputs. LCFFs are inserted when a cell driving a primary output is assigned to VDDL (as in the case of CVS). Similarly ALCs are inserted whenever a VDDL gate feeds into a

VDDH gate. ECVS is also executed in reverse topological order. Pseudo-code for the ECVS algorithm is shown in Figure 8.4.

Since ECVS subsumes CVS, it can theoretically attain a higher degree of VDDL gate assignments. ECVS, however, must consider the overheads imposed by the ALCs. Although ECVS has clear advantages over CVS, its policy of determining the VDD assignments is still constrained by levelization. An approach that avoids this constraint is described next.

```
ECVS ( ) {
    minimum power found = power of initial VDDH circuit.
    Best configuration = all VDDH assignment.

    L = gates that only drive circuit primary outputs.
    While L is non-empty {
        STEP: "SET VDDL" –
        Select candidate A from L.
        Remove A from L.
        Set the supply voltage of A to VDDL.

        If A drives a primary output, insert an LCFF.

        For each gate B ∈ fanouts(A) {
            If (supply of B = VDDH)
            Insert an ALC on the path from A to B.
        }

        Check timing.
        If circuit still meets timing constraints {
            STEP: "LEVELIZED" –
            Add to L gates that fan into A and only into other gates
            that have already been considered or primary outputs.

            Check power consumption.
            If power < minimum power found {
                minimum power found = power.
                Best configuration = current VDDL assignment.
            }
        } else {
            Remove any added LCFFs or ALCs.
            Set the supply voltage of A back to VDDH.
        }
    }
}
```

Figure 8.4 Pseudo-code for the ECVS algorithm. Note the differences from Figure 8.3.

8.3   GREEDY ECVS: A NEW DUAL–VDD ASSIGNMENT ALGORITHM

ECVS-style approaches are most effective when they are able to find 'groups' or 'clusters' of connected gates that can be assigned to the lower supply. This is so since such a grouped assignment will require fewer ALCs and minimize their resulting overhead. A sensitivity measure that uses the information available in the slack distribution of the circuit and the power savings attainable before finalizing *each VDDL assignment move* can be used for this purpose. This avoids the problems inherent in ECVS, which merely traverses the circuit (after levelization) and makes the earliest *seen* feasible move.

At each stage of the new algorithm, a sensitivity measure for all cells that are potential candidates for VDDL assignment is evaluated. Every VDDL assignment may call for either the insertion or removal of ALCs in the vicinity (at the inputs/output of the gate under consideration). This is because an ALC is required only when a VDDL driven gate needs to supply a VDDH driven gate. Since a levelized VDDL assignment (as in ECVS) is not followed here, this ALC removal is frequently required and is accomplished by the *update_vicinity( )* sub-routine in the pseudo-code below. As a result of a move, the arrival time at the output of the gate being assigned to VDDL will change (arrival time at the output includes the arrival time at the output of any added level converters, if the move requires ALC/LCFF insertion). This changes the slack of various paths in the circuit. The overall power dissipation of the circuit will also change as a result of the move. A move assigning VDDL to a gate feeding a primary output requires inserting an LCFF and the LCFF delay must be included in the arrival time calculation in this case. The LCFF data used was obtained from [1][11].

The sensitivity for a move is determined from the change in total power ΔP, change in arrival time at the gate output ΔD (summing the worst changes in rise and fall delay), and the sum of the worst rise and fall slacks of timing arcs through the gate (*slack*). The slack for a given rise or fall timing arc ij through a gate, $Slack_{arc\ ij}$, is calculated as the difference of the delay of the arc ($d_{arc\ ij}$) and the difference of the required arrival time at the output node and the arrival time at the input node of the gate:

$$Slack_{arc\ ij} = (t_{required\ at\ output\ j} - t_{arrival\ at\ input\ i}) - d_{arc\ ij}, \text{ where}$$

$$\text{(8.1)}$$

timing *arc ij* is from gate input i to gate output j.

The slack reflects the maximum possible increase in delay of the timing arc that still satisfies the timing characteristics of the overall design. The sensitivity for the gate is defined as follows:

$$\text{Sensitivity (set VDDL)} = \frac{-\Delta P \times slack}{\Delta D} \qquad \text{(8.2)}$$

where it is assumed that $\Delta D > 0$. The code handles the case where $\Delta D \leq 0$, but this does not occur in practice for a VDDH to VDDL change with our benchmarks and libraries.

Sensitivities for all gates that can undergo VDDL assignment are evaluated at every iteration of the algorithm, and the gate with the maximum sensitivity is selected. The state of the circuit is saved at this point and the algorithm proceeds to the next iteration. From the definition of sensitivity in Equation (8.2), observe that this algorithm allows negative moves to be taken, thus opening the possibility of uncovering better solutions in the long run. Essentially this sensitivity measure enables us to choose the move giving *the best power savings per unit delay penalty*. The slack term in the sensitivity computation acts as a weighting factor to encourage VDDL assignment for gates with more slack. Evaluating this sensitivity for a gate only requires the rise/fall transition and arrival times at the inputs of the gates that feed it. This sensitivity can thus be evaluated efficiently as a constant-time operation.

This algorithm, designated GECVS (Greedy-ECVS) [16], tends to group VDDL gates together inherently due to the nature of the sensitivity function. Since the ΔD and ΔP terms consider the ALC overheads associated with a particular VDDL assignment, the algorithm automatically guides itself towards building groups or clusters of VDDL gates. What is unique about GECVS is that these clusters can form *at the beginning of a path,* just as easily as they can at the end of a chain of combinational logic. This makes GECVS fundamentally more flexible than CVS or ECVS, which proceed with VDDL assignment using a backwards traversal. Since CVS and ECVS follow a backward traversal for VDDL assignment, they naturally tend to assign more gates near the primary outputs to the lower supply. As gates near primary outputs typically have low switching activity (on average) [20], this can also lead to degraded savings. GECVS avoids this levelized approach and does not suffer from this drawback. Pseudo-code for GECVS is shown in Figure 8.5.

Various approaches for the physical design of dual-VDD circuits have been discussed earlier. In [15] and [19], the authors proposed the use of macro voltage islands where entire functional units operate at different supplies. Reference [23] proposed the use of a somewhat more fine-grained strategy that used voltage islands interfaced through level converters. A more fine-grained approach is presented in [33] that employs alternating rows of VDDH and VDDL cells – this is an ideal approach for designs that are highly performance critical as well as severely power constrained. The technique presented in [33] attempts to minimize the wire length between the VDDL cell and the ALC that it drives, for delay reduction. An opposite approach that places the ALCs directly at the input of the fanout gates rather than at the output of the driving gate improves the dynamic power consumed in switching the wire at the expense of additional delay.

The *update_vicinity( )* step occurs over the local neighborhood of the gate: on the fanout side of the gate, the gate itself and any ALCs it might have induced; and on the fanin side, the fanin gates and any required ALCs.

Determining feasible moves is done as follows. Before calculating the sensitivities of all gates, all timing information, such as arrival times and slacks, is calculated with static timing analysis and stored. Then when calculating the sensitivities for each gate, these stored values are used and the new arrival times are found using only the gates in the immediate neighborhood, as described for *update_vicinity( )* above – if the slack can accommodate this increased delay, this gate is remembered. The vicinity calculations are not exact (as compared to static timing analysis over the transitive fanout from the gate's fanins), but this filtering greatly reduces the number of gates to be tried in the final timing check. The final timing checks are with static timing analysis over the full affected region, to confirm that the delay target is met.

```
GECVS ( ) {
      minimum power found = power of initial VDDH circuit.
      Best configuration = all VDDH assignment.

      Do {
            For each VDDH gate 'A' {
                  Set A to VDDL.
                  If A drives a primary output, insert an LCFF.
                  update_vicinity ( ) // insert or remove ALCs as necessary
                  Calculate sensitivity for A using Equation (8.2).
                  Set A back to VDDH
                  update_vicinity ( ) // insert or remove ALCs as necessary
            }

            Select the maximum sensitivity gate 'B' that meets timing.

            Check power consumption.
            If power < minimum power found {
                  minimum power found = power.
                  Best configuration = current VDDL assignment.
            }
      }
      While there are feasible moves (i.e., moves meeting timing)
}
```

Figure 8.5 Pseudo-code for the GECVS algorithm.

Table 8.1 Comparison of power savings using CVS and GECVS versus the original design with all gates being at VDDH and low Vth.

Circuit	VDDL = 0.6V		VDDL = 0.8V	
	CVS	GECVS	CVS	GECVS
c432	1.0%	1.5%	0.8%	0.8%
c880	8.2%	10.3%	15.0%	21.3%
c1355	0.0%	0.0%	0.0%	1.0%
c1908	4.3%	7.7%	3.4%	8.4%
c2670	21.1%	25.5%	16.5%	25.0%
c3540	3.2%	8.3%	2.9%	9.7%
c5315	7.6%	19.0%	8.3%	22.0%
c7552	14.9%	20.2%	22.0%	28.8%
Huffman	6.6%	12.7%	6.7%	14.4%
Average	**7.4%**	**11.7%**	**8.4%**	**14.6%**

8.4 POWER SAVINGS WITH CVS AND GECVS

The underlying process assumed is a 0.13um dual-Vth CMOS process. The higher (nominal) power supply VDDH is 1.2V and VDDL was either 0.6V or 0.8V. Table 8.1 summarizes the dynamic power savings achieved by CVS and GECVS for the various benchmark circuits with VDDL = 0.6V and 0.8V. The delay target was set to be 10% slower than the fastest possible all-VDDH/low Vth design as found by a TILOS [8] based gate sizing algorithm. The algorithms were evaluated on circuits from the ISCAS'85 benchmark set [3]. The data for LCFFs was adopted from [1]. This work shows that LCFFs impose a delay overhead which is about two FO4 delays in the target technology, or about 80ps in our studies. The ALC used was the circuit STR6 from [17] (shown in Figure 13.5(d)). This ALC has a delay of 84ps and consumes 6.3fJ of internal energy per transition. The switching activity at each node was computed by simulating the design for 10,000 cycles with independent random inputs with equal probabilities of logical value 0 and 1. The switching activity at each input was then scaled to so that approximately 20% of the overall power dissipation was due to leakage power. The wire capacitance is approximated by

$$C_{wire} = 3 + 2 \times fanouts_{wire} \text{ fF} \qquad (8.3)$$

where *fanouts*$_{wire}$ is the number of gates to which the wire connects, excluding its driver. Equation (8.3) is based on the model used in [29] and provides a wire capacitance of 5fF for a gate with one fanout, corresponding to a wire length of approximately 25um in our technology.

GECVS performs significantly better than CVS and provides approximately twice the power savings or more in some circuits. On average, circuits optimized using GECVS have about 6% lower power than those optimized with CVS (Table 8.1). The percentage of power consumed by ALCs under

GECVS is shown in Table 8.2. On average, ALCs consume 6% of the total power. Figure 8.6 compares the fraction of the total gates assigned to VDDL of 0.8V in the final design. GECVS enables more VDDL assignments compared to CVS thereby reducing power.

Table 8.2 Percentage of total power consumed by asynchronous level converters and their percentage gate count for the optimized GECVS results shown in Table 8.1.

		VDDL = 0.6V		VDDL = 0.8V	
Circuit	**Logic depth**	**ALC power**	**ALC count**	**ALC power**	**ALC count**
c432	23	1.1%	1.8%	0.0%	0.0%
c880	26	1.9%	2.3%	6.7%	7.2%
c1355	28	0.0%	0.0%	0.7%	1.6%
c1908	40	4.8%	5.1%	6.5%	9.3%
c2670	25	5.5%	7.2%	7.4%	9.0%
c3540	42	6.9%	12.6%	7.9%	12.7%
c5315	41	6.8%	10.3%	7.7%	9.1%
c7552	44	8.7%	10.1%	9.0%	8.5%
Huffman	47	9.0%	10.2%	12.8%	17.5%
Average		**5.0%**	**6.6%**	**6.5%**	**8.3%**

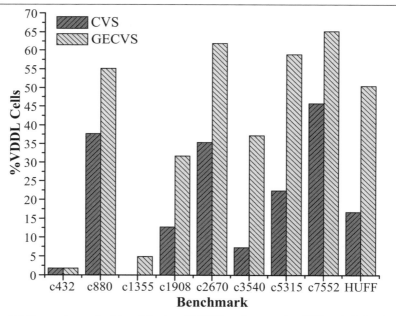

Figure 8.6 Comparison of achieved levels of VDDL assignment by CVS and GECVS, with VDDL of 0.8V.

8.5   GATE SIZING AND DUAL-VTH ASSIGNMENT

All algorithms discussed to this point achieve power savings by simply changing the supply voltage of the gates in the design to the lower supply (VDDL) as frequently as possible. Techniques such as gate sizing and dual-Vth assignment can be used to further reduce the power. In this section, we describe extensions to CVS and GECVS that incorporate these techniques.

8.5.1   CVS Based Power Optimization Using Dual-Vth Assignment and Gate Sizing

In this section we discuss a two stage sensitivity-based heuristic approach to minimize total power using dual-VDD assignment, gate sizing, and dual-Vth assignment for a standard cell library mapped design. All the gates in the design are initially assumed to be operating at the higher supply voltage and lower threshold voltage. Throughout the flow of the algorithm (which we refer to as VVS [26]) a "wave-front" is maintained located at the interface between the VDDL and VDDH gates (e.g. see Figure 8.8). Similar to CVS, level conversion within the logic itself is not allowed, and therefore we must strictly observe the topological constraint imposed in dual-VDD designs. The timing constraints on the design remain fixed throughout the flow of the algorithm.

Initially, all gates are low Vth and with VDDH supply voltage. In the first stage of the VVS algorithm, called the backward pass, VDD assignment and sizing are combined to minimize total power while we move the front from the primary outputs towards the primary inputs. Threshold voltages are kept at low Vth in the backward pass. The second stage, or the forward pass, uses the optimal location of the front found in the first stage as the starting point for the optimization and then relies on both VDD and Vth assignment along with gate sizing to further reduce total power while the front is moved back towards the primary outputs. Thus all three design variables are used to perform concurrent VDD assignment, Vth assignment and gate sizing in the forward pass.

8.5.1.1   Backward Pass

To adhere to the topological constraint imposed by dual-VDD we define the *backward front*, the list L as defined in *CVS( )*, which consists of all gates operating at VDDH that do not fanout to any gate operating at VDDH. Thus, assigning any gate on the backward front to VDDL will not violate the topological constraint since all its fanout gates also operate at VDDL. This front is initialized to be the set of gates that drive the primary outputs of the design. A simple *CVS( )* procedure is first used to assign gates on the front to VDDL as long as the circuit meets timing.

```
Backward Pass( ) {
    CVS( )

    L = backward front (VDDH gates not fanning out to VDDH gates).
    While list L is non-empty {
        // Candidate selection based on predictive metric
        STEP: "SET VDDL" –
        Calculate predictive metric for all gates in backward front.
        Select candidate A from L.
        Remove A from L.
        Set the supply voltage of A to VDDL.

        If A drives a primary output, insert an LCFF.

        STEP: "UPSIZING" –
        While circuit fails timing and number of upsizing moves
        is < 10% of total number of gates in the circuit) {
            Calculate sensitivity of all gates to upsizing with Equation (8.4).
            Upsize gate with maximum sensitivity to the next higher
            size available in the library.
        }

        Check timing.
        If circuit meets timing constraints {
            STEP: "CONSTRAINED TOPOLOGY" –
            Add to L gates that fan into A but not into any VDDH gate.

            Check power consumption.
            If power < minimum power found {
                minimum power found = power.
                Best configuration = current VDDL & sizing assignment.
            }
        } else {
            Undo upsizing moves.
            Remove any added LCFFs.
            Set the supply voltage of A back to VDDH and flag A.
        }
    }
}
```

Figure 8.7 Pseudo-code for the backward pass of the VVS algorithm.

At the end of *CVS( )*, none of the gates on the backward front can be assigned to VDDL without violating the timing constraints. Figure 8.8 shows the scenario at this stage. Gates 1 to 3 have been set to VDDL by *CVS( )* and gates 4, 5, and 8 now form the backward front. Gate sizing is then employed to compensate for the delay increase arising from the assignment of a gate to VDDL. The pseudo-code for this stage is shown in Figure 8.7.

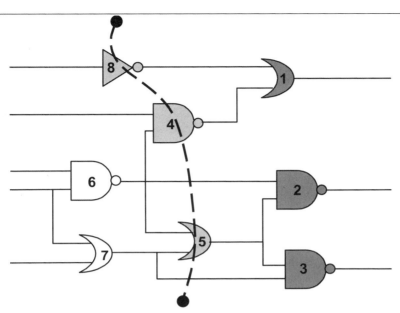

Figure 8.8 Backward "wave front" for an example circuit at the end of CVS. The gates in dark grey have been assigned to VDDL. Gates in light grey are candidates for being assigned to VDDL – the backward front.

Specifically, after a candidate gate on the backward front is assigned to VDDL, a sensitivity measure to upsize gates to the next available size in the standard-cell library for all the gates in the circuit is calculated. This is used to identify gates to be upsized. Let ΔD_{arc} represent the change in delay of a timing arc of the gate and ΔP the change in power dissipation due to upsizing a gate to the next higher size in the library. The sensitivity of each gate in the circuit to upsizing is computed as

$$\text{Sensitivity} = \frac{1}{\Delta P} \sum_{arcs} \frac{\Delta D_{arc}}{Slack_{arc} - Slack_{min} + K} \qquad (8.4)$$

where $Slack_{min}$ is the worst slack of a timing arc seen in the circuit, and K is a small positive quantity for numerical stability purposes. $Slack_{arc}$ represents the slack associated with the particular timing arc of the gate as defined in Equation (8.1).

The form of the sensitivity measure gives a higher value to gates lying on the critical paths of the circuit. The arcs represent the falling and rising arcs associated with each of the inputs of the gate. Thus, for a 3-input NAND gate the sensitivity measure will be obtained by summing over all six possible arcs. The ΔD_{arc} computation for a gate (say G) is performed by upsizing the gate to the next higher size in the library. Since only gates which are the immediate inputs of G see a different load capacitance, only these gates need to be re-simulated during sensitivity computation to calculate the new arrival time at the inputs of G which is used to calculate ΔD_{arc} at the output of G. Any change in delay due to slew changes in the fanout cone of G is considered to be second order and neglected for sensitivity computation. Similarly ΔP can also be easily computed by considering only the immediate fanin of the gate, the gate itself, and the load capacitance (with switching voltage changing from VDDH to VDDL). The gate with the maximum sensitivity is then selected and sized up to the next available size in the library. It is important to note that the sensitivity calculation for a gate does not require a full circuit timing analysis or an incremental timing analysis (which would propagate the impact of the change through the fanout cone), which would otherwise make the runtime prohibitively large. Complete timing analysis is performed and this process is repeated until all slacks in the circuit become positive. While performing gate upsizing, delays and slacks of the gates that form the fanin and fanout cone are modified and hence we need to re-compute the timing information and sensitivities only for these gates.

The number of upsizing moves allowed to meet timing is fixed to a constant large number (10% of the number of gates in the circuit) to avoid pursuing bad solutions that could also possibly result in overly large area increases. The choice of a 10% limit on the number of gates to be upsized is based on the observation that varying this percentage from 8% to 50% results in a very small change in the power dissipation achieved. The power dissipation for values less than 8% gradually increases as we reduce the maximum number of gates that can be upsized. This limit on the percentage of upsizing moves that provides the maximum reduction was not found to increase with circuit size (i.e. the absolute number increases linearly with circuit size). If the circuit fails to meet timing after the maximum number of upsizing moves, then the VDDL assignment and the associated upsizing moves are reversed and the gate assigned to VDDL is flagged so that it is not reconsidered for VDDL assignment.

We do allow moves that result in a net increase of total power in an attempt to allow the flow of the algorithm to escape local minima. Due to the topological constraints imposed on VDDL assignment, if a gate is not assigned to VDDL then none of the gates in its input cone can be assigned to VDDL. Otherwise a steepest decent only approach is likely to get stuck in a local minimum that may be far from the global minimum. Consider the case where the path that goes through gates 7, 5 and 2 forms the critical path of the circuit in Figure 8.8. If gate 5 is not assigned to VDDL, gate 6 and other gates in its fanin cone (if present) cannot be assigned to VDDL. Thus a lower total power might be achieved if gate 6 can be assigned to VDDL, after having assigned gate 5 to VDDL with upsizing to meet delay constraints which resulted in an increase in power at that step.

Gates on the backward front are ordered using a predictive metric – a heuristic used to steer the flow of the algorithm in the right direction. The predictive metric can be used to identify the capacitance and the slack associated with the fanout cone of a gate. The sum of the product of the capacitance and timing slack at each node in the fanin cone has been used as the predictive metric.

The end of the backward pass is signaled when the list containing the gates on the backward front becomes empty or else none of the gates in the list can be assigned to VDDL without violating timing (even with the maximum allowed amount of upsizing). At all points during the backward pass the best-seen solution is saved and this solution is restored at the end of the backward pass.

8.5.1.2   Forward Pass

At the end of the backward pass, the circuit sizing and VDD assignments (the VDDH to VDDL "wave front") which best minimized total power for the dual-VDD, single low Vth is chosen. The second stage, or forward pass, is then used to move the front forward towards the primary outputs in conjunction with high Vth allocation and possible gate upsizing to minimize the total power in a dual-Vth scenario.

We now define the *forward front*, which consists of all gates that are operating at VDDL and have all of their fanins operating at VDDH. In Figure 8.8, assuming that upsizing in the backward pass allows us to further assign gates 4, 5 and 8 to VDDL, these same three gates would now form the forward front. Importantly, assigning a gate on the forward front to operate at VDDH will not lead to a violation of the topological constraint. We now calculate 1) a sensitivity measure for gates on the forward front with respect to VDDH operation, and 2) a sensitivity measure for all gates in the circuit with respect to upsizing to the next higher size in the library. Both these sensitivities are calculated as the ratio of the sum of the delay changes of all timing arcs to the change in power dissipation as a result of the corresponding operation. An expression similar to Equation (8.4) is not used since this operation is not used to identify gates that are critical and is only used to generate additional timing slack in the circuit (to enable high Vth assignment). The gate with the maximum sensitivity is then either assigned to VDDH or upsized based on the operation to which the maximum sensitivity corresponds. Note that any gate in the circuit may be upsized whereas only gates in the forward front may be re-assigned to VDDH.

Once a gate is upsized or reset to VDDH operation, timing slack has been created in the circuit. To exploit this slack and reduce total power, the next step begins by computing the sensitivity of all gates in the circuit with respect to operation at high Vth (recall that initially all gates are low Vth). This sensitivity is calculated as the ratio of the change in power to the change in delay multiplied by the slack of the gate in order to identify gates that provide the maximum decrease in power for the minimum increase in delay and is expressed as

$$\text{Sensitivity} = \Delta P \sum_{arcs} \frac{Slack_{arc}}{\Delta D_{arc}} \qquad (8.5)$$

Based on this sensitivity measure gates are assigned to high Vth as long as the timing constraints of the design are met. This set of moves (assignment to VDDH or upsizing a gate followed by the associated high Vth assignments) is then accepted if the total power is found to decrease. If the total power increases and the initial move was an upsizing move then all these moves are reversed, otherwise the moves are accepted in keeping with our approach to avoid local minima. In addition, if the initial move was an upsizing move then the gate is flagged so that it is not reconsidered for upsizing. The best-seen solution is always maintained and restored at the end of the forward pass. The pseudo-code for this stage of the algorithm is shown in Figure 8.9.

This two-stage VVS algorithm allows us to make intelligent choices to trade-off dynamic power for leakage power in order to obtain a reduction in the total power dissipation. The algorithm is effectively directed to auto-matically provide either more leakage power or dynamic power reduction based on the initial design point. The two-stage algorithm can easily quantify the impact of setting a gate to high Vth on the extent to which other gates in the circuit can be assigned to VDDL. In other words, we can independently judge the impact of Vth and VDD assignment on total power, something that is difficult to achieve in a flow that simultaneously assigns VDDL and high Vth throughout the optimization or performs VDD and Vth assignments completely separately in two independent stages. The important capability of reassigning gates to VDDL leads to a reduction in the total power dissipation of the design in low activity cases (leakage power dominated designs) and steers the algorithm towards a proper low-power solution. In such cases an optimization approach where a dual-Vth and sizing optimization is followed by a dual-VDD and sizing optimization would result in highly sub-optimal results.

8.5.2   GECVS Based Power Optimization Using Dual-Vth Assignment and Gate Sizing

The GECVS algorithm can be extended to include gate sizing and dual-Vth optimization. Two major heuristic modules are incorporated: the first module seeks to increase the VDDL assignment in the circuit (referred to as 'Assign−VDDL'), while the second seeks to increase the high Vth assignment (referred to as 'Assign−High Vth'). The overall flow containing GECVS, dual-Vth and gate sizing is referred to as GVS.

```
Forward Pass( ) {
    L = forward front (VDDL gates with only VDDH fanins).
    While list L is non-empty {
        // Sensitivities to upsizing and VVDH assignment calculated
        Calculate sensitivity of gates in L to changing to VDDH.
        Calculate sensitivity of all gates to upsizing.

        Select candidate A with maximum sensitivity.
        Upsize or assign A to VDDH based on maximum sensitivity.

        If A is changed to VDDH {
            Remove A from L.
            Add to L gates that fan out of A but are not fanouts of any
            VDDL gate.
            If A drives a primary output, remove the LCFF.
        }

        Calculate sensitivity of all low Vth gates changing to high Vth.
        While timing is not violated {
            Set low Vth gate with maximum sensitivity (Equation (8.5))
            to high Vth.
        }

        Check power consumption.
        If power < minimum power found {
            minimum power found = power.
            Best configuration = current VDDL, high Vth &
                                          sizing assignment.
        } else if upsizing initiated move and total power increases {
            Undo upsizing move.
            Flag gate A not to be considered again for upsizing.
            Undo associated high Vth assignment moves.
        }
    }
}
```

Figure 8.9 Pseudo-code for the backward pass of the VVS algorithm.

8.5.2.1   Assign-VDDL

At the end of GECVS, any slack remaining in the circuit is not sufficient to support additional VDDL assignments which provide power reductions. In other words, any further VDDL assignments will either cause the circuit to fail timing or increase power consumption. This heuristic attempts to increase the number of VDDL assignments by employing the technique of gate upsizing in order to create slack. Note that during this step, only those VDDH driven gates that *do not* have any VDDH gates in their fanouts are considered as candidates for VDDL assignment (such gates are simply referred to as '*candidates*' below). This condition is imposed since assigning such a candidate VDDH gate to VDDL will not require the insertion of ALCs and thus the overhead for changing to VDDL is less. This thinking is in line with the concept of building 'groups' or 'clusters' of VDDL cells as introduced in Section 8.3. This heuristic also serves to reduce the number of gates considered for VDDL assignment, which helps reduce the execution time. Figure 8.10 gives examples showing a gate that can be a candidate for VDDL assignment, gate A in Figure 8.10(a); and a gate that cannot be a candidate, gate E in Figure 8.10(b), in this step.

After identifying the candidates for VDDL assignment we next evaluate the sensitivities as was done in standard GECVS using Equation (8.2). Once the sensitivities of all candidates have been evaluated, the gate with the maximum sensitivity is assigned to VDDL. No ALC insertion is needed at this point as none of the fanouts are at VDDH. Once the gate with the best sensitivity has been assigned to VDDL, the circuit no longer meets timing and we now upsize gates on critical paths to meet timing. In identifying gates to upsize we evaluate the sensitivities of all gates to upsizing using the following definition. This sensitivity was also employed in Section 5.1 by Equation (8.4) and is reproduced here for convenience.

$$\text{Sensitivity} = \frac{1}{\Delta P} \sum_{arcs} \frac{\Delta D_{arc}}{Slack_{arc} - Slack_{min} + K} \qquad (8.6)$$

where ΔD_{arc} and ΔP are the change in delay and power dissipation due to upsizing (by one drive strength); $Slack_{min}$ is the worst timing slack in the circuit; and K is a small positive quantity for numerical stability. The 'arcs' in Equation (8.6) are the falling and rising arcs associated with each gate input. The sensitivity has higher values for gates on critical paths and thus guides the algorithm towards upsizing the most beneficial gates [6].

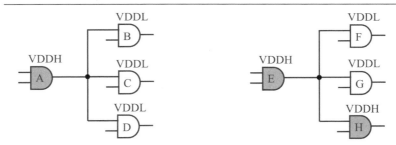

(a) Gate A is a potential candidate for VDDL assignment. Setting gate A to VDDL does not require an ALC to be inserted since all fanout gates (B,C,D) are fed by VDDL.

(b) In this case, gate E is not a potential candidate for VDDL assignment. Setting gate E to VDDL requires an ALC to be inserted since gate H is fed by VDDH.

Figure 8.10 Candidates for additional VDDL assignment during the 'Assign–VDDL' step.

Once sensitivities for all the gates are evaluated, the gate with the maximum sensitivity is selected and sized up. This sensitivity evaluation does not require a full circuit timing analysis and hence does not lead to large runtimes. This procedure is repeated until the circuit no longer fails timing. The number of upsizing moves is limited to a large number (10% of the total number of gates in the circuit) in order to stop pursuing bad moves that require too much upsizing (which would lead to a high area overhead and also smaller power improvements). We again allow moves that immediately result in increased power (after VDDL assignment and required upsizing moves) in order to include hill-climbing capability in the algorithm.

The list of candidates to assign to VDDL is updated after each accepted move, since setting a gate to VDDL may create more candidates in its fanin cone. This procedure of VDDL assignment is continued as long as there are candidates remaining that can be set to VDDL without violating timing given upsizing.

The pseudo-code for the assign VDDL step is shown in Figure 8.11.

8.5.2.2 Assign-High Vth

At the end of the 'Assign–VDDL' step, some slack may still remain in the circuit. We next attempt to convert this slack into power savings by converting gates from low Vth to high Vth. Although assigning a gate to high Vth will clearly slow it down, the gate input pin capacitances also reduce somewhat [24] (~8% in our technology), speeding up gates that fan into it. The approach used in GECVS (Figure 8.5) is again employed for the high Vth assignment – the only difference is that moves are from low Vth to high Vth here, rather than VDDH to VDDL as in GECVS. At the end of this step, no further gates can be set to high Vth without violating timing. Thus, in order to explore the possibility of increasing power savings by more high Vth assignment, we upsize certain gates or assign gates back to VDDH (thus creating slack). In order to identify the gates to be upsized or set to VDDH the following sensitivity measure is defined:

$$\text{Sensitivity (upsizing/set-VDDH)} = \frac{\Delta D}{\Delta P} \qquad (8.7)$$

where, ΔD and ΔP are the change in delay and power dissipation.

This sensitivity enables us to choose the gate giving the best delay improvement per unit power penalty. In considering gates to be set to VDDH, a heuristic analogous to the one followed in Assign-VDDL (Figure 8.11) can be followed. Specifically in the Assign-High Vth step while setting gates back to VDDH, we only consider gates that do not have any VDDL gates as their fanins. This ensures no ALC insertion will be required. Sensitivities of setting all such gates to VDDH are evaluated while sensitivities of *all* gates in the circuit with respect to upsizing (by one drive strength) are evaluated using Equation (8.7). Once all sensitivities have been computed, the gate with the largest sensitivity is set to VDDH or upsized accordingly, creating slack. Once slack is created, more gates can be assigned to high Vth following the approach outlined above.

```
Assign-VDDL( ) {
    L = Candidate gates identified in Figure 8.10, i.e., VDDH gates not
        fanning out to VDDH gates.
    While list L is non-empty {
        // VDDL assignment sensitivity calculated with Equation (8.2)
        Calculate sensitivity of gates in L to changing to VDDL.

        STEP: "SET VDDL" –
        Select candidate A with maximum sensitivity from L.
        Remove A from L.
        Set the supply voltage of A to VDDL.

        If A drives a primary output, insert an LCFF.

        STEP: "UPSIZING" –
        While circuit fails timing and number of upsizing moves
        is < 10% of total number of gates in the circuit) {
            Calculate sensitivity of all gates to upsizing with Equation (8.6).
            Upsize gate with maximum sensitivity to the next higher
            size available in the library.
        }

        Check timing.
        Check power consumption.
        If circuit meets timing and power increase < hill-climbing
        tolerance {
            STEP: "CLUSTERING" –
            Add to L gates that fan into A but not into any VDDH gate.

            If power < minimum power found {
                minimum power found = power.
                Best configuration = current VDDL, Vth &
                                            sizing assignment.
            }
        } else {
            Undo upsizing moves.
            Remove any added LCFFs.
            Set the supply voltage of A back to VDDH.
        }
    }
}
```

Figure 8.11 Pseudo-code for the Assign-VDDL algorithm.

Table 8.3 Power savings with the VVS algorithm, which adds dual-Vth and sizing to CVS. VDDL = 0.8V. High Vth = 0.23V. Low Vth = 0.12V.

Circuit	Initial Power (uW)			VVS Power (uW)			VVS Savings
	Leakage	Switching	Total	Leakage	Switching	Total	
c432	43	155	198	22	151	173	12.7%
c880	61	250	310	31	205	236	23.9%
c1355	163	537	699	106	532	638	8.8%
c1908	89	339	428	36	313	349	18.4%
c2670	118	541	658	40	435	475	27.9%
c3540	161	570	731	82	529	611	16.4%
c5315	219	1017	1235	117	892	1009	18.3%
c7552	232	983	1215	196	732	928	23.6%
Huffman	68	310	378	25	278	303	19.7%
						Average:	18.9%

8.6 POWER SAVINGS WITH VVS AND GVS

Table 8.3 summarizes our results after applying the heuristics described above to CVS and GECVS. The initial power point is from delay minimization in the manner of TILOS (per Equation (8.6)), then backing off 10% from the minimum delay point ($1.1 \times T_{min}$) with power minimization (per Equation (8.5)). The same $1.1 \times T_{min}$ delay constraints are used for the results presented in Table 8.3 and Table 8.4 – i.e. comparing sizing/dual-VDD/dual-Vth results versus a sizing only/high VDD/low Vth initial point. Only results for VDDL of 0.8V are detailed here, as these results were better than for VDDL of 0.6V. From comparing the percentage improvements in Table 8.1 with those in the most right-hand column of Table 8.3 and Table 8.4, power savings can be improved significantly when sizing, VDD and Vth assignment are utilized together for reasonably large test cases – on average 10% additional savings vs. CVS, and 7% additional savings vs. GECVS.

In contrast, approaches such as the genetic algorithm presented in [9] are expected to fail for larger benchmarks because of the increased problem size. We implemented the genetic algorithm and found about 15% power savings versus the initial power point for c17, but the genetic algorithm provided no power savings for the larger benchmarks in comparison to the initial point.

Level converter performance has an important impact on achievable power savings. The asynchronous level converter we used was the circuit referred to as STR6 in [17]. This ALC has a delay of 84ps and consumes 6.3fJ of internal energy per transition. The LCFF data was based on [1][11] which show that LCFF delay overhead is about 2 FO4 delays, or about 80ps in our technology. The LCFFs have no internal power overheads and reduced switching power due to VDDL driving the input capacitance of the register.

Table 8.4 Power savings with the GVS algorithm, which adds dual-Vth and sizing to GECVS. VDDL = 0.8V. High Vth = 0.23V. Low Vth = 0.12V.

	Initial Power (uW)			GVS Power (uW)			GVS
Circuit	Leakage	Switching	Total	Leakage	Switching	Total	Savings
c432	43	155	198	31	148	179	9.6%
c880	61	250	310	36	189	225	27.3%
c1355	163	537	699	119	514	633	9.5%
c1908	89	339	428	64	300	364	15.0%
c2670	118	541	658	74	391	466	29.2%
c3540	161	570	731	128	492	621	15.1%
c5315	219	1017	1235	147	760	907	26.5%
c7552	232	983	1215	135	662	797	34.4%
Huffman	68	310	378	41	234	275	27.3%
						Average:	**21.6%**

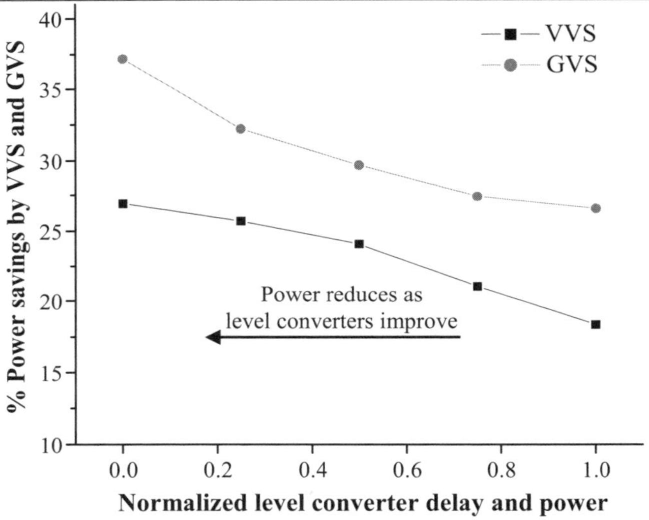

Figure 8.12 Impact of level converter performance on system level power dissipation of benchmark c5315 with VDDL of 0.8V.

By scaling the data for the level converters, we studied the possible power enhancements that can be obtained via further improved level converter circuits. Figure 8.12 shows the variation in the achieved power savings using VVS and GVS for an example case as level converter delay and power are varied, sweeping from the full power and delay (normalized to 1.0) to zero power and delay overhead. A reasonable sensitivity of power savings to level converter performance can be seen.

Table 8.5 This table compares the runtimes of CVS, GECVS, VVS, and GVS. The runtime complexity of the algorithms is summarized at the bottom.

Circuit	Number of Gates	Logic Depth	Runtime (s)			
			CVS	GECVS	VVS	GVS
c432	166	23	0.6	0.7	1.2	0.9
c880	390	26	2.7	4.6	5.4	8.4
c1355	558	28	6.1	6.4	13.6	7.4
c1908	432	40	3.6	4.0	9.1	5.3
c2670	964	25	16.8	21.9	56.1	57.4
c3540	962	42	24.5	31.8	65.1	51.4
c5315	1,627	41	68.8	119.9	229.9	386.6
c7552	1,994	44	110.6	286.7	222.9	889.9
Huffman	509	47	3.7	4.7	15.7	33.0
Runtime Complexity			$O(n^2)$	$O(n^3)$	$O(n^3)$	$O(n^3)$

8.6.1 Runtime and Complexity of the Multi-Vdd Algorithms

The final runtimes of the CVS, GECVS, VVS and GVS algorithms are compared in Table 8.5, as measured on a computer with a 3GHz Pentium microprocessor and 2GB RAM. The worst case complexities of these algorithms in terms of the number n of gates are as follows.

CVS can have at most all n gates assigned to VDDL. Each VDDL assignment needs one execution of the static timing analyzer (STA) to check if timing is met. Since the complexity of the STA is $O(n)$, the worst case complexity of CVS is $O(n^2)$.

In the case of GECVS, each potential VDDL assignment begins with the sensitivity calculation for all gates (Equation (8.2)). Since each sensitivity calculation takes $O(1)$ time, this takes $O(n)$ time. Once all sensitivities are calculated, finding the timing feasible gate with maximum sensitivity has a worst case complexity of $O(n^2)$ (since in the worst case an STA run is needed for each gate). Overall, since all gates can potentially be assigned to VDDL, we get a worst case complexity of $O(n^3)$ for GECVS.

The worst-case run-time complexity of the VVS algorithm is $O(n^3)$. Static timing analysis has a run time complexity of $O(n)$ and in the worst case we can make $O(n^2)$ moves in both the backward and forward passes. In the backward pass we can potentially attempt to assign $O(n)$ gates to low Vdd and for each of these possible assignments we can maximally have $O(n)$ upsizing moves in the circuit. The total number of upsizing moves (due to the size of the standard-cell library) is $O(n)$, therefore the worst-case occurs when all upsizing moves are reversed. Thus the possible number of upsizing moves is $O(n^2)$, making the overall worst-case complexity of the backward pass $O(n^3)$. Similarly we can find the worst-case complexity of the forward pass and hence of the overall approach to be $O(n^3)$. For the forward pass the worst-case complexity occurs only when we revert back to the original circuit after assigning $O(n)$ gates in the circuit to high Vth. However, since the amount of slack generated in the circuit due to a single high Vdd assignment or gate upsizing is small, the number of possible high Vth assignments due to upsizing or high Vdd assignment of a single gate can be expected to be $O(1)$, hence the average complexity of the forward pass can be expected to be $O(n^2)$. For the backward pass the complexity is actually given by $O(n^2s)$ where s is the number of gates on the boundary of low and high Vdd gates at the end of the backward pass. This boundary forms the cutset of the acyclic graph which represents the circuit network. The cutset size is relevant since we only undo the up-sizing associated with the gates that form the cutset. The number of upsizing moves associated with gates other than the ones forming the cutset is $O(n)$ since we only have a fixed number of drive strengths for a given logic gate. In the worst-case s can be $O(n)$ and this gives us the worst-case complexity of $O(n^3)$.

The complexities of the Assign-VDDL and Assign High-Vth modules of GVS can be found as follows. In case of Assign-VDDL, we first need $O(n)$ time to calculate the sensitivities (Equation (8.2)) of all gates. After choosing the gate with the maximum sensitivity, we need another $O(n)$ time for calculating sensitivities to upsizing (Equation (8.6)) and an STA run for checking timing after the gate with maximum sensitivity is sized up. Since we allow $O(n)$ number of upsizing moves per VDDL assignment, this takes $O(n^2)$ time. Finally, since all gates can potentially go to VDDL, the overall complexity for Assign-VDDL becomes $O(n^3)$. In case of Assign High-Vth, we first calculate the sensitivities of all gates to upsizing and VDDH assignment (Equation (8.7)). This takes $O(n)$ time. Once the gate with maximum sensitivity is upsized or assigned back to VDDH, we calculate sensitivities of all gates for high Vth assignment. As in the case of GECVS, selecting the timing feasible move with maximum sensitivity takes $O(n^2)$ time. Since every iteration of Assign-High Vth creates slack through one VDDH assignment or one drive strength upsizing, only a handful (typically less than 10) of gates can go to High Vth per iteration. And since we can have at most $O(n)$ number of such iterations, the overall complexity of Assign-High Vth becomes $O(n^3)$. Hence, the complexity of GVS which includes GECVS, Assign-VDDL and Assign-High Vth is $O(n^3)$.

8.7   SUMMARY

This chapter overviewed some algorithms for supply voltage assignment in multi-VDD circuits. Heuristics for combining the three optimization techniques of gate sizing, multi VDD assignment and multi Vth assignment were also presented. We quantified the impact of level converters on system-level power dissipation, motivating further work in the development of fast and low-energy asynchronous level converters.

The approaches discussed in this chapter are sensitivity based techniques which are able to accurately consider the impact of signal slews et al. on the timing of the design. Thus, these approaches are guaranteed to meet timing at the end of the optimization which is extremely important in an ASIC methodology. Moreover, the inherently discrete nature of Vth and VDD assignment problems does not allow direct application of traditional continuous optimization techniques, and combinatorial optimization techniques are extremely costly. Thus, efficient and intelligent heuristics that can consider the impact of a single Vth/VDD assignment on the final power savings are very useful.

Other important considerations such as the physical design and power delivery also arise when implementing multi-VDD circuits. The problem of physical design can be handled by dividing the floorplan into islands of VDDL and VDDH cells. Alternatively, modifying the standard cell layouts to accommodate the multiple power supplies allows complete freedom in choosing which gates to operate at each of the supplies [2].

Robust power distribution grids need to be designed as cells supplied by lower supplies are very susceptible to power supply variations. This can be accomplished at no area or wire congestion overheads by recognizing that multi-VDD circuits operate at lower supply currents [18]. Efficient DC-DC converters for delivering power to multi-VDD chips are discussed in [13].

8.8   ACKNOWLEDGMENTS

This work was supported by the Semiconductor Research Corporation, the MARCO/DARPA Gigascale Systems Research Center, and Intel Corporation. The authors thank Y. Kim from the University of Michigan, Ann Arbor, for the characterization of the standard cell libraries used in this work. The authors also thank Professor B. Nikolic from the University of California, Berkeley for providing the Huffman benchmark circuit.

8.9   REFERENCES

[1] Bai, M., and Sylvester, D., "Analysis and design of level converting flip-flops for dual-Vdd/Vth integrated circuits," *Proc. Int. Symp. System-on-Chip*, 2003, pp. 151-154.

[2] Bai, M., Kulkarni, S., Kwong, W., Srivastava, A., Sylvester, D., and Blaauw, D., "An implementation of a 32-bit ARM processor using dual power supplies and dual threshold voltages," *Proc. Ann. Symp. VLSI*, 2003, pp. 149-154.

[3] Brglez, F., and Fujiwara, H., "A neutral netlist of 10 combinational benchmark circuits and a target translator in Fortran," *Proc. Int. Symp. Circuits and Systems*, 1985, pp. 695-698.

[4] Chen, C., and Sarrafzadeh, M., "Simultaneous voltage scaling and gate sizing for low-power design," *IEEE Trans. Circuits and Systems II: Analog and Digital Signal Processing*, Jun. 2002, pp. 400-408.

[5]   Chen, C., Srivastava, A., and Sarrafzadeh, M., "On gate level power optimization using dual-supply voltages," *IEEE Trans. VLSI Syst.*, vol. 9, Oct. 2001, pp. 616-629.

[6]   Dharchoudhury, A., Blaauw, D., Norton, J., Pullela, S., and Dunning, J., "Transistor-level sizing and timing verification of domino circuits in the Power PC microprocessor," *Proc. Int. Conf. Computer Design*, 1997, pp. 143-148.

[7]   Dhillon, Y., Diril, A., Chatterjee, A., and Lee, H., "Algorithm for achieving minimum energy consumption in CMOS circuits using multiple supply and threshold voltages at the module level," *Proc. Int. Conf. Computer Aided Design*, 2003, pp. 693-700.

[8]   Fishburn, J., and Dunlop, A., "TILOS: a posynomial programming approach to transistor sizing," *Proc. Int. Conf. Computer Aided Design*, 1985, pp. 326-328.

[9]   Hung, W., Xie, Y., Vijaykrishnan, N., Kandemir, M., Irwin, M., and Tsai, Y., "Total power optimization through simultaneously multiple-VDD multiple-VTH assignment and device sizing," *Proc. Int. Symp. Low-Power Electronics Design*, 2004, pp. 144-149.

[10] Igarashi, M., Usami, K., Nogami, K., Minami, F., Kawasaki, Y., Aoki, T., Takano, M., Mizuno, C., Ishikawa, T., Kanazawa, M., Sonoda, S., Ichida, M., and Hatanaka, N., "A low-power design method using multiple supply voltages," *Proc. Int. Symp. Low-Power Electronics Design*, 1997, pp. 36-41.

[11] Ishihara, F., Sheikh, F., and Nikolic, B., "Level conversion for dual supply systems," *Proc. Int. Symp. Low-Power Electronics Design*, 2003, pp. 164-167.

[12] Karnik, T., Ye, Y., Tschanz, J., Wei, L., Burns, S., Govindarajulu, V., De, V., and Borkar, S., "Total power optimization by simultaneous dual-Vt allocation and device sizing in high performance microprocessors," *Proc. Design Automation Conf.*, 2002, pp. 486-491.

[13] Hazucha, P., Schrom, G., Hahn, J., Bloechel, B., Hack, P., Dermer, G., Narendra, S., Gardner, D., Karnik, T., De, V., and Borkar, S., "A 233-MHz 80%-87% efficient four-phase DC-DC converter utilizing air-core inductors on package," *IEEE J. Solid-State Circuits*, Apr. 2005, pp. 838-845.

[14] Ketkar, M., and Sapatnekar, S., "Standby power optimization via transistor sizing and dual threshold voltage assignment," *Proc. Int. Conf. Computer Aided Design*, 2002, pp. 375-378.

[15] Kosonocky, S., Bhavnagarwala, A., Chin, K., Gristede, G., Haen, A., Hwang, W., Ketchen, M., Kim, S., Knebel, D., Warren, K., and Zyuban, V., "Low power circuits and technology for wireless digital systems," *IBM J. R&D*, vol. 47, no. 2/3, 2003.

[16] Kulkarni, S., Srivastava, A., and Sylvester, D., "A new algorithm for improved VDD assignment in low power dual VDD systems," *Proc. Int. Symp. Low-Power Electronics Design*, 2004, pp. 200-205.

[17] Kulkarni, S., and Sylvester, D., "High performance level conversion for dual VDD design," *IEEE Trans. VLSI Syst.*, Sep. 2004, pp. 926-936.

[18] Kulkarni, S., and Sylvester, D., "Power distribution techniques for dual VDD circuits," *Proc. Asia-South Pacific Design Automation Conf.*, 2006, pp. 838-843.

[19] Lackey, D., Zuchowski, P., Bednar, T., Stout, D., Gould, S., and Cohn, J., "Managing power and performance for SOC designs using voltage islands," Proc. Int. Conf. Computer *Aided Design*, 2002, pp. 195-202.

[20] Nemani, M., and Najm, F., "Toward a high level power estimation capability," *IEEE Trans. Computer Aided Design*, vol. 15, Jun. 1996, pp. 588-598.

[21] Nguyen, D., Davare, A., Orshansky, M., Chinnery, D., Thompson, B., and Keutzer, K., "Minimization of dynamic and static power through joint assignment of threshold voltages and sizing optimization," *Proc. Int. Symp. Low-Power Electronics Design*, 2003, pp. 158-163.

[22] Pant, P., Roy, R., and Chatterjee, A., "Dual-threshold voltage assignment with transistor sizing for low power CMOS circuits," *IEEE Trans. VLSI Syst.*, 2001, pp. 390-394.

[23] Puri, R., Stok, L., Cohn, J., Kung, D., Pan, D., Sylvester, D., Srivastava, A., and Kulkarni, S., "Pushing ASIC performance in a power envelope," *Proc. Design Automation Conf.*, 2003, pp. 788-793.

[24] Sirichotiyakul, S., Edwards, T., Oh, C., Zuo, J., Dharchoudhury, A., Panda, R., and Blaauw, D., "Stand-by power minimization through simultaneous threshold voltage selection and circuit sizing," *Proc. Design Automation Conf.*, 1999, pp. 436-441.

[25] Srivastava, A., "Simultaneous Vt selection and assignment for leakage optimization," *Proc. Int. Symp. Low-Power Electronics Design*, 2003, pp. 146-151.

[26] Srivastava, A., Sylvester, D., and Blaauw, D., "Power minimization using simultaneous gate sizing, dual-Vdd, and dual-Vth assignment," *Proc. Design Automation Conf.*, 2004, pp. 783-787.

[27] Sundararajan, V., and Parhi, K., "Low power synthesis of dual threshold voltage CMOS VLSI circuits," *Proc. Int. Symp. Low-Power Electronics Design*, 1999, pp. 139-144.

[28] Sundararajan, V., and Parhi, K., "Synthesis of low power CMOS VLSI circuits using dual supply voltages," *Proc. Design Automation Conf.*, 1999, pp. 72-75.

[29] Sylvester, D., and Keutzer, K., "System-level performance modeling with BACPAC – Berkeley advanced chip performance calculator," *Int. Workshop System-Level Interconnect Prediction* (workshop notes), 1999, pp. 109-114.

[30] Usami, K., and Horowitz, M., "Clustered voltage scaling technique for low-power design," *Proc. Int. Symp. Low-Power Electronics Design*, 1995, pp. 3-8.

[31] Usami, K., and Igarashi, M., "Low-power design methodology and applications utilizing dual supply voltages," *Proc. Asia South Pacific Design Automation Conf.*, 2000, pp. 123-128.

[32] Usami, K., Igarashi, M., Ishikawa, T., Kanazawa, M., Takahashi, M., Hamada, M., Arakida, H., Terazawa, T., and Kuroda, T., "Design methodology of ultra low-power MPEG4 codec core exploiting multiple voltage scaling techniques," *Proc. Design Automation Conf.*, 1998, pp. 483-488.

[33] Usami, K., Igarashi, M., Minami, F., Ishikawa, M., Ichida, M., and Nogami, K., "Automated low-power technique exploiting multiple supply voltages applied to a media processor," *IEEE J. Solid-State Circuits*, Mar. 1998, pp. 463-472.

[34] Wei, L., et al., "Mixed-Vth (MVT) CMOS Circuit Design Methodology for Low Power Applications," in *Proceedings of the Design Automation Conference*, 1999, pp. 430-435.

[35] Wei, L., Roy, K., and Koh, C., "Power minimization by simultaneous dual-Vth assignment and gate sizing," *Proc. Custom Integrated Circuits Conf.*, 2000, pp. 413-416.

[36] Yeh, C., Chang, M., Chang, S., and Jone, W., "Gate-level design exploiting dual supply voltages for power-driven applications," *Proc. Design Automation Conf.*, 1999, pp. 68-71.

Chapter 9

PLACEMENT FOR POWER OPTIMIZATION
Physical Synthesis

Ameya R. Agnihotri, Satoshi Ono, Patrick H. Madden
Computer Science Department
T. J. Watson School of Engineering
State University of New York at Binghamton
P.O. Box 6000
Binghamton, NY 13902

9.1   INTRODUCTION

Circuit placement is a well studied area of VLSI design. The logic elements in a circuit design must be transferred onto the silicon substrate – transistors are not allowed to overlap, and there are a variety of spacing and size constraints. In this chapter, we survey techniques to minimize power within a placement context. By "placement", we mean a mapping of each logic element to a physical location.

Minimization of power during circuit placement requires a delicate balance of constraints. There is always a trade-off between power and speed. If speed is not an issue, power can be reduced by operating at a low frequency, increasing device threshold voltages, and down-sizing devices which reduces the layout area. However for modern circuits, it's rare to have such low performance objectives. Rather, the challenge is to design a circuit that is both fast and low power.

Early place and route power minimization methods focused on reducing the lengths of interconnect wires with higher switching activity. Minimization of the length of delay critical nets has also been a traditional concern. Recent focus has been on the integration of multiple supply voltages and multiple threshold voltages, so that circuit performance can be finely tuned.

The traditional "wire length" objective in placement addresses power minimization by reducing the switching capacitance of the interconnect wires. Each wire has a capacitive load; charging and discharging this load consumes power. In one recent study [41], for example, it was estimated that 50% of the dynamic power consumption of a microprocessor designed in 0.13um technology could be attributed to the switching capacitance of circuit interconnect. This percentage is from a highly optimized design – a poor quality circuit placement would have longer interconnect lengths, contributing an even greater share of the dynamic power.

Reducing the length of delay critical nets is beneficial to power consumption, as this makes achieving timing closure easier. If a long interconnect wire is on a delay critical path, the only way to meet a timing constraint may be to either insert buffers and/or to increase the size of gates along the path; each of these will increase power consumption.

The physical locations of each circuit element determine the lengths of interconnecting wires, which in turn influences the need for gate sizing and buffer insertion. While interconnect length reduction is not the only objective in placement, it is an extremely important one

There are three dominant algorithmic techniques in use for circuit placement: analytic methods, partitioning based placement, and simulated annealing. Most commercial placement tools rely heavily on analytic methods, but hybrids are common – each technique has its strengths and weaknesses. Circuit placements produced by current methods are known to be significantly suboptimal for even simple metrics [10]; for timing and power optimization, the suboptimality is likely even greater. Many placement researchers believe that significant improvements are possible, and that an algorithmic breakthrough would have tremendous benefit.

The circuit placement is frequently broken into global and detailed placement steps. During global placement, only coarse estimates of wire length and wire delay are known. While there are techniques to gain rough estimates of interconnect lengths before placement [16][50], these methods lack the accuracy needed for fine tuning [49]. For modern high-performance design, a great deal of circuit optimization must be performed *after* global placement has been completed.

In global placement, power minimization can be handled by biasing the solution such that there is a preference for shorter lengths on the most active signal nets. We discuss this biasing in the section on "net weighting". As a placement algorithm converges towards a final configuration, estimates of individual net lengths become more accurate, and the true delay critical paths of the circuit begin to emerge. Biasing the nets such that those on the critical paths become shorter also benefits power: shorter nets require smaller drivers to meet performance constraints.

During detailed placement, the circuit can be fine-tuned, with combinations of gate sizing, buffer insertion, logic resynthesis, and small-scale placement modifications. Combinations of logic optimization and physical layout (both placement and routing) are generically known as *physical synthesis*.

Recent efforts to further minimize power consumption by using multiple supply voltages have presented new challenges. In particular, cells must be placed in rows (or portions of a row) with appropriate supply voltage and that level converters be placed such that they can access both supply voltage levels. This complicates placement legalization.

It is important to note that the placement problem has been changing over the past few years. The number of available transistors has been increasing exponentially, but the ability of designers to utilize them has lagged behind. In most cases, modern chips are "power and speed limited", and not "device limited". It is not uncommon for large designs to have a great deal of open space between blocks and logic elements – the capacity of each die exceeds the needs for the circuit. Thus, the "packing" nature of the problem that was a significant challenge in previous generations is now almost irrelevant – handling the abundance of "white space" has emerged as a new concern [2][6].

The placement optimization problem and objectives for this are discussed in Section 9.2. Placement approaches and physical synthesis are detailed in Section 9.3, then Section 9.4 examines placement issues with multiple supply voltages. State of the art results for placement tools are examined in Section 9.5. We conclude in Section 9.6.

9.2 PLACEMENT BASICS

On the surface, circuit placement may appear to be a relatively simple problem. In Figure 9.1, we illustrate three types of common placement problems: standard cell placement, mixed size placement, and floorplanning. In standard cell placement, large numbers of relatively simple logic elements are arranged in horizontal rows – the elements frequently have the functionality of basic Boolean operators. Mixed size placement problems contain both standard cells, and also larger and more complex circuit blocks; this is frequently referred to as the "boulders and dust" problem. At the highest level is floorplanning: large blocks of circuitry must be arranged such that they fit together, while minimizing wire length (or a variety of other objectives).

The input to a placement tool is generally a circuit net list, coupled with some constraints on the "core area" into which the circuit is to be embedded. A circuit net list contains a large number of logic elements – almost always rectilinear, and for standard cell design, rectangular with uniform height. These logic elements must be arranged within the core area such that they do not overlap.

The set of logic elements in a circuit is typically denoted as $C=\{c_1,c_2,...,c_n\}$. C is the entire set of logic elements, while each c_i corresponds to a single device (ranging from a simple logic gate to a macro block). Connecting the logic elements together are the signal nets $N=\{n_1,n_2,...,n_m\}$. Often, it is useful to treat the circuit as a hypergraph, with the logic elements as vertices, and the signal nets as hyperedges.

With a small amount of excess area, fitting logic elements into the space provided is usually relatively easy. Complicating the placement problem are wire length, power, and delay considerations. We will consider these briefly; detailed discussions of placement objectives can be found in [3][34][40].

9.2.1   Wire Length

The most studied placement objective is wire length minimization. Consider a simplified placement problem: we will assume that each logic element is square, and that they are to be placed onto a two dimensional grid. Packing the elements in such a way that they do not overlap is trivial – so long as the number of grid spaces is at least equal to the number of logic elements.

When one considers the interconnecting signal nets, however, the problem becomes NP-Complete [47]. If we have n logic elements, and an equal number of grid spaces, there are $O(n!)$ different ways to embed them into the grid.

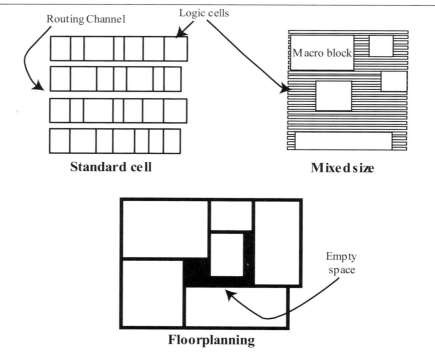

Figure 9.1 Standard cell placement, mixed size placement, and floorplanning problems. We focus primarily on standard cell and mixed size placement.

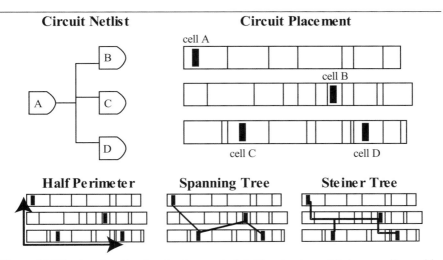

Figure 9.2 Wire length estimation in placement. A half perimeter metric is normally used in global placement. For gate sizing and buffer insertion, however, it is preferable to have more accurate measures.

For each signal net, the half perimeter of the bounding box is a rough estimate of the interconnect wiring needed for the net – for two and three pin nets, this is exact. Thus, a common objective is to minimize the sum of the half perimeters; HPWL (half perimeter wire length) is well known.

Improving somewhat over the perimeter objective are minimum spanning tree (MST) and Steiner minimal tree (SMT) objectives. In some respects, these objectives are better than HPWL – but as they ignore the actual topologies found by global and detail routing, they are also inaccurate. We illustrate different methods for estimating the wire length of a net in Figure 9.2.

9.2.2 Power

The impact of placement on dynamic power consumption has been studied extensively. A typical formulation [41] to capture this is

$$P = \sum_{j \,\in\, \text{signal nets } N} \frac{1}{2} a_j C_j V_{dd}^{\,2} f \qquad (9.1)$$

where P is the total dynamic power consumption due to switching wire capacitances; a_j is the switching activity factor for signal net j; C_j is the switching capacitance of the net; V_{dd} is the supply voltage; and f is the clock frequency. While the notation varies slightly, the basic equations remain the same; [44] utilized a similar formula in an early survey of power minimization.

Switching capacitance, C_j, depends heavily on the interconnect length. To accommodate this in placement optimization, it is possible to perform *net weighting* to bias the solution such that nets that switch frequently have reduced lengths. How net weighting for power is handled in the different placement approaches is discussed in detail in Section 9.3.

9.2.3   Delay

Delay minimization is the most elusive objective in placement, for a number of reasons. Even with a fixed placement, determining the longest delay path through a circuit is nontrivial.

During placement, the locations of individual logic elements may change repeatedly. As wire lengths between logic elements change, the critical paths in a circuit change. If one uses an accurate delay analysis method during circuit placement, run times are unacceptable. Fast delay analysis methods are inaccurate, which can result in over optimization of non-critical paths, or in performance objectives being missed.

Note that some "false" paths can't actually affect the circuit delay; determining false paths is in principle a difficult problem. To impact the delay of a circuit, a path must be sensitized; this is effectively circuit satisfiability, a classic NP-Complete problem. Fortunately, circuit designers frequently have good insight into the nature of the true critical paths, and many false paths can be eliminated by heuristic methods.

9.2.4   Routability

While we consider this only briefly, it should be clear that the actual routing of a circuit has a great impact on interconnect length, and thus the power consumption of a circuit. In portions of many designs, the routing demand can be close to the available resources; these regions are "congested". Routing congestion can be reduced through the introduction of routing detours; congested designs have increased numbers of vias between metal layers (which add to the capacitance of the nets), and an overall increase in interconnect length.

There has been extensive work in routability-driven placement in recent years [37][59]. Generally, excess area in the design, the "white space", is distributed within the placement region. In general, this has the effect of increasing half-perimeter wire length estimates; the routed wire lengths, however, are reduced due to reduced congestion with less routing detours.

A primary challenge to routability-driven placement is accurately estimating routing demand [38][39][58]. If the routing estimates do not match the actual behavior of the routing tools, space may be inserted where it is not required (increasing wire length), while areas that need space are overlooked (resulting in detours, or even routing failure).

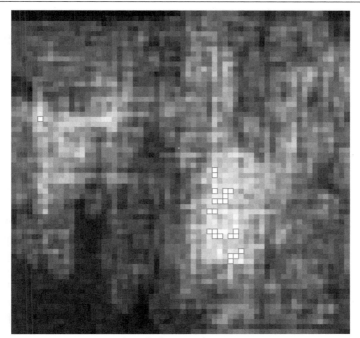

A typical congestion map

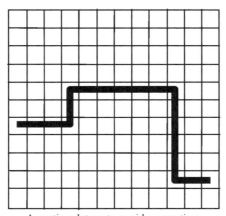

A routing detour to avoid congestion

Figure 9.3 A "congestion map" normally shows densely occupied regions with light colors. A routing detour may erase any improvements made by careful placement. If the design is congested, wire detours will increase the capacitance of interconnect nets, increasing switching power. Congested regions also have more net-to-net coupling capacitance, and thus wider variation in delay.

Routing detours, as illustrated in Figure 9.3, may eliminate any gains that may have been expected from power and delay optimization – wire lengths are unexpectedly higher than the estimates used by the optimization tools. For this reason, some have advocated a return to the "variable die" routing model [55]. Most modern design flows use a "fixed-die" routing model [59]. Variable die routing allows expansion between rows of standard cells to increase routing space, avoiding the introduction of routing detours.

9.2.5   Problem Complexity

Evaluating the "quality" of a given placement is difficult; compounding this is the inherent NP-completeness of placement. For n logic elements, there can be $O(n!)$ different arrangements. There are no known optimal algorithms for even the simplest of metrics.

While there is disagreement regarding the degree of suboptimality of current placement methods, there is general agreement that the degree is quite significant. It should be stressed that the methods discussed in the next section are all heuristic in nature. Current tools based on analytic methods, bisection, and annealing, produce similar results on some benchmarks, and widely differing results on others. On a set of synthetic benchmarks with known optimal configurations [10], wire lengths produced by modern tools were anywhere from 30% to 150% away from optimal—with some tools exhibiting pathological behavior. We speculate that in terms of average wire length, most methods are at least 50% away from optimality on "real" circuits. For delay optimization, we would speculate that results could be a factor of two or more away from optimality.

9.3   PHYSICAL SYNTHESIS

In this section, we first discuss general methods for net weighting and global placement. Most placement approaches first find a rough distribution of logic elements across the layout area, while addressing power and performance objectives by weighting individual signal nets.

Following global placement are legalization, gate sizing, and buffer insertion. It is during this phase that "physical synthesis" has departed most significantly from traditional placement. In early fabrication technologies, circuit delay was relatively independent of interconnect delay; device sizes could be fixed at an early stage, and nearly any reasonable placement would produce performance results close to those expected. With modern fabrication, interconnect delay is far more significant, and it is only at the last stages of physical design that performance can be accurately estimated.

The amount of optimization to be performed may make design closure difficult. For example in one recent study [48], large numbers of repeaters were needed to meet performance targets. The insertion of repeaters into the design caused changes to the overall structure of the placement, making some optimizations ineffective.

Significant changes in area due to gate sizing and buffer insertion cause wire estimates to be inaccurate, which can require layout to be redone. Layout and sizing iterations may fail to converge on an acceptable solution and results can be unpredictable. This is essentially why traditional placement flows failed – wire load models were so inaccurate that the results were quite suboptimal in terms of area and power due to oversized cells to conservatively try and drive what might be long wires, but in the typical case were over-estimated, also limiting the minimum delay that could be achieved. While an underestimate in the less common case of longer wires would lead to failure to satisfy delay constraints for paths with long wires after routing.

The "stability" of a placement in this context is a key concern [7]. Some designs now contain a great deal of internal white space, so that insertion of buffers and gate sizing does not disrupt the placement structure. Logical effort [51] based optimization can be extremely effective, but can require a large overhead in terms of total wire length.

9.3.1 Net Weighting

At the core of almost all performance-driven placement techniques is a *net weighting* scheme. In a circuit, some nets are delay critical, or transition very frequently. By increasing the weight of a net in something as simple as a half perimeter wire length calculation, the results of a placement algorithm can be tuned towards better performance. Figure 9.4 shows a simplified example; if power is the only objective, one of first two arrangements might be acceptable. The third arrangement will minimize delay.

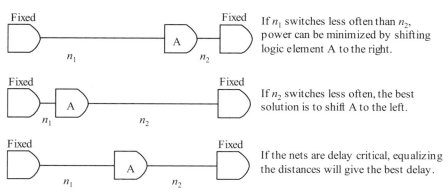

Figure 9.4 If nets have different switching activity, it may be beneficial to weight the nets so that their lengths are different after placement. In this figure, the first two arrangements optimize the length of n_1 or n_2; which is better depends on the switching activity. In practice, neither may be desirable: interconnect delay is roughly quadratic with net length, and a balanced arrangement may give better delay, and require less sizing of drivers.

Net weighting methods are not new. An early method was developed by Dunlop [19], and most current placement tools use something similar. Weights for individual nets are frequently based on the "slack allocation" methods of Frankle [22]. By traversing the circuit with a longest-path algorithm (easily done in a directed graph), it is possible to find long paths that may limit performance. Increasing the weight of nets along the path will result in the placement algorithm pulling the logic elements along the path together, reducing wire length. Decreasing the weight of non-critical nets also achieves this. Integrating switching activity into this approach is trivial.

A common criticism of net weighting is that it addresses the nets individually, but not the paths. As the placement changes, or gate sizing and buffer insertion are performed, the critical path can change repeatedly. Without frequent recomputation of net weights, it is likely that the placement tool will optimize non-critical portions of the circuit.

It is important to note again that the longest path may not necessarily be relevant; it is not uncommon for a long path to be "false". In practice, performance driven design is done by either focusing on a set of paths provided by the circuit designer, or by having a set of false paths to explicitly ignore. As placement and physical synthesis operations are performed, the timing of a circuit is updated repeatedly (usually with an incremental method to minimize computation overheads); net weights are recomputed, and optimization continues.

9.3.2   Global Placement

In current practice, there are three dominant placement techniques: analytic, recursive partitioning, and simulated annealing.

9.3.2.1   Analytic Placement

Analytic placement [21][33][46][54][56] is a generic term for methods that formulate the placement problem as a set of equations; the objective is to minimize the sum of the distances between connected logic elements. Figure 9.4 provides a simple example; the optimal position for element "A" can be formulated such that we minimize the distance, or the square of the distances. The position of the logic element would typically be represented as a pair of variables for the x and y location. Linear and quadratic programs can then be formulated, and individual connections can be weighted.

The circuit in Figure 9.4 is trivial; the optimal solution can be found easily with a pencil and paper. For large circuits, one might expect the formulation to have tens of thousands of variables; there are many algorithmic techniques to solve problems of this size quickly. Figure 9.5 illustrates pseudocode for a generic analytic approach; there are many variations (for example [21][33][54][56]).

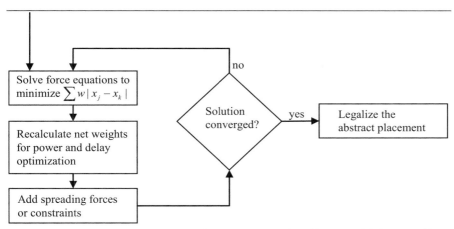

Figure 9.5 Flowchart for an analytic placement approach, with net weighting to address power and performance. There are many different ways to formulate the optimization objectives; the x and y optimization objectives are frequently computed independently. There are a variety of methods to remove overlap, and this is an active research area.

In typical formulations, the signal nets n_i connect the circuit elements c_jc_k. The objective in this formulation is to minimize the sum of distances between circuit elements. Many analytic placement tools minimize the square of the distance between connected elements: this formulation is differentiable, which makes it relatively easy to solve. The quadratic formulation also captures a useful aspect of the placement problem somewhat naturally: if interconnect wiring is unbuffered, delay is approximately proportional to the square of the net length. In some sense, quadratic formulations can be viewed as minimizing the sum of net delays – not necessarily a bad objective.

While squared distance formulations are easier to solve, linear objective functions are also common. Linear wire length more accurately models routed wire length, and a great deal of effort has gone into the development of efficient solution methods.

With analytic methods, there is an "obvious" optimal solution to the set of equations; one in which all circuit elements are directly overlapping. While there are differences in how the equations are formulated or solved, it is in the handling of overlap where one sees the greatest variation between approaches. Methods based on partitioning [33][56], the introduction of additional forces [21], and cell shifting [54] have all been investigated.

The natural integration of power and delay objectives, coupled with efficient mathematical solvers, has made analytic placement extremely popular. There are many variations on this theme, and a majority of commercial placement tools utilize some form of analytic placement.

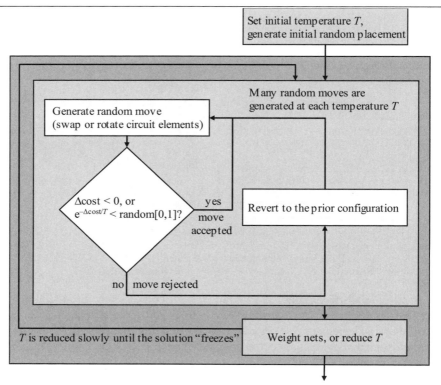

Figure 9.6 Flowchart for an annealing based placement approach. There are a variety of different cooling schedules and move strategies. To adjust the weights of interconnect nets, there are periodic calls to an algorithm that finds a set of long paths.

9.3.2.2   Simulated Annealing

A second common placement approach is *simulated annealing*, based on methods first described in [32]. In terms of wire length minimization, annealing can produce excellent results, but at the expense of high run times. The current academic tool *Dragon* [57], for example, produces leading results in terms of length, but has comparatively high run times. For industrial placement tools, annealing has fallen out of favor.

As is done with analytic placement, power and delay optimization is integrated into annealing tools through net weighting. An early work to perform timing-driven placement was the academic tool *TimberWolf* [52]; we show pseudocode in Figure 9.6.

In [52], a method by Dreyfus [18] was used to find a fixed number of the longest paths in the circuit. Slack-based methods were used to weight the nets. Throughout the annealing process, the set of long paths was repeatedly computed, and weights were reassigned.

Recently, performance driven placement within an annealing framework was revisited, with a surprising result [55]. Rather than attempting to optimize the circuit through weighting of nets, the objective was simply to minimize routed wire length. Delay and power considerations were then addressed through extensive gate sizing, and custom cell generation. Compared to commercial tools, the approach produced far better results.

The results in [55] pose an interesting question: namely, is timing driven placement (with net weighting) an essential part of timing driven design? The placement method used, while based on [52], simply optimized wire length – net weighting was entirely ignored. That superior results (in terms of both circuit delay and power consumption) were obtained by ignoring net weights is somewhat counter-intuitive. One way to interpret this is that a more complex (and perhaps accurate) formulation is also more difficult to optimize; the solution quality obtained for the simplified problem is thus better than the solution for the complex version.

A second interesting outcome of the work is a reconsideration of the "fixed-die" routing model – the authors used a "variable die" formulation, which eliminated routing detours while also allowing very dense placement. In most current design methodologies, the spacing between rows of logic elements is fixed; the total area is also fixed. The variable-die methodology allows increased space between rows of logic elements, which provides needed routing resources and allows detours to be eliminated.

9.3.2.3 Partitioning Based Placement

A third popular placement approach is recursive partitioning (and most frequently, recursive bisection). The advent of strong multi-level partitioning algorithms [13][30] has made the basic methods outlined by Breuer [9] quite effective. With the terminal propagation techniques of Dunlop and Kernighan [20], modern bisection based placement tools can produce leading results on both standard cell [4] and mixed size [31] placements. We show a flowchart for a typical recursive bisection approach in Figure 9.7.

However, in terms of performance optimization for power and delay, partitioning methods are at a bit of a disadvantage when compared to analytic or annealing methods. Partitioning methods approach the placement problem with a top-down perspective; subcircuits are treated as generic clusters of logic until fairly late in the placement process. Because net lengths within the cluster are not known, delay and power consumption estimates cannot be made accurately until fairly late in the placement process – frequently too late to make effective changes.

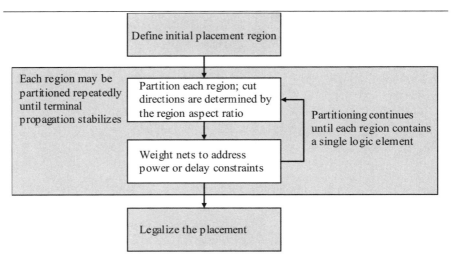

Figure 9.7 Flowchart for a recursive bisection placement approach. The fractional cut formulation greatly simplifies cut line insertion, and results in improved wire lengths. Most current tools are based around multi-level partitioning algorithms.

One method to have some success was the approach of Ou and Pedram [43]. In bisection based methods, nets that are cut early in the placement process generally have higher length; if a net is cut repeatedly, it can be very long. To avoid having long nets, and in particular, long nets along a critical path, nets were weighted based on if they had been previously cut.

9.3.3   Legalization and Detailed Placement

We discuss placement legalization and detailed placement extensively, as these are key components of an effective physical synthesis flow. Even if the placement tool provides a legal solution initially (both bisection and annealing frequently can do this), it may become illegal – gate sizing and buffer insertion may change the size of logic elements, or introduce new ones. Thus, legalization must be an essential part of any successful optimization strategy.

Traditionally, legalization has been most closely associated with analytic methods. Tools such as *Kraftwerk* [21], *APlace* [28][29], and *FastPlace* [53] are known to produce high quality results. Recently, partitioning based placement tools have made a great deal of improvement [4], and are making significant use of legalization techniques. Dynamic programming based legalization has been used [4][26], and a simple "tetris" method patented by Hill [23] has been adapted to handle mixed size designs [31].

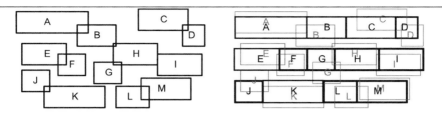

| Logic elements are globally placed, but are not row aligned and can have overlaps. | Legalization shifts each element slightly, to remove overlap and align with rows. |

Figure 9.8 An example of placement legalization. A common objective is to minimize displacement.

9.3.3.1 Traditional Legalization Methods

An early approach to placement legalization was developed in the well known tool *Domino* [14], which uses a network flow approach to move cells from over congested areas to less congested regions. Flow-like techniques have been used in a number of other works (for example [15][25]).

A recent version of the placement tool *Capo* has a single row dynamic programming based legalization approach [26]. It uses "cell juggling" to adjust the density of cells within a single row. The method uses a number of cost functions based on minimum perturbation, minimizing half perimeter wire length (HPWL), minimum maximum movements (legalizing a row by minimizing the maximum movement from the original locations), and an iterative modification of the minimum HPWL cost function. The approach needs prior assignment of cells to rows. [4] also uses dynamic programming legalization.

For mixed size placement, [8] used both flow based techniques and dynamic programming. The method was effective for large industrial designs.

We illustrate the legalization problem in Figure 9.8. The general objective is to move logic elements that are not row-aligned, or are overlapping, to new positions that are both aligned and overlap free.

While there are many complex methods, a remarkably simple "tetris" [23] based method has gained popularity. We show pseudocode for the method in Figure 9.9. In this method, all cells are first sorted by their horizontal positions. Each cell, in sorted order, is then placed into a legal position that minimizes displacement from the abstract position. *APlace* [28] [29] uses a variant of the tetris method. *Feng Shui* [31] extended it to handle mixed size placements. In most cases, the method produces excellent results.

```
TetrisLegalization( ) {
    Sort all elements by their left edge location
    Initialize all cell rows as empty
    Initialize the right edge of each row

    For each element in order {
        Find a legal position in a row to minimize displacement
        Move the element to that position
        Update the right edge of the row
    }
}
```

Figure 9.9 Pseudocode for the "tetris" legalization method by Hill. This is a simple greedy algorithm that processes the logic elements from left to right. Each cell is placed in the row that minimizes displacement. The method is extremely fast, and for uniformly distributed abstract placements, surprisingly effective.

The tetris method does have some shortcomings, however. In placements produced by *Feng Shui*, the cells are closely packed. When cells are distributed more widely (as is done with fixed-die placement methods), not all cells are properly legalized by this tool. The "tetris" method also has no way to handle out of core cells effectively, and "stacking" of cells or macro blocks can degrade results.

In [1], it was observed that the placements for *Feng Shui* were illegal on some industrial benchmarks. When the abstract placement contains areas with significant overlap, solution quality of placements legalized by the Tetris method can degrade abruptly. Study of these placements showed that the increase in wire length came from only a subset of the nets – those connected to cells that had not only been displaced during legalization, but in particular to those in "pyramid" shaped areas of the legal placement. Figure 9.10 shows such an instance – in the center of the placement, there is a dark triangular shaped area of cells, with empty space in the surrounding regions.

Considering the operation of the greedy legalizer reveals how the pyramids are constructed. If there is little overlap, and the cells can be placed into legal positions with only small amounts of displacement, the process works extremely well. When there is overlap, however, cells must be displaced – and this displacement can occur horizontally or vertically. As an extreme case, consider a sample placement in which all logic elements are stacked on top of each other: each time an element is moved to a legal location, the position with minimum distance is at the perimeter of a growing "Manhattan circle."

The tetris method is attractive due to its simplicity. However for it to be effective, it is critical for the placement to not have areas with excessive demand. Methods to eliminate dense regions are discussed below.

Nine overlapping logic
elements prior to
placement legalization.

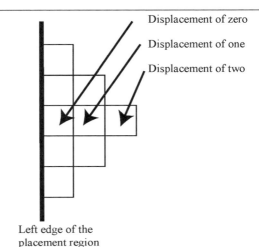

Displacement of zero

Displacement of one

Displacement of two

Left edge of the
placement region

In mixed size placement, large blocks
can introduce significant overlap,
resulting in a great deal of displacement.

The pyramid configurations produced by
the tetris legalization method are easy to
recognize, and increase wire lengths by
10% or more in some cases.

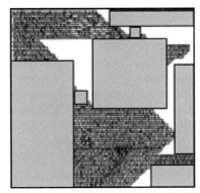

Figure 9.10 The "pyramid" effect in that occurs in Tetris-based legalization where significant overlap is present.

9.3.3.2   Detailed Placement

After legalization, many tools commonly apply single and multiple row branch-and-bound optimizations to improve wire lengths.

By simply passing a "sliding window" over the placement region, and enumerating the different permutations of cells within the window, wire length improvements can normally be obtained. While the number of permutations can be exponential, by keeping the window small – in most cases, from six to eight cells – this can be done with acceptable run times.

A variety of other techniques are also available. For example, the "optimal interleaving" work of Hur and Lillis [25] has many applications. With minor modifications, it can be used to distribute open space within a row, and the legalization method of [4] has a number of similarities.

9.3.4   Integration with Logic Synthesis

Many significant changes in the traditional design flow have occurred at the transition from detailed placement to routing. In earlier design flows, the typical sequence was *logical synthesis, placement*, and then *routing*. With the increased impact of interconnect on overall performance, logic synthesis optimizations are now frequently performed after detailed placement – and these optimizations must be incorporated into the layout.

Other chapters consider in depth the types of optimizations normally performed. For simplicity, we will focus here on gate sizing – optimizations such as buffer and repeater insertion, or the wholesale modification of portions of the circuit net list, are handled in a similar manner.

After the completion of placement, and possibly routing, one might find that the performance of a circuit could be improved by changing the size of a subset of gates. If sizes increase, cell overlaps can occur; we discuss methods to remove overlap here. After overlap removal is performed, the design can be made legal again, and the optimization process continues.

9.3.4.1   White Space

Many industrial designs contain a great deal of excess white space. For example, a recent set of benchmarks released by IBM [42] has a mixture of fixed macro blocks and standard cells. The space available for placement of the standard cells can be twice as large as the area of the cells themselves – there is a great deal of open space available.

There can be many reasons for having large amounts of open space. For the example benchmarks, this space allows for the "logical effort" [51] approach to circuit optimization to be performed with relative ease. Logic gates can be sized extensively – with abundant space, overlaps are relatively small and can be removed easily. Furthermore, there is space available for the insertion of buffers and repeaters.

9.3.4.2   Placement Transformation

If there is abundant open space, cell sizing and buffer insertion can be done without disrupting the overall placement. Without extra space, the integration of logic synthesis and placement can be much more difficult and "straight-forward" legalization is likely to produce unacceptable results. As described above, the greedy legalization method by Hill exhibits a "pyramid effect", and other legalization methods can also perform poorly. If the relative positions of logic changes significantly during legalization, the wire lengths anticipated during cell sizing don't match the final placement, making the cell sizing suboptimal in terms of delay, area, and power. In general, a design with a great deal of overlap, or areas with high utilization, poses a significant challenge to legalization methods.

```
CutLineShifting( ) {
    STEP: 'SET INITIAL PLACEMENT REGION' –
    r₁ = entire placement area
    R = {r₁}
    Assign all moveable objects to r₁

    While (R contains a region with more than one element) {
        For each rᵢ in R {
            // Aspect ratio determines cut direction
            If (tall) {
                Split the region at 50% horizontally
                If the cell areas of the two subregions do not match {
                    Compress the larger region and expand the smaller
                    region vertically

                }
            } else if (wide) {
                Split the region at 50% vertically
                If the cell areas of the two subregions do not match {
                    Compress the larger region and expand the smaller
                    region horizontally

                }
            }
            Remove rᵢ from the set of regions
            Add new smaller regions
        }
    }
    // Now legalize the placement
}
```

Figure 9.11 Pseudocode for the cut line shifting method; the overall approach can be thought of as "fractional cut bisection" in reverse. If a portion of the placement is too dense, logic elements can be moved in a relatively stable and uniform manner.

To simplify the legalization problem, recent research has focused on "placement transformation" techniques [36]. The objective of placement transformation is relatively simple: logic elements should be spread out by some combination of horizontal or vertical shifting, while avoiding large disruptions in interconnect lengths.

Note that this is not in any sense a "minimum displacement" objective: if a group of logic elements shift in the same direction, their interconnecting nets do not change in length. A solution that has a great deal of displacement [7] may be perfectly acceptable. Stable net lengths will lead to small changes in circuit delay, and easier timing and power convergence.

The "cut line shifting" method developed by Li [37] was shown to be effective in removal of overlaps, without introducing large changes in net lengths. The method is remarkably simple, and effective on placements produced by a variety of tools. The algorithm is outlined in Figure 9.11 and the operation of the approach is shown in Figure 9.12.

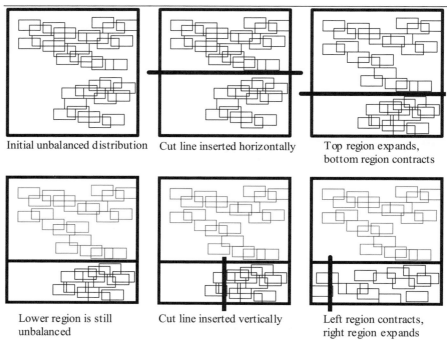

| Initial unbalanced distribution | Cut line inserted horizontally | Top region expands, bottom region contracts |
| Lower region is still unbalanced | Cut line inserted vertically | Left region contracts, right region expands |

Figure 9.12 An example of cut line shifting.

We should note that there can be many ways of overlap removal, and methods developed as part of analytic placement can also be applied. For example, consider the "cell shifting" method used in the placement tool *FastPlace* [54]. The circuit is divided into horizontal or vertical strips; each strip is then divided into a set of bins. By adjusting the height or width of a band to adjust to cell area constraints, the degree of cell overlap can be minimized.

The cell shifting technique, shown in Figure 9.13, is similar in spirit to cut line shifting. As such, it can achieve a similar effect. Both techniques are extremely fast, and can remove overlap while preserving the basic structure of the placement.

9.3.4.3   Stability of New Placements

Placement transformation is a relatively new development in physical design. To enable physical synthesis, "stability" of a placement algorithm is essential.

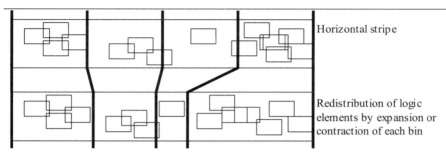

Horizontal stripe

Redistribution of logic elements by expansion or contraction of each bin

Figure 9.13 An example of cell shifting. Rather than the alternating cut lines of cut line shifting, this approach divides the circuit into horizontal or vertical stripes, and then adjusts the positions of elements within each stripe.

For analytic methods, slight changes to a circuit net list (through the insertion of repeaters and buffers, or through the sizing of logic elements) have a modest impact on the overall structure of the placement solution [7]. By contrast, recursive bisection placement and annealing methods can produce wildly different placements from two different runs. The stability of analytic solutions is yet another reason the approach is preferred for industrial tools. In an industrial flow, a circuit net list may change repeatedly – if each new placement solution is fundamentally different than the prior one, effort spent on gate sizing and buffer insertion will have little effect.

With the introduction of placement transformation, one can obtain stability within any global placement flow – provided that the degree of change to the circuit is relatively modest. We anticipate that there will be a great deal of progress in this area over the next few years, and that transformation will alter how many industrial groups perform logic synthesis. Rather than running a placement engine "from scratch" with each circuit modifycation, an existing placement may be adjusted with transformation, to meet the space requirements for gate sizing, buffer insertion, or small scale logic changes.

9.4 MULTIPLE SUPPLY VOLTAGE PLACEMENT

The techniques discussed in the previous sections can be viewed to a large degree as "enhancements" to traditional placement objectives. To minimize power dissipation of high activity nets, or to shorten wire length of critical path nets, a simple net-weighting approach can be applied. These modifications do not make fundamental changes to the basic placement algorithms.

Even the integration of gate sizing and buffer insertion has a relatively modest impact. If abundant white space is available, the "new" circuit can be legalized easily. Using recently developed techniques for placement transformation, space can be made available without significantly disrupting the overall structure.

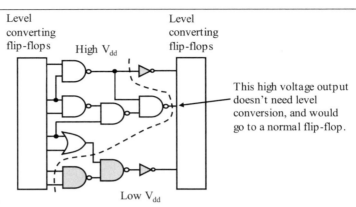

Figure 9.14 A simple diagram illustrating clustered voltage scaling. Clustered voltage scaling integrates level conversion within the latches. Downstream gates can switch to lower voltage, as long as this transition is monotonic because no additional voltage level converters are allowed between combinational logic in a clustered voltage scaling approach. (Low V_{dd} gates are shaded.)

In this section, we focus on a recent trend in low power circuitry: the utilization of multiple supply voltages as a method to reduce total power consumption. Methods to determine appropriate supply and threshold voltages are covered in other chapters. Here we focus on methods to place and legalize a multiple voltage circuit netlist. In general, it is acceptable for a high-V_{dd} gate to drive a low-V_{dd} gate, but not vice-versa – a logic 1 low V_{dd} output is unable to fully turn off the PMOS transistors in the high V_{dd} gate, resulting in considerable leakage current.

Multiple supply voltages impact the placement problem in a fundamental way. Firstly, construction of the power grid must be considered. If different voltages are scattered throughout the design, two complete power grids must be constructed, consuming valuable routing resources. The preferred method is to have logic with the same supply voltage clumped together spatially to some degree; the extent of aggregation required is an area of active research. Secondly, in bulk CMOS there are spacing requirements between regions with different supply voltages; the transistor wells must be separated[1], and this again is a motivation for aggregation. Finally, when transitioning from low V_{dd} to high V_{dd}, voltage level converters must be inserted into the design, and they typically require access to both power levels, adding yet another placement constraint.

---

[1] Otherwise the low V_{dd} gate PMOS n-wells must be connected to high V_{dd}. This reverse biases the well (vs. low V_{dd}), raising the PMOS transistor threshold voltage and reducing the pull-up drive strength, which is already substantially less due to operating at low V_{dd} (see Figure 7.2).

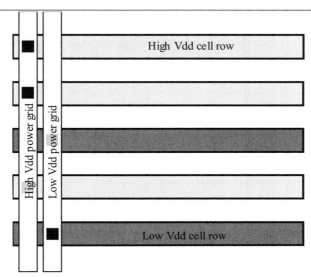

Figure 9.15 In some cases, supply voltages may be selected on a row-by-row basis. If a row is extremely long, this may constrain the solution significantly. The row-based restriction constrains placement, which may result in increased wire length and lower area utilization.

For voltage assignments at the macro block level, the problem is relatively simple; the blocks are large enough that power routing is easy, and the locations of level converters can be planned. The situation with gate level assignment is more interesting, and we focus on that here.

9.4.1   Clustered Voltage Scaling

The first approach to fine-grained voltage assignment was clustered voltage scaling [53]. With clustered voltage scaling, level converters are integrated with latches. The low voltage input is stepped up within the latch, which has a high output voltage. Cells connected directly to the latch can then be at either high or low V_{dd}. Different transition points between high-V_{dd} and low-V_{dd} can be examined. At the circuit level, the voltage assignment problem can be viewed as one of finding an appropriate logic "layer" to transition from high V_{dd} to low, as illustrated in Figure 9.14. A number of other methods have also been explored (e.g. [11][12] [17]).

Placement constraints have been addressed by restricting entire standard cell rows to use only a single power level, as illustrated in Figure 9.15. The supply voltages for a given row may be determined in an iterative manner; first a rough placement is performed, and then supply voltages are assigned to logic elements. Once the total area of logic elements at a given voltage level is known, the number of rows needed to perform legalization can be determined. This is normally done in a fairly simple manner, with an attempt to make legalization possible without major disruptions in the placement.

After rows have been assigned power levels, the placement may be improved by (for example) low temperature simulated annealing. During optimization, there may be high voltage logic elements assigned to low voltage rows, and vice versa. This mismatch can be penalized; the annealing process can then move these elements to rows with appropriate power levels, while also optimizing wire length.

9.4.2   Voltage Islands

A concern with the "row-based" constraint is that it normally increases interconnect lengths significantly. In many cases, circuit elements must be moved a great distance to obtain a legal placement.

There is growing interest in "voltage island" [24][35][45] configurations to address this problem. This approach is illustrated in Figure 9.16. Portions of a row may have different power levels. This requires increased spacing between some elements in a row, and can have an overhead in terms of the power grid wiring. The benefit is in reduced constraints on the placement, resulting in improved interconnect lengths and better area utilization.

The minimum size of a power island depends a great deal on the circuit structure, performance requirements and so on. Block-level power assignments can be viewed as one end of the spectrum. How finely grained power should be is currently being investigated.

We conclude this section by noting that there is active research on methods to legalize multiple V_{dd} designs. The added constraint can be integrated into the method by Hill [23] by simply restricting the rows (or portions of a row) that are considered. However, the restrictions may cause displacement, which can result in increased wire lengths.

9.5   STATE OF THE ART

In this section, we present experimental results of current academic placement tools on standard benchmarks. For some classes of problems, results of leading tools are similar; for others, results differ widely.

The nature of public benchmarks illustrates many of the difficulties faced by both academic and industry research groups. Most leading-edge circuit designs contain valuable intellectual property. Thus, circuits that have been released to the public are commonly several years old, and are also relatively small. Furthermore, in almost all cases, the logical behavior of the circuit has been stripped. Without knowing the functionality of each logic element, it is impossible to perform timing or power analysis.

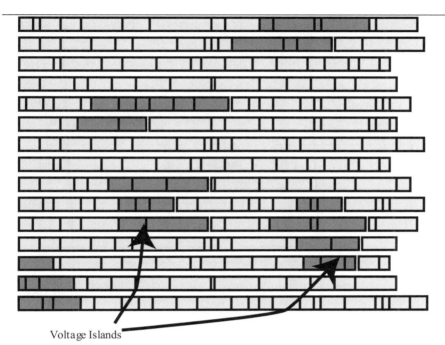

Voltage Islands

Figure 9.16 A "voltage island" approach allows for fine-grained selection of voltages, with variation within a row. There are minimum lengths of a row to maintain a uniform power level, and there can be a spacing requirement between voltages.

For academic research groups, these limitations prevent almost all meaningful comparisons *except* for half perimeter wire length. Many commercial tools have restrictions against benchmarking as part of their license agreements. One can assume that comparisons of tools are made within industry groups, but these results are not made public.

9.5.1 Standard Cell Placement and Routability

For the traditional standard cell placement problem, there has been something of a convergence of results for half perimeter wire length. The "Version 2 IBM Place" benchmarks are widely used [57]. These are based on hypergraph partitioning benchmarks [5], which were mapped to a commercial standard cell library. The partitioning benchmarks were in fact derived from IBM circuits, with logical functionality stripped off to protect intellectual property.

Note that only a subset of the eighteen hypergraph benchmarks has been converted into placement benchmarks; the numbering of these benchmarks corresponds to the hypergraphs that they are based on.

Table 9.1 Routed wire length results on IBM Place benchmarks; these designs use a 0.18um standard cell library. Half perimeter wire lengths are normally within a few percent of each other, with *mPL-R* frequently having the highest wire length. There is wide variation in the routed wire length, however; routing congestion results in wire length increases or outright routing failure. Successful routing results are shown in **bold**. The average routed wire length relative to mPL-R is shown at the bottom of the table.

Benchmark	# Cells	# Nets	Routed Wire Length		
			Dragon	Feng Shui	mPL-R
ibm01	12,282	11,507	0.93	0.85	**0.77**
ibm02	19,321	18,429	**2.18**	2.37	**1.89**
ibm07	45,135	44,394	4.55	4.49	**4.29**
ibm08	50,977	47,944	**4.78**	5.19	**4.58**
ibm09	51,746	50,393	**3.81**	3.56	**3.50**
ibm10	67,692	64,227	7.46	7.02	**6.84**
ibm11	68,525	67,016	5.68	5.41	**5.16**
ibm12	69,663	67,739	10.61	10.47	**10.52**
Comparison			×1.09	×1.08	×1.00

These benchmarks contain between 12,000 and 70,000 movable objects. This is extremely small in comparison to typical industrial designs, but representative of small blocks within a larger design.

In terms of wire length, the analytic placement tool *mPL-R*, the annealer *Dragon*, and the bisection based tool *Feng Shui*, all produce results within a few percent of each other. However, the results differ significantly after routing by a commercial tool: congestion results in routing detours, and the different placement tools have significantly different results.

In Table 9.1, we show routed wire length results for each tool on each of the benchmarks. If routing is successful, the result is listed in bold face; despite having good half perimeter wire length results, both *Dragon* and *Feng Shui* frequently fail during routing. *mPL-R*, which frequently has the highest half perimeter wire length results, produces successful routings on all benchmarks – it utilizes the cut line shifting method to insert space into the design, thereby eliminating congestion.

9.5.2   Mixed Size Benchmarks

While there has been something of a convergence of results on standard cell half perimeter wire length results, there is greater variation in mixed size placement. Two main approaches are used. One is to first place the macro blocks, and then place the standard cells around them. The second and more effective approach is to place both large and small objects simultaneously.

Table 9.2 Half perimeter wire length comparisons of *Capo*, *mPG*, and *Feng Shui* on IBM Place benchmarks. By using a fractional cut representation and a relatively simple legalizer, *Feng Shui* obtained large improvements over prior methods for mixed size designs. The average half perimeter wire length relative to Feng Shui is shown at the bottom of the table.

Benchmark	#Cells	#Nets	Capo	mPG	Feng Shui
ibm01	12,282	11,507	3.1	3.0	2.4
ibm02	19,321	18,429	6.8	7.4	5.3
ibm03	22,207	21,621	10.4	11.2	7.5
ibm04	26,633	26,163	10.1	10.5	8.0
ibm05	29,347	28,446	11.1	10.9	10.1
ibm06	321,825	33,354	9.9	9.2	6.8
ibm07	45,135	44,394	15.3	13.7	11.7
ibm08	50,977	47,944	17.9	16.4	13.6
ibm09	51,746	50,393	19.9	18.6	13.8
ibm10	67,692	64,227	45.5	43.6	37.5
ibm11	68,525	67,016	29.4	26.5	20.0
ibm12	69,663	67,739	55.8	44.3	35.6
ibm13	81,508	83,806	37.7	37.7	25.0
ibm14	146,009	143,202	50.3	43.5	38.5
ibm15	158,244	161,196	65.0	65.5	52.1
ibm16	182,137	181,188	90.0	72.4	61.3
ibm17	183,102	180,684	89.2	78.5	70.6
ibm18	210,323	200,565	51.8	50.7	45.1
Comparison			×1.29	×1.26	×1.00

Our bisection based placement tool *Feng Shui*, using the fractional cut approach and a legalization method based on Hill's work, obtained improvements of 26% or more *on average* over prior works. Results of experiments on the mixed size benchmarks (also derived from the partitioning benchmarks), are shown in Table 9.2. Very recently, the analytic tool *APlace* [28] was able to match the results of *Feng Shui*, using the same legalization and detailed placement methods.

9.5.3 Abundant White Space

The most recently released set of benchmarks is the ISPD2005 placement contest suite [43]. This set is unusual in many respects. Firstly, some circuits contain more than two million movable objects, substantially larger than other public benchmarks; these sizes are typical for current industry designs. Secondly, there is abundant white space: 50% to 85% of the space is *open*, making the effective handling of white space essential for good results. Finally, the macro blocks are fixed within the main placement area, creating obstacles that must be avoided.

Table 9.3 Results of the ISPD2005 placement contest. Large amounts of white space and fixed macro blocks caused poor behavior in some tools, resulting in a large gap in results. The average half perimeter wire length relative to APlace is shown at the bottom of the table.

Benchmark	# Cells	# Nets	APlace	Dragon	Feng Shui	Kraftwerk
adaptec2	254,457	266,009	87	95	123	158
adaptec4	494,716	515,951	188	201	337	352
bigblue1	277,604	284,479	95	103	115	149
bigblue2	534,782	577,235	144	160	285	322
bigblue3	10,995,519	1,123,170	358	380	471	656
bigblue4	2,169,183	2,229,886	833	904	1040	1404
Comparison			×1.00	×1.08	×1.50	×1.84

From the experimental results shown in Table 9.3, it should be clear that some placement tools handle the constraints better than others. The best and worst performing tools were the analytic placers *APlace* [28] and *Kraftwerk* [21]. The tool *Dragon* [59], which is a hybrid of annealing and bisection, produced results that were on average 8% higher than *APlace*. The bisection based tool *Feng Shui* [31], which performs well on designs with movable macro blocks, produced results with 50% higher wire lengths due to inadequate handling of open space.

9.6   SUMMARY

˙ There has been an upswing in academic placement research, illustrating the rising importance of the problem. Circuit designers are under intense pressure to minimize both power and circuit delay. The quality of a placement can make or break a design. The semiconductor industry is extremely competitive, and there is very little margin for error.

Poor placement can be very costly. Excess area increases the cost per die and reduces the yield. High power reduces battery life and can cause chip failure, or power may exceed design specifications rendering the chip unusable for its given task. Likewise worse delay may fail to meet design specifications, or cause incorrect functioning of the chip – which can also be caused by layout errors. Any of these problems increase time-to-market, and a product that is late or misses the market window can be extremely costly. Consequently, engineers need tight control of the design flow from architecture to synthesis to placement and routing, including verification at all steps; and the design flow must be predictable and converge.

While the problem has been studied for many years, the general consensus of the research community is that placement results are significantly suboptimal with respect to wire length objectives [10][27]. In [10], a set of synthetic benchmarks with known optimal configurations were created; placements produced by leading academic and commercial tools were commonly 50% or more away from optimal, and in some cases more than a factor of two away. While many disagree with the analysis, and argue that the benchmarks are not representative of "real" circuitry, the magnitude of sub-optimality was startling.

Considerable improvement has been obtained. In a recent placement competition [42], the wire lengths produced by the older analytic placement tool *Kraftwerk* were almost a factor of two times the results from *APlace*. One can conclude that there has in fact been a great deal of progress, and that more improvement is to be expected.

Wire length can be considered a relatively simple objective to capture. Power and delay objectives are much more complex, making it reasonable to assume that the magnitude of suboptimality may be even greater.

9.7 REFERENCES

[1] Adya, S., et al., "Unification of partitioning, placement, and floorplanning," in *Proc. Int. Conf. on Computer Aided Design*, 2004, pp. 550–557.

[2] Adya, S., Markov, I., and Villarrubia, P., "On whitespace in mixed-size placement and physical synthesis," in *Proc. Int. Conf. on Computer Aided Design*, 2003, pp. 311–318.

[3] Adya, S., et al., "Benchmarking for large-scale placement and beyond," *IEEE Trans. on Computer-Aided Design of Integrated Circuits and Systems*, vol. 23, no. 4, 2004, pp. 472–487.

[4] Agnihotri, A., et al., "Fractional cut: Improved recursive bisection placement," in *Proc. Int. Conf. on Computer Aided Design*, 2003, pp. 307–310.

[5] Alpert, C., et al., "The ISPD98 circuit benchmark suite," in *Proc. Int. Symp. on Physical Design*, 1998, pp. 80–85.

[6] Alpert, C., Nam, G., and Villarrubia, P., "Free space management for cut-based placement," in *Proc. Int. Conf. on Computer Aided Design*, 2002, pp. 746–751.

[7] Alpert, C., et al., "Placement stability metrics," in *Proc. Asia South Pacific Design Automation Conf.*, 2005, pp. 1144–1147.

[8] Brenner, U., Pauli, A., and Vygen, J., "Almost optimum placement legalization by minimum cost flow and dynamic programming," in *Proc. Int. Symp. on Physical Design*, 2004, pp. 2–9.

[9] Breuer, M., "A class of min-cut placement algorithms," in *Proc. Design Automation Conf.*, 1977, pp. 284–290.

[10] Chang, C., Cong, J., and Xie, M., "Optimality and scalability study of existing placement algorithms," in *Proc. Asia South Pacific Design Automation Conf.*, 2003, pp. 621–627.

[11] Chang, J., and Pedram, M., "Power minimization using multiple supply voltages," in *Proc. Int. Symp. on Low Power Electronic Design*, 1996, pp. 157–162.

[12] Chen, C., and Sarrafzadeh, M., "An effective algorithm for gate-level power-delay tradeoff using two voltages," in *Proc. Int. Conf. on Computer Aided Design*, 1999, pp. 222–227.

[13] Cong, J., and Smith, M., "A parallel bottom-up clustering algorithm with applications to circuit partitioning in VLSI design," in *Proc. Design Automation Conf.*, 1993, pp. 755–780.

[14] Doll, K., Johannes, F., and Antreich, K., "Iterative placement improvement by network flow methods," *IEEE Trans. on Computer-Aided Design of Integrated Circuits and Systems*, vol. 13, no. 10, 1994, pp. 1189–1200.

[15] Donath, W., et al., "Transformational placement and synthesis," in *Proc. Design, Automation and Test in Europe Conf.*, 2000, pp. 194–201.

[16] Donath, W., "Placement and average interconnection lengths of computer logic," *IEEE Trans. on Circuits and Systems*, vol. CAS-26, no. 4, 1979, pp. 272–277.

[17] Donno, M., et al., "Enhanced clustered voltage scaling for low power," in *Proc. Great Lakes Symposium on VLSI*, 2002, pp. 18–23.

[18] Dreyfus, S., "An appraisal of some shortest-path algorithms," *Operations Research*, vol. 17, 1969, pp. 395–412.

[19] Dunlop, A., et al., "Chip layout optimization using critical path weighting," in *Proc. Design Automation Conf.*, 1984, pp. 133–136.

[20] Dunlop, A., and Kernighan, B., "A procedure for placement of standard-cell VLSI circuits," *IEEE Trans. on Computer-Aided Design of Integrated Circuits and Systems*, vol. CAD-4, no. 1, January 1985, pp. 92–98.

[21] Eisenmann, H., and Johannes, F., "Generic global placement and floorplanning," in *Proc. Design Automation Conf.*, 1998, pp. 269–274.

[22] Frankle, J., "Iterative and adaptive slack allocation for performance-driven layout," in *Proc. Design Automation Conf.*, 1992, pp. 536–542.

[23] Hill, D., "Method and system for high speed detailed placement of cells within an integrated circuit design," U.S. Patent No. 6,370,673, Apr. 9, 2002.

[24] Hu, J., et al., "Architecting voltage islands in core-based system-on-a-chip designs," in *Proc. Int. Symp. on Low Power Electronic Design*, 2004, pp. 180–185.

[25] Hur, S., and Lillis, J., "Mongrel: Hybrid techniques for standard cell placement," in *Proc. Int. Conf. on Computer Aided Design*, 2000, pp. 165–170.

[26] Kahng, A., Markov, I., and Reda, S., "On legalization of row-based placements," in *Proc. Great Lakes Symposium on VLSI*, 2004, pp. 214–219.

[27] Kahng, A., and Reda, S., "Evaluation of placer suboptimality through zero-change netlist transformations," in *Proc. Int. Symp. on Physical Design*, 2005, pp. 208–215.

[28] Kahng, A., and Wang, Q., "An analytic placer for mixed-size placement and timing-driven placement," in *Proc. Int. Conf. on Computer Aided Design*, 2004, pp. 565–572.

[29] Kahng, A., and Wang, Q., "Implementation and extensibility of an analytic placer," in *Proc. Int. Symp. on Physical Design*, 2004, pp. 18–25.

[30] Karypis, G., "Multilevel hypergraph partitioning: Application in VLSI domain," in *Proc. Design Automation Conf.*, 1997, pp. 526–529.

[31] Khatkhate, A., et al., "Recursive bisection based mixed block placement," in *Proc. Int. Symp. on Physical Design*, 2004, pp. 84–89.

[32] Kirkpatrick, S., "Optimization by simulated annealing: Quantitative studies," *J. Statistical Physics*, vol. 34, 1984, pp. 975–986.

[33] Kleinhans, J., et al., "GORDIAN: VLSI placement by quadratic programming and slicing optimization," *IEEE Trans. on Computer-Aided Design of Integrated Circuits and Systems*, vol. 10, no. 3, 1991, pp. 356–365.

[34] Koźmiński, K., "Benchmarks for layout synthesis – evolution and current status," in *Proc. Design Automation Conf.*, 1991, pp. 265–270.

[35] Lackey, D., et al., "Managing power and performance for System-on-Chip designs using voltage islands," in *Proc. Int. Conf. on Computer Aided Design*, 2002, pp. 195–202.

[36] Li, C., Koh, C., and Madden, P., "Floorplan management: Incremental placement for gate sizing and buffer insertion," in *Proc. Asia South Pacific Design Automation Conf.*, 2005, pp. 349–354.

[37] Li, C., et al., "Routability-driven placement and white space allocation," in *Proc. Int. Conf. on Computer Aided Design*, 2004, pp. 394–401.

[38] Liu, Q., and Marek-Sadowska, M., "Pre-layout wire length and congestion estimation," in *Proc. Design Automation Conf.*, 2004, pp. 582–588.

[39] Lou, J., Krishanmoorthy, S., and Sheng, H., "Estimating routing congestion using probabilistic analysis," in *Proc. Int. Symp. on Physical Design*, 2001, pp. 112–117.

[40] Madden, P., "Reporting of standard cell placement results," *IEEE Trans. on Computer-Aided Design of Integrated Circuits and Systems*, vol. 21, no. 2, February 2002, pp. 240–247.

[41] Magen, N., et al., "Interconnect-power dissipation in a microprocessor," in *Proc. System Level Interconnect Prediction Workshop*, 2004, pp. 7–13.

[42] Nam, G., et al., "The ISPD2005 placement contest and benchmark suite," in *Proc. Int. Symp. on Physical Design*, 2005, pp. 216–220.

[43] Ou, S., and Pedram, M., "Timing-driven placement based on partitioning with dynamic cut-net control," in *Proc. Design Automation Conf.*, 2000, pp. 472–476.

[44] Pedram, M. "Power minimization in IC design: Principles and applications," *ACM Trans. on Design Automation of Electronics Systems*, vol. 1, no. 1, 1996, pp. 3–56.

[45] Puri, R., et al., "Pushing ASIC performance in a power envelope," in *Proc. Design Automation Conf.*, 2003, pp. 788–793.

[46] Rohe, A., and Brenner, U., "An effective congestion driven placement framework," in *Proc. Int. Symp. on Physical Design*, 2002, pp. 1–6.

[47] Sahni, S., and Bhatt, A., "The complexity of design automation problems," *Proc. Design Automation Conference*, 1980, pp. 402—411.

[48] Saxena, P., et al., "Repeater scaling and its impact on CAD," *IEEE Trans. on Computer-Aided Design of Integrated Circuits and Systems*, vol. 23, no. 4, 2004, pp. 451–463.

[49] Scheffer, L., and Nequist, E., "Why interconnect prediction doesn't work," in *Proc. System Level Interconnect Prediction Workshop*, 2000, pp. 139–144.

[50] Stroobandt, D., "A priori system-level interconnect prediction: Rent's rule and wire length distribution models," in *Proc. System Level Interconnect Prediction Workshop*, 2001, pp. 3–21.

[51] Sutherland, I., Sproull, R., and Harris, D., *Logical Effort: Designing Fast CMOS Circuits*. Morgan Kaufmann, 1999.

[52] Swartz, W., and Sechen, C., "Timing driven placement for large standard cell circuits," in *Proc. Design Automation Conf.*, 1995, pp. 211–215.

[53] Usami, K., and Horowitz, M., "Clustered voltage scaling technique for low-power design," in *Proc. Int. Symp. on Low Power Electronic Design*, 1995, pp. 3–9.

[54] Viswanathan, N., and Chu, C., "Fastplace: Efficient analytical placement using cell shifting, iterative local refinement and a hybrid net model," in *Proc. Int. Symp. on Physical Design*, 2004, pp. 26–33.

[55] Vujkovic, M., et al., "Efficient timing closure without timing driven placement and routing," in *Proc. Design Automation Conf.*, 2004, pp. 268–273.

[56] Vygen, J., "Algorithms for large-scale flat placement," in *Proc. Design Automation Conf.*, 1997, pp. 746–751.

[57] Wang, M., Yang, X., and Sarrafzadeh, M., "Dragon2000: Standard-cell placement tool for large industry circuits," in *Proc. Int. Conf. on Computer Aided Design*, 2000, pp. 260–263.

[58] Westra, J., and Groeneveld, P., "Is probabilistic congestion estimation worthwhile?" in *Proc. System Level Interconnect Prediction Workshop*, 2005, pp. 99–106.

[59] Yang, X., Choi, B., and Sarrafzadeh, M., "Routability driven white space allocation for fixed-die standard cell placement," in *Proc. Int. Symp. on Physical Design*, 2002, pp. 42–50.

Chapter 10

POWER GATING DESIGN AUTOMATION

Jerry Frenkil, Srini Venkatraman
Sequence Design, Inc.
Westford, MA 01886

10.1   INTRODUCTION

The demand for portable electronic devices is growing rapidly and, due in large part to the development of wireless communications, is expected to continue to grow. This demand has generated great interest in low power design, which initially focused on controlling dynamic power consumption. While this focus resulted in significant improvements in dynamic power efficiency, two issues subsequently arose which rendered this initial focus inadequate. The combination of these two issues has motivated the development of leakage reduction techniques and related design automation.

The first issue pertains to the operational characteristics of wireless devices – basically, their operation tends to be bursty. That is, relatively short periods of activity are followed by relatively lengthy periods of inactivity, and while the power consumption during the active period is dominated by dynamic power, the power consumption during the inactive period (known as standby or sleep mode) is dominated by leakage power.

The second issue pertains to leakage power itself. Leakage is increasing exponentially with each new process generation due to the scaling of transistor threshold voltages [19].

This chapter will describe in detail the use of power gating for leakage reduction along with cell-based design automation methods employed by the CoolPower™ design tool, and is organized as follows. The next section briefly surveys different leakage reduction techniques, providing the motivation for power gating. The subsequent sections describe design issues, CoolPower automation methods including analysis and optimization techniques, and two different power gating application flows as well as results from using those flows. This chapter then concludes with a view to the future and likely new developments in power gating design.

10.2   LEAKAGE CONTROL TECHNIQUES

This section briefly presents and compares several different leakage control techniques to enable the reader to understand the motivations for the development and deployment of MTCMOS power gating.[1]

Leakage has several different components, however the largest components are sub-threshold related [11]. The equation for sub-threshold leakage current is

$$I_{leakage} = I_{s0}\, e^{(V_{gs} - V_{th})/nV_T}\, (1 - e^{-V_{ds}/V_T})$$ (10.1)

where

$$I_{s0} = K(W_{eff}/L_{eff})V_T^2$$ (10.2)

$$V_{th} = V_{th0} - \gamma V_{bs} - \eta V_{ds}$$ (10.3)

and V_{gs} is the transistor-gate to source voltage; V_{ds} is the drain to source voltage; V_{th0} is the zero bias threshold voltage; γ is the linearized body effect coefficient; V_{bs} is the source to body voltage; η is the DIBL (drain induced barrier lowering) coefficient; n is the subthreshold swing coefficient; V_T is the thermal voltage; K is a process constant; W_{eff} is the effective transistor width; and L_{eff} is the effective transistor channel length. [7][15]

Leakage control techniques focus on controlling one or more terms in these equations. The most prevalent techniques can be categorized as reducing V_{gs}, increasing V_{th0}, lowering V_{bs}, and reducing V_{ds}. Several different methods for controlling these terms are described below along with how they relate to equations (10.1) to (10.3).

10.2.1   Reverse Body Bias (RBB)

Since leakage currents are a function of the device thresholds, one method for controlling leakage is to control V_{th} through the use of substrate, or body, bias. In this case, the substrate or the appropriate well is biased so as to raise the transistor thresholds thus reducing leakage. Since raising V_{th} also affects performance, the bias can be applied adaptively such that during active mode the reverse bias is small while in standby mode the reverse bias is more negative. Thus, reverse body bias reduces leakage by increasing V_{th} due to decreasing the γV_{bs} term in Equation (10.3).

---

[1] Multi-Threshold CMOS (MTCMOS) is commonly used as a synonym for power gating, since the most prevalent power gating implementations utilize multiple transistor thresholds. However, it has also been used to refer to the use of non-power gated CMOS circuits designed utilizing multiple transistor thresholds. In this chapter, MTCMOS will be used synonymously with power-gating, while multi-V_{th} denotes the use of multiple transistor thresholds in otherwise conventional circuit design.

Use of body bias requires a substrate bias-generator to generate the bias voltage. This generator will consume some dynamic power, partially offsetting any gain from reduced leakage.

However, a more significant issue with the use of substrate biasing for leakage reduction is that it is generally less effective in advanced technologies [10]. The body effect factor γ decreases with advanced technologies [2], reducing the extent of the leakage control. Consequently, reductions of 4× at 90nm and only 2× at 65nm have been reported [21].

10.2.2 Dynamic Voltage Scaling (DVS)

One technique for reducing dynamic power, dynamic voltage scaling, can also be used for reducing leakage power. DVS works by reducing the power supply voltage V_{dd} when the work load does not require maximal performance.

DVS can also be applied to inactive circuits for leakage reduction. In Equation (10.1), the reduction in V_{dd} is reflected in a smaller value for V_{ds} which has an exponential effect on leakage. Power savings of 8× to 16× have been reported when scaling the voltage to the 300mV range, the lowest voltage at which state can be maintained [3].

However, DVS requires additional circuitry to monitor and predict the workload as well as a dynamic voltage regulator to dynamically adjust the supply voltage. Also, the timing analysis of DVS circuitry is complicated since proper operation must be validated over a number of additional voltage points. Nevertheless, DVS has been combined with RBB for even greater leakage reduction than either technique can achieve alone [16].

10.2.3 Multi-Vth Cell Swapping

The most prevalent technique used to date for leakage reduction is multi-V_{th} cell swapping, most commonly deployed with two different transistor thresholds (and hence known as dual-V_{th} cell optimization) [20][26]. In this technique, low-V_{th} cells are used on critical paths while high-V_{th} cells are used on non-critical paths. The low-V_{th} cells are fast but leaky, while the high-V_{th} cells are just the opposite. Thus, the multi-V_{th} technique can reduce leakage power without any performance penalty.

A significant advantage of multi-V_{th} cell swapping is that it is generally footprint neutral. That is, no floorplanning or layout changes are required for implementation. High-V_{th} cells replace their low-V_{th} equivalents in exact positions in the layout, thus effectively changing only the implant mask.

Leakage can typically be reduced by about 50% compared to a circuit implemented with all low-V_{th} cells although the reduction is heavily dependent upon the amount of available slack in the original circuit [25]. However, the remaining low-V_{th} cells still consume significant amounts of leakage power.

Thus, this technique is usually insufficient for achieving large reductions in standby mode leakage power. For this reason, designers have turned to more aggressive leakage control design techniques such as MTCMOS power gating [23][24].

10.2.4   MTCMOS Power Gating

MTCMOS power gating is a design technique in which a power gating transistor is inserted in the stack between the logic transistors and either power or ground, thus creating a virtual supply rail or a virtual ground rail, respectively. (In order to simplify the descriptions of power gated circuitry, the following text will refer to virtual grounds only, except in those cases where header switches or virtual supplies present issues that are different from those related to footers and virtual grounds.)

Such configurations are shown in Figure 10.1. The logic block contains all low-V_{th} transistors for fastest switching speeds while the switch transistors, header or footer, are built using high-V_{th} transistors to minimize the leakage. Power gating can be implemented without using multiple thresholds, but it will not reduce leakage as much as if implemented with multiple thresholds.

MTCMOS refers to the mixture of the transistor thresholds in power gating circuits. The most common implementations of power gating use a footer switch alone to limit the switch area overhead. High-V_{th} NMOS footer switches are about half the size of equivalent-resistance high-V_{th} PMOS header switches due to differences in majority carrier mobilities.

Power gating reduces leakage by reducing the gate-to-source voltage which in turn drives the logic transistors deeper into the cutoff region. This occurs because of the stack effect. The source terminal of the bottom-most transistor in the logic stack is no longer at ground, but rather at a voltage somewhat above ground due to the presence of the power gating transistor. Leakage is reduced due to the reduction of the V_{gs} term in Equation (10.1).

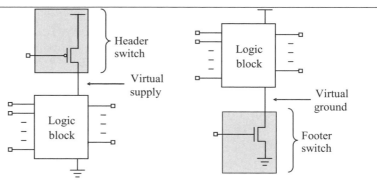

Figure 10.1 MTCMOS power-gating circuit topology

Power gating itself has several variants, such as Super Cut-off CMOS [9] and Zigzag Super Cut-off CMOS [17]. In Super Cut-off CMOS, instead of using high-V_{th} NMOS or PMOS switch transistors, low-V_{th} switch transistors are used. In standby mode, the switches are driven deeper into cut-off by applying a gate voltage below V_{ss} for NMOS switches and above V_{dd} for PMOS switches, thus decreasing V_{gs} beyond what can be achieved with conventional gate voltages. In Zigzag Super Cut-off CMOS, both header and footer switches are used in an alternating fashion along logic paths in combination with Super Cutoff CMOS to reduce the amount of time required for the virtual rails to settle after turning on the switch transistors.

Power gating can be combined with other leakage reduction techniques, such as those described above, to achieve even greater leakage reduction. When implemented alone, power gating can achieve 10 to 100× reduction in leakage. When implemented in combination with other techniques, such as reverse body bias on the switch, the reduction can be even larger. [13]

While power gating can be implemented in either a custom design style or an ASIC cell based design style, the following section will describe issues and automation techniques for the ASIC cell based design style.

10.3 POWER GATING DESIGN ISSUES

The design of power gated circuits presents the designer with a number of issues that are not usually encountered in designing non-power gated circuits. This section briefly describes some of these issues to give some perspective on the design automation presented in subsequent sections.

10.3.1 Power Gating Topologies

Power gating can be implemented using several different topologies, such as global power gating, local power gating, and switch-in-cell power gating. Each of these topologies is primarily distinguished by the connections between the switches and the logic and, as can be expected, each has its own advantages and disadvantages.

10.3.1.1 Global power gating

Global power gating refers to a logical topology in which multiple switches are connected to one or more blocks of logic, and a single virtual ground is shared in common among all the power gated logic blocks. In this arrangement, illustrated in Figure 10.2, there is a single virtual ground for each sleep domain (a group of logic controlled by a particular sleep enable signal). This topology is effective for large blocks in which all the logic is power gated, but is less effective, for physical design reasons, when the logic blocks are small. It does not apply when there are many different power gated blocks, each controlled by a different sleep enable.

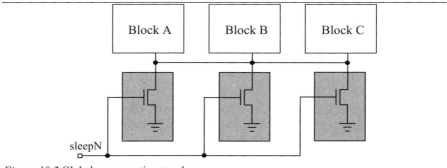

Figure 10.2 Global power gating topology

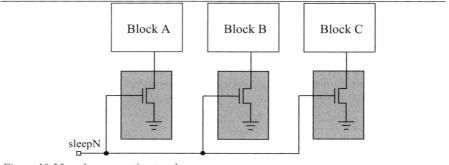

Figure 10.3 Local power gating topology

10.3.1.2 Local power gating

Local power gating refers to a logical topology in which each switch singularly gates its own virtual ground connected to its own group of logic. The key issue here is that a single switch cell is used for each logic group (with the single switch being shared among all cells in that group of logic), as opposed to using multiple, arrayed switch cells. This arrangement results in multiple segmented virtual grounds for a single sleep domain. Figure 10.3 illustrates the connections for local power gating. Compared to global power gating (as illustrated in Figure 10.2), local power gating provides more flexibility in floorplanning since the various power gated blocks within a given sleep domain need not be physically contiguous.

10.3.1.3 Switch-in-cell

Switch-in-cell may be thought of as an extreme form of local power gating implementation. In this topology, each logic cell contains its own switch transistor, as illustrated with an inverter in Figure 10.4.

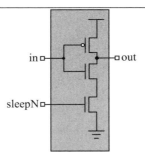

Figure 10.4 Switch-in-cell: a switch is included in each individual logic cell

The switch-in-cell approach has several notable advantages and disadvantages. Its primary advantages are that delay calculation is very straightforward (since each cell is timing characterized with its own, dedicated internal switch) and that it can be placed, generally without restriction, like any other standard cell. However, its disadvantages are significant, chief among them being that the area overhead is substantial (due to an additional transistor in the pulldown stack, and the need to size up the previously existing logic transistors to compensate for the additional device in the stack). And, given that each individual instance has its own switch, the aggregate input capacitance presented to the sleep signal is much larger than needed for shared switches requiring a larger than necessary amount of dynamic energy to open and close the switches. Additionally, since the size of the switch transistor is set during the design of each of the individual cells, the performance impact of the switch is also set at the time of the cell design, thus potentially limiting the applicability of the cells to either low-performance (small switches with a large performance impact) or high-performance (large switches with a small performance impact), but not both, unless of course two (or more) complete sets of logic cells are designed with each set utilizing different switch transistor sizes.

10.3.2   Switch Sizing Tradeoffs

All power gating topologies face the challenging tradeoff of switch sizing. A common switch sizing goal is to minimize the switch area, but this results in a larger virtual ground voltage which degrades switching performance but produces a larger reduction in leakage currents.

Sizing must respect one fundamental constraint: switches must be large enough to hold their virtual grounds sufficiently close to ground potential. That is, switches must limit "ground bounce" – the smaller the switch resistance, the smaller the voltage on the virtual ground.

However, achieving smaller switch resistance requires a physically larger switch. Unfortunately, the larger the switch the smaller the leakage reduction [4][13] since a larger switch, with smaller on resistance, reduces the body effect on the logic transistors – the larger switch produces a smaller virtual ground voltage (logic transistor source voltage) which in turn results in a less negative γV_{bs} term in Equation (10.3).

On the other hand, the virtual ground voltage must be minimized to minimize its impact on performance. The larger the virtual ground voltage, the smaller the gate drive on the logic transistors and the slower the logic transistors will switch. Also, the logic transistors will not pull down as far, thereby slowing transitions on gates that they drive.

Thus, we have a classic tradeoff: minimizing the performance impact of the virtual ground results in more area overhead and lesser leakage reduction due to larger switches. One method of reducing the area consumed by the switches is to share them, since using the switch-in-cell approach consumes relatively significant amounts of area. However, the use of shared switches complicates delay calculation and timing analysis. In any case, it is clear that switch sizing has a major impact on critical circuit characteristics and thus deserves careful attention.

10.3.3   Delay Calculation and Timing Analysis

Given that the size of the switches affects the voltage drop on the virtual grounds, and that the voltage drop impacts timing performance, we must consider how power gating affects delay calculation and timing analysis.

There are two general methods for the timing analysis of power gated circuits. The first method uses conventional delay calculation and relies upon tightly controlling the virtual ground voltage drop. The second method uses back-annotated virtual ground voltages in the delay calculator to compute a set of instance-specific voltage-sensitive delay values.

The first method is identical to the existing non-power-gated delay calculation method. All variations in supply voltages are assumed to lie within the voltage range for the cell library timing characterization. During timing characterization of the library logic cells, a non-zero voltage is asserted on the ground line to approximate the effects of the cell being connected to a non-ideal rail. If the voltage drop seen by the cell *in-situ* is less than the value used during characterization, then the cell is considered to be operating within the characterization limits, or guardbands. This common practice for the timing analysis of non-power gated circuits also applies to power gated circuits provided that the voltage drop on the virtual ground is constrained to be within the cell-library characterization limits.

The second method, by contrast, places no *a priori* constraints on virtual ground voltages. Instead, post-route voltage drop analysis determines the virtual ground voltages. The delay calculator then computes the delays through each instance based on the particular virtual ground voltages seen by each instance. This flow requires a set of library timing models that accurately model voltage effects upon delay.

10.3.4   Power Gating Granularity

Granularity refers to the size of each logic block controlled by a single switch or single sleep domain. This section describes the basic choices.

10.3.4.1   Coarse Grained and Fine Grained Power Gating

Consider two different chips, one that has a single sleep control that can power down the entire chip and a second chip that has multiple sleep control signals, each of which separately controls different logic functions such as an execution unit, memory controller, instruction decoder, etc. The former design is said to use coarse-grained power gating, since power is gated very coarsely, in this case either all or nothing. The latter design is said to use fine-grained power gating since power can be shut off to individual units without shutting off the power to other units at that time.

The choice of granularity has both logical and physical implications. A *power domain* refers to a group of logic with a logically unique sleep signal. Each power domain must be physically arranged to share the virtual ground common to that particular group (except for the boundary case of the switch-in-cell topology in which there are no shared virtual grounds).

The motivation for fine-grained power gating is to reduce active mode leakage power, that is, the leakage power consumed during normal operation. While the coarse-grained example above will reduce leakage during standby, it will not affect active leakage since with a single sleep domain the power supply is either completely connected (active mode) or completely disconnected (standby mode). However, with fine-grained switching, portions of the design may be switched off while the other portions continue to operate. For example, in a VLIW processor with four execution units, if only three of the execution units are active, the fourth may be put to sleep until such time as it is scheduled to resume computation.

10.3.4.2   Full and Selective Power Gating

Chips, or modules, may be completely or partially power gated. With *full* power gating, all logic instances are power gated. With *selective* power gating, only a subset of the logic instances are power gated.

Selective power gating typically combines fine-grained power gating with multi-V_{th} cell swapping. In this case, selective power gating refers specifically to the power gating of individual instances along a critical path. In this implementation style, all instances along the non-timing critical paths use high-V_{th} cells, while those instances along the critical paths use low-V_{th} cells to maintain performance but are power-gated to minimize leakage while not operating.

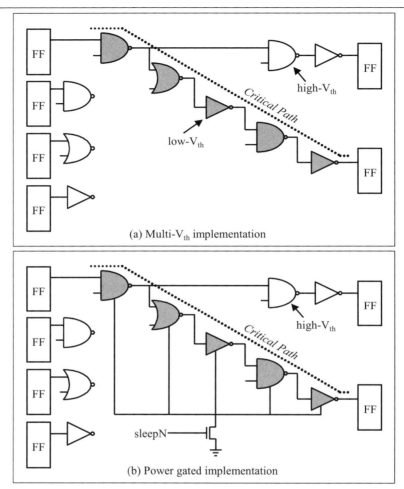

(a) Multi-V_{th} implementation

(b) Power gated implementation

Figure 10.5 Selective power gating

Figure 10.5 illustrates an example of selective power gating. The top schematic shows the results of a multi-V_{th} optimization in which only the instances along the critical path utilize low-V_{th} cells while all other instances are high-V_{th} cells. The bottom schematic shows the same circuit after power gating the low-V_{th} cells.

Selective power gating minimizes the area overhead of the switches while maintaining fast switching speeds. Since only the low-V_{th} fraction of the logic instances are power gated, fewer switches are needed. However, placement and clustering issues present physical implementation challenges. Switch-in-cell libraries are often used in selective power gating applications since they eliminate the problem of sharing virtual grounds, although they still require routing of the sleep signals.

10.3.5 State Retention

When state registers are power gated, they will lose their state unless particular measures are taken to prevent the loss of memory. Multiple methods exist for retaining state, including saving state to off-chip memory by scanning out the internal state prior to power-down and subsequently scanning it back in upon power-up, utilizing specially designed state retention registers that remember their state even when power gated [3], and not power gating state registers (only power gating combinational logic).

Each of these methods has its own advantages and disadvantages. Scanning state in and out is relatively straightforward but takes time and consumes dynamic power in the process. Use of state retention registers simplifies the logic design, but requires complex circuit designs for the state retention registers and often impacts both area and performance. Power gating only the combinational logic resolves the issue of state retention, but requires circuitry to prevent the interface nodes from floating.

10.3.6 Power Domain Interfacing

When only a portion of a chip is power gated, the power-gated logic will drive some signals that are received by non-power gated logic. These signals are called interface or fence nodes and require special attention, as they will float when the driving logic is disconnected from the power rails. [14][27]

When an interface node floats to an intermediate voltage, approximately $V_{dd}/2$, both the p-channel and the n-channel transistors in the receiving logic will conduct, drawing large amounts of current from the power supply. Not only does this negate the power savings from power gating, it can also cause reliability problems or outright failure due to electromigration, as the interconnect is not sized to support these types of currents.

To prevent floating, the interface nodes can simply use float-prevention mechanisms, such as an isolation cell, as illustrated in Figure 10.6. Note that in the case of fine-grained power gating, outputs of power gated logic that drive other power gated logic must also be prevented from floating if the receiving logic is power gated by a logically different sleep signal.

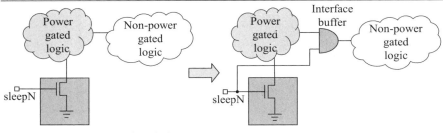

Figure 10.6 Power domain interfacing

10.4   COOLPOWER DESIGN AUTOMATION

To date, design automation for power gating has been narrowly focused and has not adequately addressed the issues of area overhead, performance impact, or overall cell-based design flows. However, it is essential that design automation address these issues holistically, since the use of switches complicates so many design facets. These facets include mixing power-gated and non-power-gated logic, current flow analysis, switch and virtual ground optimization, and worst case design, among others. For example, one approach simplified the design automation requirements by employing the switch-in-cell structure described above [23]; since each cell contains its own dedicated virtual ground, no virtual ground sharing or routing is required. However, the area overhead of this approach is substantial, as much as 80% additional area per power-gated instance [4][23], driving up the per-die cost. Other approaches used shared virtual grounds [1][8], reducing the area overhead compared to the switch-in-cell approach but complicating the design automation.

In all cases, the issue of switch sizing is central to the overall solution. Our overall approach embodied in CoolPower is similar to that of [1] and [8] in that we use shared virtual grounds and dynamic currents to size switches. However we overcome significant limitations of those approaches both in current calculation as well as in optimization. The switch network employs local power gating with a shared-switch architecture. Our current calculation solution computes current waveforms based upon an all-events static timing analysis. During optimization, we size the switches based not only on sink currents but also on virtual ground parasitic resistance. These optimizations are performed subject to user-specified constraints for peak transient voltage on the virtual ground, maximal distance between switches, and electromigration limits.

CoolPower includes internal delay calculation, timing and signal integrity analysis and optimization, and incremental placement capabilities, but relies upon external routers to route signals, power and ground, as well as the virtual grounds. Operating at the cell level, CoolPower requires cell-level models for non power gated cells, power gated cells, and switch cells (LEF for placement, Liberty for timing and power analysis), and netlist and placement files to describe the design (Verilog and DEF formats, respectively). CoolPower relies on a conventional standard-cell placement architecture which, coupled with its current calculation and switch optimization, enables completely automatic design of electrically robust power-gated MTCMOS circuits. It produces as output power-gated netlist and placement files (again, in Verilog and DEF formats, respectively).

The following sections will describe CoolPower's operation in more detail.

10.4.1 Design Transformation

CoolPower begins implementation of a power-gated circuit by loading the design, including a list of modules to be power gated, and the names of the sleep signals for those modules. CoolPower inserts a virtual ground into the circuit and logically gates the specified logic module or group of cells. First, CoolPower replaces all specified non-power gated instances (without a virtual ground connection) with power gated instances (with a virtual ground connection). Next, CoolPower inserts switches to connect the power-gated instances, through the virtual ground connection, to the real ground and the switch input is wired to the specified sleep control signal. Additionally, CoolPower inserts interface cells on power-gated block outputs that drive non-power gated logic to prevent any floating inputs. Interface cells are also inserted on outputs from power-gated blocks that drive power-gated logic that is controlled by different sleep signals, as there is no guarantee that the two power-gated blocks are put to sleep at the same time.

State retention is handled during transformation by one of two methods. The first method gates the power to only the combinational logic and not the registers; the registers thus maintain state since they would be continuously powered. The second method uses specially designed state-retention registers in the input netlist such that even if the entire design is power gated, state is maintained in the retention registers.

Finally, after these transformations are completed, CoolPower performs a timing analysis to ensure that no critical paths or timing parameters were violated during the transformation process. If any violations are found, CoolPower's timing closure optimizations are run to repair the violations.

These capabilities enable the pre-synthesis design phase to proceed without modification, exactly as it would for non-power-gated circuits, since CoolPower performs all transformations, insertions, and connections needed for virtual grounds, switches, and interface cells.

10.4.2 Analysis

The analysis of current flow is an essential element in the design, optimization, and verification of virtual grounds and switch networks. Not only is it necessary to analyze the current flow, but its accuracy affects the optimization results in terms of both area and electrical integrity.

10.4.2.1 Types of Current Analysis

In early power gating implementations, an average current was computed under the assumption that the ratio of peak current to average current is approximately constant, thus enabling design decisions to be based on average currents, [27] where the computation of the average currents could be based upon average power values determined from simulation traces. While easy to implement, this approach is problematic in that peak currents can deviate substantially from simple multiples of average current due to issues such as clock skew, decoupling capacitance, and package inductance; each of these issues affects the peak current and voltage spikes but does not alter the average values. Thus the use of average currents to size virtual grounds and switches is risky, as the peak value of the dynamic voltage drop may be significantly underestimated.

More recent work has suggested the use of dynamic, or time varying, currents for analyzing and optimizing virtual ground networks. [1] proposed the use of probabilistic analysis to produce *expected* dynamic discharge currents, which are calculated to be the product of the peak discharge current of the cell and the probability of its occurrence. However, this approach also has a very serious issue in that it is not a worst case calculation, and thus presents similar risks to the use of average currents since it cannot deterministically account for worst case switching scenarios wherein multiple cells switch simultaneously.

10.4.2.2   Static timing analysis based current analysis

CoolPower addresses all of these problems by producing a worst case dynamic current waveform. A vectorless static timing analysis (STA) computes the entire set of potential switching events [6] which is subsequently used to compute the dynamic currents. This set of switching events contains all of the potential events, both rising and falling, scheduled in time. This set is then filtered to remove redundant and don't care events, such as those that cannot occur due to modal operation. This filtered set of switching events is used to compute a set of current events, from which a *composite current waveform* is created for each instance. The composite current waveform represents the maximum current consumed, at each point in time, by that particular instance. It includes current consumed by rising and falling output transitions, internal crowbar currents, as well as currents consumed by input only events. Thus, the composite current waveform may have numerous peaks, with each peak occurring at the time at which that cell is scheduled to switch. In this way, neither switching events nor current peaks are neglected as is the case for average current methods or probabilistic dynamic current methods.

It should be noted that min-max timing analyses, which produce a range of switching times, are much too conservative. It is often the case that the range between minimum and maximum is quite lengthy such as for a two-input NAND that has one very early arriving input and one very late arriving input. In our approach we use exact switching times to compute current events, thus avoiding the need to smear the current event from the earliest switching time to the latest switching time as in [1]. Variation is considered by widening the calculated exact switching time to a small range for each event, as opposed to the overly generalized min-max approach.

An example of a composite current waveform for one cell is shown in Figure 10.7. The multiple peaks in the current waveform correspond to different current events that could occur at different times due to different stimuli. For example, events 1 and 3 are output rising events while events 2 and 4 are output falling events. Thus the waveform represents the composite of all the cell's current consuming events.

10.4.3 Optimization

The goals of switch optimization are to minimize area while meeting performance and electrical constraints in reasonable computation time.

10.4.3.1 Optimization Constraints

Switch transistor optimization has previously been treated primarily as that of a sizing problem [1][8] with the proper sizing being determined only by the voltage drop across the switch. The motivation for controlling the voltage drop is to control the impact of the virtual ground on the delay of the switching circuits connected to the virtual ground.

One solution, switch-in-cell [23], involves no switch optimization as the sizing is handled during the design of each individual cell, so no sizing is required during chip assembly. However, for shared switches, block- or chip-level virtual ground currents must be computed in order to properly size the switches. In [27], average currents are used for sizing, while [1][8] employ dynamic currents; [8] sizes the switches based on mutually exclusive switching and then merges the sizes into a single equivalent switch for a global power gating structure, whereas [1] sizes the switches for clustered groups of logic using local power gating.

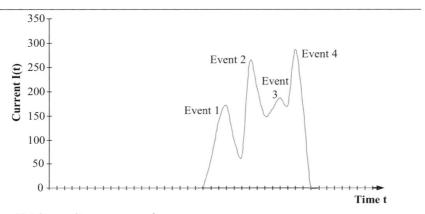

Figure 10.7 Composite current waveform

While sizing is clearly a first order concern, it is not the only issue that optimization should address in the design of robust power gated circuits. In particular, capacitive coupling and electromigration effects can render unreliable an otherwise voltage compliant virtual ground design.

Since the primary goal of optimization is to minimize area, switch resistances are maximized subject to a voltage constraint. This resistance, along with the potentially lengthy route of the virtual ground net, makes the virtual ground subject to substantial coupling capacitance. Thus potential aggressor signals can couple to the virtual ground causing temporary, albeit substantial, voltage bounces.

Electromigration becomes a concern because usually the area occupied by the virtual ground route is minimized by making the route width as narrow as possible. Since a route's electromigration limit is proportional to width, minimizing the width reduces the electromigration limit.

10.4.3.2 Current Scheduling

Our solution for switch optimization has two components, the first of which is *current scheduling* – the waveshaping of virtual ground currents through the assignment of power gated logic instances to particular switches [5]. The goal of current scheduling is to assign logic instances to particular switches such that the currents due to those instances do not cause a voltage violation; that is, we schedule the currents for each switch subject to when the currents occur and how large they are. This is similar to the approach in [8] in that we do the assignments based on when the cells switch. However, by contrast, our implementation utilizes detailed timing information as opposed to unit-delays. Additionally, our approach is not limited to mutually exclusive switching; non-mutually-exclusive switching is allowed subject to a user specified constraint on overlapping current waveforms. This implementation is similar to that of [1], however that approach's use of simulation based activities makes it susceptible to missing a worst case switching scenario. By contrast, our use of STA based composite current waveforms in optimization enables us to avoid that problem and produce switch networks suitable for worst case design.

The assignments of logic instances to switch cells are made using a Bin Packing algorithm in which each bin is filled according to the amount of current consumed by each logic instance per time bin. The assignments are also subject to distance constraints and electromigration constraints. If a logic instance is beyond the specified distance limit for a particular switch, it is not considered as a candidate to be connected to that particular switch. If the logic instance is close enough to the switch, and its currents can be successfully bin packed but in so doing it would violate the specified electromigration limit, then that particular connection is discarded and another connection is evaluated.

10.4.3.3 Switch Sizing

Once the logic instance to switch assignments have been made, the switches are individually sized according to the specified voltage constraint. The most appropriate switch size for each instance is determined by evaluating the various switch sizes available in the library. CoolPower chooses the physically smallest switch that still meets the voltage constraint given the expected current flow established during the prior current scheduling optimization.

In contrast to previous approaches, our sizing is subject to the voltage drop across the virtual ground route in addition to the voltage drop across the switch itself. This is particularly significant given the aforementioned motivation to minimize the width of the virtual ground route. Thus the critical evaluation metric during optimization is not the voltage drop across the switch, but rather the voltage seen by each individual logic instance's virtual ground connection.

The routing of the virtual ground affects not only the voltage seen by the power gated logic cells, but also the total area occupied by the switches – the larger the IR drop along the virtual ground route, the larger the switch must be in order to meet the voltage constraint at the connection to the logic instance. Thus the precise placement of the switches relative to the switching logic is critical as the sizing is dependent upon the placement. To enable correct simultaneous switch sizing and placement, CoolPower can move logic instances to create open space in the desired switch location. However, to prevent timing closure problems due to logic instance movement, Cool-Power performs an internal *trial* timing analysis for each potential move. If the move under consideration would result in a timing or signal integrity violation, the move is discarded, the logic instance is left in its original location, and another instance movement is considered in order to free up placement space. However, if the trial timing analysis indicates that the contemplated instance movement would not adversely affect timing or signal integrity characteristics then the placement change is committed to the internal design database along with the switch insertion. These cell movement capabilities enable CoolPower to not only control the voltage drop on the virtual ground, but also to minimize the area overhead of the switches as the logic cell movement creates placement space for the switches in the precisely desired locations.

The effects of virtual ground route length upon switch sizing is illustrated in Figure 10.8 and Figure 10.9 which plot data for a particular 130nm module optimized by CoolPower.

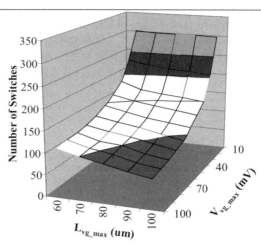

Figure 10.8 Number of inserted switch cells as a function of voltage V_{vg_max} and virtual ground length L_{vg_max} constraints

Figure 10.8 shows the effects of the virtual ground route length and voltage constraints upon the number of switch cells inserted. As can be seen, the longer the route length, the fewer switches are needed – more logic cells can share a single switch than when the route length is constrained to be shorter. Similarly, when the dynamic voltage constraint is larger, fewer switches are needed because more logic cells can share a given switch without violating the voltage constraint.

Figure 10.9 shows similar data, however here total switch cell area is plotted as a function of the virtual ground route length and voltage constraints. The overall shape of the surface is similar to Figure 10.8, but the switch cell area is a weaker function of the route length, because when switches must be placed closer together, less voltage drop builds up along the virtual ground route. Thus each of the switches may be sized smaller even though more overall switches may be required.

10.4.3.4 Optimization in the Design Flow

The optimizations described above are performed both prior to and after routing. Prior to routing, estimated parasitics are used for both timing analysis and power analysis, and the results of those two analyses are in turn used by the current scheduling and sizing routines. The objective of the pre-route optimizations is to not only size the switches, but also to place them effectively, while the objective of the post-route optimizations is to adjust the design to account for any deviation from the pre-route estimated performance.

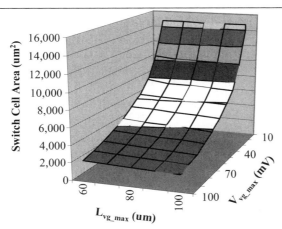

Figure 10.9 Switch cell area vs. voltage V_{vg_max} and virtual ground length L_{vg_max} constraints

10.5 APPLICATION FLOWS

Power gating can generally be implemented in two different applications. The first, Full Power Gating, refers to power gating all the logic elements in a block or design. The second, Selective Power Gating, is used to power gate only portions of a block. Implementation flows using CoolPower for both types of power gating are described in this section. User inputs include the logic to be power gated along with the associated sleep control signal names, type of interface buffer to be used if any, and optimization constraints for maximum dynamic voltage drop and current density. In each case, the flow is completely automated and the switches are sized for worst case operation.

10.5.1 Full Power Gating

Full, or complete, power gating is just what the name implies – all logic instances are connected to power (ground) through a header (footer) switch. The implementation of full power gating suggests that state is not retained unless a particular mechanism, such as the use of state retention registers, is employed to save state.

The design flow to generate a fully power gated chip is shown in Figure 10.10. This flow is notable in that the up-front design of the logic generally need not consider any implications of the physical level power gating.

The flow begins with the design of the register transfer level (RTL) code in the conventional manner; that is, no special considerations need be given to the fact that power gating circuitry will be inserted during a later step. One possible exception here is the sleep control signal – if the synthesized RTL code does not contain an explicit sleep control signal, then the RTL code must be modified such that the resulting synthesized netlist contains one.

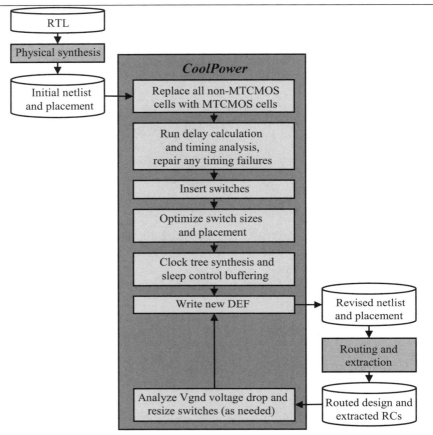

Figure 10.10 Full power gating design flow

As illustrated in Figure 10.10, CoolPower loads the initial netlist and placement and, as a first step, replaces all non-power gated logic instances with power gated versions. Next, unsized switches are inserted into the netlist and floorplan followed by an optimization step, during which the switches are clustered with associated logic and sized according to the algorithms described above in the current scheduling and sizing sections. The sizing optimization considers the availability of switch placement locations; if the desired switch size cannot be placed in the desired location, logic cells are moved in order to free up available space. At this point all the switch sizes and locations are known. The next operation buffers the fanout tree for the sleep signal, since the aggregate switch cell input capacitance can be significant. Clock tree synthesis is also performed at this time, after which a new DEF file is produced containing the modified placement including the inserted switches and interface cells.

The design is now routed, including signal, real supply, and virtual supply routing. After routing and subsequent parasitic extraction, the design is analyzed post-route to verify the electrical characteristics (voltage drop and electromigration), similar to post-route timing verification. If any violations are found, then post-route sizing operations can be employed to repair the violations. An ECO route would be needed to route any changes introduced by the post-route repair operations, however the number of changes introduced at this step is minimal requiring few reroutes and no floor plan changes.

10.5.2   Selective Power Gating

In selective power gating only the selected instances are power gated and the un-selected logic remains ungated. While conceptually any portion of a design could be power gated while not gating the other portions, in practice selective power gating specifically refers to gating only those logic elements that are implemented with low-V_{th} logic cells. Thus, selective power gating is an extended version of the multi-V_{th} cell swapping technique described above [12][23] – logic on non-critical paths utilize high-V_{th} cells while the low-V_{th} cells on the critical paths are replaced with power gated cells.

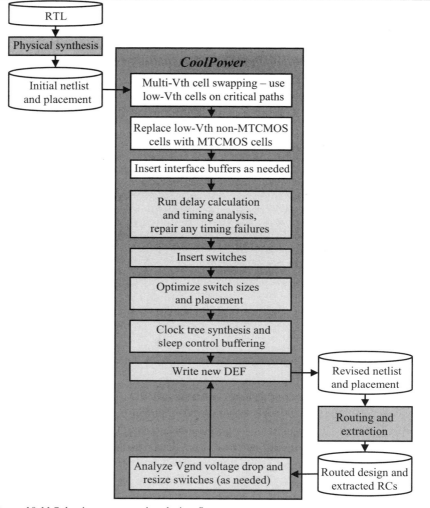

Figure 10.11 Selective power gating design flow

The Selective Power Gating flow is shown in Figure 10.11. This flow is a modification of the full power gating flow shown in Figure 10.10. More specifically, there are a couple of additional steps in the selective flow during which the low-V_{th} logic instances are identified for power-gating.

10.6   RESULTS

CoolPower has been used to implement and optimize both full and selective power gating in several different designs. Some results of these efforts are briefly described below.

Table 10.1 Full power gating results

Parameter	Design		
	A	D	E
Process technology	90nm	130nm	90nm
Supply voltage	1.5V	1.5V	1.2V
Logic function	32-bit ALU	8-bit datapath	multi-processor
Retain state in registers?	yes	yes	no
# of instances	1,852	118	182,225
# of power-gated logic instances	1,388	80	181,809
# of switch instances	104	3	15,872
# of interface instances	206	10	0
Logic cell to switch cell ratio	13.3	26.7	11.5
Power-gated logic cell area (um^2)	15,259	886	1,457,391
Switch cell area (um^2)	2,565	114	136,545
Switch area overhead (%)	16.8%	12.9%	9.4%
Interface cell area (um^2)	791	38	0
Interface cell area overhead (%)	5.2%	4.3%	0.0%
Original bounding-box area (um^2)	977,725	3,483	22,156,698
New bounding-box area (um^2)	977,725	3,483	22,156,698
Bounding-box area increase (%)	0.0%	0.0%	0.0%

10.6.1 Full Power Gating

Full power gating was implemented and optimized by CoolPower using footer switches in several different blocks of varying sizes at both 130nm and 90nm. The results of these implementations are shown in Table 10.1.

For designs A and D, all combinational logic was power gated; state retention was implemented by *not* power gating any of the registers, which necessitated the insertion of interface cells. However, for design E, all logic elements, including registers, were power gated; state was retained in on-chip memories which were left continuously powered. In all cases, the maximum virtual ground dynamic voltage target was 100mV.

The area overhead of the inserted power gating switches varied from 9% to 17%, with interface cells adding another 4% to 5% where utilized. The latter area could be reduced significantly by adding registers to the standard cell library that include the interface structure within the register cell. Nevertheless, CoolPower in all cases was able to physically insert all of these structures, using the cell movement facilities to precisely position logic cells and switches without increasing the bounding box of the design, thus achieving zero net area overhead.

Table 10.2 Selective power gating results

Parameter	Design			
	A	B	C	E
Process technology	90nm	90nm	90nm	90nm
Supply voltage	1.5V	1.5V	1.5V	1.2V
Logic function	32 bit ALU	32 bit DSP	32 bit DSP	multi-processor
Retain state in registers?	yes	yes	yes	yes
# of instances	1,808	148,879	226,259	182,225
# of power-gated logic instances	359	14,418	55,479	19,639
# of switch instances	55	1,005	2,057	4,060
# of interface instances	206	9,213	29,140	12,259
Logic cell to switch cell ratio	6.5	14.3	27.0	4.8
Power-gated logic cell area (um^2)	6,136	248,173	218,846	143,563
Switch cell area (um^2)	1,192	46,954	23,303	17,923
Switch area overhead (%)	19.4%	18.9%	10.6%	12.5%
Interface cell area (um^2)	791	35,378	54,820	43,249
Interface cell area overhead (%)	12.9%	14.3%	25.0%	30.1%
Original bounding-box area (um^2)	977,725	5,651,221	34,552,882	22,156,698
New bounding-box area (um^2)	977,725	5,651,221	34,552,882	22,156,698
Bounding-box area increase (%)	0.0%	0.0%	0.0%	0.0%

10.6.2 Selective Power Gating

Selective power gating was implemented and optimized by CoolPower using footer switches in several different 90nm designs. The results for these designs are tabulated in Table 10.2. Designs A and E are the same designs used for full power gating in the preceding section. As above, the maximum virtual ground dynamic voltage target was 100mV.

In each of these cases, selective power gating was implemented by first performing a multi-V_{th} optimization and then replacing the low-V_{th} instances with power gated instances. Interface cells were inserted on all nets driven by power gated instances and received by non-power gated logic.

The area occupied by the switch cells varies from 11% to 19%; this variance is a function primarily of two factors. First, the number of power gated instances depends upon how many low-V_{th} instances were needed to meet timing in the multi-V_{th} optimized design, which in turn is determined by how much performance margin existed in the original design. Second, the number of switch instances is determined in large part by the degree of sharing achieved by the current scheduling algorithm; sharing is affected by the logic function and connectivity, the geographic distribution of the power gated logic instances, and the voltage constraint used for the optimization. For example, compared to the full power gating results, the area overhead for selective power gating is generally larger – with full power gating the geographic clustering is usually more effective since all logic instances in the physical vicinity will be power gated leading to greater degree of sharing. Nevertheless, in all cases, the final area overhead was again zero since the logic cells were moved during the optimization process so as to insert the switch cells in the precisely desired positions. This capability, in the end, enabled CoolPower to avoid bloating the bounding box of the placed logic.

10.6.3 Performance

As described earlier, one of the issues with power gating is understanding and managing the performance impact of the virtual grounds. In order to determine the specific effects on a particular design, we studied the impact of switch sizing upon performance using a logic block power gated by Cool-Power.

In this case, we optimized the D block using the full power gating flow described above for a particular virtual ground voltage constraint. The design was then routed after which the critical path was extracted and simulated using HSPICE, for both rising and falling transitions under worst case conditions (slow-slow process, 1.4V, 100°C). This simulation included the effects of the switches and virtual ground parasitics. The switches were then resized for different voltage constraints and the resulting circuits were reanalyzed. The data from this study is presented below in Table 10.3. In this table, the Avg Vssv voltage and Max Vssv voltage columns reflect the average and maximum measured Vssv values, respectively. The top table reflects the performance of a rising edge propagating through the critical path, the bottom table a falling edge. Each row corresponds to a different set of switch sizings, however the sizings are identical for the two tables.

As can be expected, the delay along the path increases as the maximum virtual ground voltage target is relaxed, although the magnitude of the delay increase is relatively small. Here the maximum virtual ground voltage reflects an instantaneous or dynamic effect, implying that not all of the cells along the path "see" or experience that maximum voltage when they switch. This effect can be confirmed by considering the relationship between the virtual ground average and maximum dynamic voltages. In this case the maximum dynamic voltage is six to eight times larger than the average voltage. This indicates that instances switching at times other than that of the peak voltage must experience a much lower dynamic voltage.

The falling edge path delay increases more than the rising edge path delay, as expected. The falling edges in the path are directly affected by the virtual ground since the increased virtual ground voltage reduces the effective gate-to-source driving voltage on the n-channel pull-downs. However, the output rising edges of the power gated instances in the path are less affected since the virtual ground is not directly connected to the p-channel devices supplying the charging currents.

Table 10.3 Power gating delay effects

Design	Rising Edge			
	Avg Vssv voltage (mV)	Max Vssv voltage (mV)	Critical path (ns)	Delay change (%)
Non-Power Gated	n/a	0.0	0.994	n/a
Power Gated	2.89	22.9	1.003	0.92%
	5.88	42.0	1.012	1.81%
	11.81	74.8	1.020	2.67%
	15.51	91.7	1.034	4.00%

Design	Falling Edge			
	Avg Vssv voltage (mV)	Max Vssv voltage (mV)	Critical path (ns)	Delay change (%)
Non-Power Gated	n/a	0.0	1.073	n/a
Power Gated	4.55	22.7	1.087	1.33%
	9.19	41.1	1.106	3.04%
	18.86	74.5	1.137	5.97%
	25.06	91.3	1.148	6.96%

Table 10.4 Leakage reduction results

Design	Max Vssv voltage (mV)	Sleep Mode Leakage (uA)	Leakage reduction factor (X)
Non-Power Gated	0.0	2.170	n/a
Power Gated	22.9	0.035	62
	42.0	0.021	102
	74.8	0.011	195
	91.7	0.009	235

Leakage currents were also measured for each of the sizing sets used in Table 10.3. This data is presented in Table 10.4. It illustrates the tradeoff between virtual ground voltages and leakage reduction – the larger the allowed peak voltage on the virtual ground, the smaller the standby leakage current. Figure 10.12 overlays the delay results from Table 10.3 with the leakage results from Table 10.4 enabling us to see that small increases in delay are accompanied by large reductions in standby leakage. In round numbers, a two orders of magnitude reduction in leakage currents can be achieved with only a 3% delay push out due to virtual ground effects. Thus we can clearly see that the central design issue for the leakage reduction – area overhead – performance impact tradeoff is the careful sizing of the switch cells (the Vssv peak voltage acting as proxy for the switch sizing).

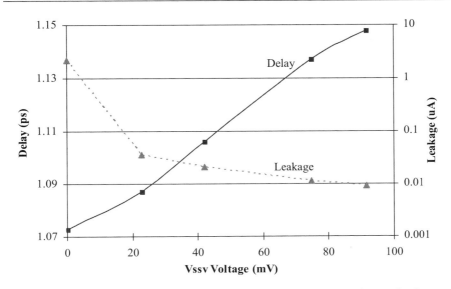

Figure 10.12 Critical path delay and standby leakage as a function of virtual ground voltage

10.7 FUTURE WORK

While power gating as a leakage control technique has been known for several years [18], few production designs have utilized it due to numerous design issues. This situation will change with the growing need to control the larger leakage currents in advanced technologies. Additionally, the development of power gating specific design tools such as CoolPower will reduce the difficulty and time required to implement power gating, thus enhancing its adoption.

Nevertheless, both the complexity of power gating implementations and associated design automation will increase. For example, to reduce leakage currents during a chip's active mode, dynamic fine grained power gating will be deployed wherein many relatively small blocks will be power gated independently of each other, much in the same manner as clock gating is implemented today. This will require more complex control logic as well as more attention to the transient characteristics of turning the switches on and off. In particular, the rush currents that flow when a switch is closed will need to be carefully controlled so as to induce only a minimal voltage bounce on the real and virtual rails. And, in a quest for even greater levels of leakage reduction, power gating will be combined more often with other techniques, such as RBB and DVS, although RBB is unlikely to be used with extreme fine grained power gating due to the mismatch in recovery times (RBB requires a much longer wake up time than power gating [22]).

As the logical and physical design issues become automated, more attention will be paid to run-time and compile-time software control of the switches. In the former case, the switches' control logic will be designed for operating system control much in the way that high level clock gating is controlled today. Additionally, for programmable applications, compilers will optimize not only for execution speed and code density, but also for the length of time the power gated logic blocks can be kept continuously inactive so as to maximize the amount of leakage reduction.

10.8   SUMMARY

Leakage has become one of the most critical challenges facing integrated circuit designers and threatens to become even more so. Since part of the challenge is that advanced processes exacerbate leakage instead of mitigating it (as occurred in the past for other issues), the leakage challenge must be addressed largely in the design domain.

MTCMOS power gating has emerged as an effective design technique for controlling leakage, however it has not yet been widely deployed due to a variety of unique design issues and a lack of effective design automation.

We have presented in this chapter an answer for those issues – CoolPower, a fully automated solution for the efficient implementation of MTCMOS power gated circuits. We have outlined a number of critical design issues, such as switch sizing for worst case design and sleep domain interfacing, and described how CoolPower addresses those issues automatically. Finally, we presented detailed results of CoolPower's automation demonstrating its viability and effectiveness.

10.9   REFERENCES

[1] Anis, M., Areibi, S., and Elmasry, M., "Design and optimization of multithreshold (MTCMOS) Circuits," *IEEE Transactions on Computer-Aided Design of Integrated Circuits and Systems*, vol. 22, no. 10, October 2002, pp. 1324-1342.

[2] von Arnim, K., et. al., "Efficiency of body biasing in 90-nm CMOS for low-power digital circuits," *IEEE Journal of Solid State Circuits*, vol. 40, no. 7, July 2005, pp. 1549-1556.

[3] Calhoun, B., et. al., "Power gating and dynamic voltage scaling," *Leakage in Nanometer CMOS Technologies*, S. Narendra and A. Chandraksan, editors, Springer, 2005.

[4] Choi, K., Xu, Y., and Sakurai, T., "Optimal zigzag (OZ): an effective yet feasible power-gating scheme achieving two orders of magnitude lower standby leakage," *proceedings of the Symposium on VLSI Circuits*, 2005, pp. 312-315.

[5] Frenkil, J., "Current scheduling system and method for optimizing multi-threshold CMOS designs," U. S. Patent No. 7117457, Oct. 3, 2006.

[6] Frenkil, J., "Vectorless instantaneous current estimation," U. S. Patent No. 6807660, Oct. 19, 2004.

[7] Kao, J., Miyazaki, M., and Chandrakasan, A., "A 175-mV multiply-accumulate unit using an adaptive supply voltage and body bias architecture," *IEEE Journal of Solid State Circuits*, vol. 37, no. 11, November 2002, pp. 1545-1554.

[8] Kao, J., Narendra, S., Chandrakasan, A., "MTCMOS hierarchical sizing based on mutual exclusive discharge patterns," *proceedings of the Design Automation Conference*, 1998, pp. 495-500.

[9] Kawaguchi, H., Nose, K., and Sakurai, T., "A super cut-off CMOS (SCCMOS) scheme for 0.5-V supply voltage with picoampere stand-by current," *IEEE Journal of Solid State Circuits*, vol. 35, no. 10, October 2000, pp. 1498-1501.

[10] Keshavarzi, A., et. al., "Effectiveness of reverse body bias for leakage control in scaled dual Vt CMOS ICs," *proceedings of the International Symposium on Low Power Electronics and Design*, 2001, pp. 207-212.

[11] Keshavarzi, A., Roy, K., and Hawkins, C., "Intrinsic leakage in low power deep submicron CMOS ICs," *International Test Conference Proceedings*, 1997, pp. 146-155.

[12] Kitahara, T., et. al., "Area-efficient Selective Multi-Threshold CMOS Design Methodology for Standby Leakage Power Reduction," *proceedings of the Design Automation and Test in Europe Conference*, 2005, pp. 646-647.

[13] Kosonocky, S., et. al., "Enhanced multi-threshold (MTCMOS) circuits using variable well bias," *proceedings of the International Symposium on Low Power Electronics and Design*, 2001, pp. 165-169.

[14] Lackey, D., et. al., "Managing Power and Performance for System-on-Chip Designs using Voltage Islands," *proceedings of the International Conference on Computer-Aided Design*, 2002, pp. 192-202.

[15] Liu, W., et. al., "BSIM3v3.2.2 MOSFET Model User's Manual," Department of Electrical Engineering and Computer Sciences, University of California at Berkeley, technical report no. UCB/ERL M99/18.

[16] Martin, S., et al., "Combined dynamic voltage scaling and adaptive body biasing for lower power microprocessors under dynamic workloads," *proceedings of the International Conference on Computer-Aided Design*, 2002, pp. 721-725.

[17] Min, K., Kawaguchi, H., and Sakurai, T., "Zigzag super cut-off CMOS (ZSCCMOS) block activation with self-adaptive voltage level controller: an alternative to clock-gating scheme in leakage dominant era," *proceedings of the International Solid State Circuits Conference*, 2003, pp. 400-401.

[18] Mutoh, S., et. al., "1-V power supply high-speed digital circuit technology with multi-threshold-voltage CMOS," *IEEE Journal of Solid State Circuits*, vol. 30, no. 8, August 1995, pp. 847-853.

[19] Semiconductor Industry Association, *The International Technology Roadmap for Semiconductors*, 2003.

[20] Sirichotiyakul, S., et al., "Stand-by power minimization through simultaneous threshold voltage selection and circuit sizing," *proceedings of the Design Automation Conference*, 1999, pp. 436-441.

[21] Taiwan Semiconductor Manufacturing Company, "Fine Grain MTCMOS Design Methodology," *TSMC Reference Flow Release 6.0*, 2005.

[22] Tschanz, J., et. al., "Dynamic sleep transistor and body bias for active leakage power control of microprocessors," *IEEE Journal of Solid State Circuits*, vol. 38, no. 11, November 2003, pp. 1838-1845.

[23] Usami, K., et. al., "Automated selective multi-threshold design for ultra-low standby applications," *proceedings of the International Symposium on Low Power Electronics and Design*, 2002, pp. 202-206.

[24] Uvieghara, G., et al., "A highly-integrated 3G CDMA2000 1X cellular baseband chip with GSM/AMPS/GPS/Bluetooth/multimedia capabilities and ZIF RF support," *proceedings of the International Solid State Circuits Conference*, 2004, pp. 422-423.

[25] Wang, Q., and Vrudhula, S., "Algorithms for minimizing standby power in deep submicrometer, dual-Vt CMOS circuits," *IEEE Transactions on Computer-Aided Design of Integrated Circuits and Systems*, vol. 21, no. 3, March 2002, pp 306-318.

[26] Wei, L., et al., "Design and optimization of low voltage high performance dual threshold CMOS circuits," *proceedings of the Design Automation Conference*, 1998, pp. 489-494.

[27] Won, H., et al., "An MTMCOS design methodology and its application to mobile computing," *proceedings of the International Symposium on Low Power Electronics and Design*, 2003, pp. 110-115.

Chapter 11

VERIFICATION FOR MULTIPLE SUPPLY VOLTAGE DESIGNS

Barry Pangrle, Srikanth Jadcherla
ArchPro Design Automation, Inc.
San Jose, CA 95124, USA

11.1 INTRODUCTION

Power management is an increasingly important aspect of system and integrated circuit (IC) design. As designers are learning more about power management techniques and incorporating them into their designs, they are left wondering how to verify that these new energy saving strategies haven't created a flaw in the final implementation. Designs, which previously appeared functionally correct based upon assumptions that voltage levels were held constant across the logic portion of the design, may break when voltages in the design vary or during voltage shutdown or wakeup. New tools are needed to verify designs where voltages are no longer constant.

The current state of electronic design automation (EDA) tools is largely built upon an underlying logical representation. The typical register transfer level (RTL) to GDSII flow starts with an RTL description. This RTL is processed by a *logic* synthesis tool that produces a *logical* netlist that is then simulated by a *logic* simulator. This flow has worked across a number of CMOS generations based on the previously mentioned assumption that the voltages were held constant. As designers incorporate new voltage varying power management techniques in their designs, today's tools fall short of meeting the challenge to adequately represent and handle the impact of designed-in voltage variance.

The verification of a multi-voltage system on a chip (SOC) is analogous to a multi-level video game. Level 1 is just getting through the flow and making sure that all the islands and their interconnections are properly handled. Diagnostics written at this level ensure that each chip function is supplied with the appropriate voltage rails and that the correct protection devices are in place. In a sense, this is a verification of spatial partitioning.

Of course, this is not possible if you cannot express your voltage based partitions (islands) at the electronic system level (ESL), register transfer level (RTL), or gate-level netlist description.

The second level entails ensuring that the whole sequencing of power management happens as desired. This is the verification of the temporal variation in voltage rails. It must be verified that the SOC can circulate through all the desired states and transitions. It must also be verified that the SOC does not get stuck in deadlock, transition into any illegal states, or perform a disallowed transition between legal states. This is indeed quite complicated. SOC designers need a lot of assistance from software/driver authors to come up with these vectors and vice versa. Typically, the SOC itself has finite state machines (FSMs) that induce transitions between states. It also has logic to monitor power, performance, temperature, or other metrics that feed into this FSM. Often these signals come from domains that are shut down leading to a "chicken and egg" situation. This is what makes verifying the sequence of state transitions so complicated. Sequences also need to be verified for the handshake between devices such as the SOC, voltage regulator module (VRM), battery, temperature sensor and software.

Here is a simple rule to follow: All voltage related events must be visible to software and/or system electronics. Spatial elements (groups of cells) in a chip are resources to accomplish functions. Voltage control alters the availability, performance and load characteristics of these functions. Thus, it is not possible to do any meaningful voltage based control without coordinating with at least the software and the VRM. Hence the need to carefully step through power management states in a coordinated manner across devices and software.

The third level involves ensuring that the desired power/energy savings are happening and better yet, that they are cost effective. Power management neither comes easily nor cheaply. Power management schemes typically impose new costs on layout area, performance and design effort. You may have to choose between multiple schemes. Table 11.1 presents a simple example of the flavor of some of the tradeoffs. After initial design estimates, these trade-offs need to tracked and updated as a chip is designed.

Some of these costs may be surprising. The package cost of a five island AVS chip may rise due to the complicated routing of multiple rails and the steps needed to put in the necessary decoupling etc. This is highly case sensitive, which means that the IC design team must keep an eye on the impact of power management on schedule and cost overall. You may also be surprised by five island AVS possibly having very little battery life benefit over three island AVS, for a given class of applications.

True verification of a power management system achieves all of these levels, as neglecting any level can cause a product to fail. Designers need to do this at every design step: There are so many tools and scripts that modify the design. Any of these can break a multi-voltage design at some level.

Table 11.1 A simple example of the relative power savings, design time, and cost trade-offs for different power schemes. Adaptive voltage scaling (AVS) refers to when island voltages are dynamically adjusted to meet performance requirements.

	Single Voltage	3 Island AVS	5 Island AVS
Average power	1.0	0.6	0.4
Package cost	1.0	0.6	0.7
Chip design time (man months)	1.0	1.5	3.0
Voltage regulator modules	1×	2×	6×
Heat sink	Yes	Yes, lower cost	No
Fan with fan driver IC	Yes	No	No
Die cost	1.0	?	?
Test Cost	1.0	3.0	5.0
Debug cycle	1.0	3.0	5.0
Software development time (man months)	1.0	1.5	3.0

For example after detailed routing, you might find that one or more islands have no space for decoupling capacitance. This will mean either an accommodation in the package or an external decoupling capacitor, both of which increase the cost. On the other hand, you may also decide to revisit the class of applications for which the decoupling capacitance is calculated, or even choose to transition voltages on the affected islands in a slower manner to reduce the decoupling capacitance needed.

This chapter examines using multiple voltages and the impact it has on the functionality and correctness of a design from architecture to implementation. Logical netlists in EDA tools have long represented the logical connectivity of signals between gates with the underlying assumption being that supply voltage V_{DD} and ground voltage V_{SS} were non-varying and "always on". Section 11.2 discusses different types of voltage techniques used to improve the energy efficiency of designs. Section 11.3 contains examples to illustrate several important multiple voltage issues and ways to address these issues. We conclude with a summary in Section 11.4.

11.2   MULTIPLE VOLTAGE DEFINITIONS AND SCENARIOS

One aim of power management is to increase the energy efficiency for a given design. Energy within an implementation is typically consumed by the charging and discharging of load capacitances and by leakage paths that cause unwanted electrical current to flow. The power P due to the charging and discharging of a capacitance C is

$$P = \frac{1}{2}\alpha f C (V_{DD} - V_{SS})^2 \qquad (11.1)$$

where α is the switching activity factor, f is the clock frequency, and $(V_{DD} - V_{SS})$ is the voltage across the capacitor when it is charged.

Since the voltage term is squared, it is a popular target for reducing dynamic power. Studies have shown that the operating frequency for a given design scales approximately linearly with the voltage [8]. Therefore, if the voltage is scaled to match the operating frequency, there is roughly a V_{DD}^3 impact on the switching power (which can be derived from Equation (11.1), assuming $V_{SS} = 0V$). This relationship has led designers to partition their designs into separate areas that run at the lowest necessary voltage to ensure proper operation of the design. This section will look at design partitioning to reduce power and establish a set of common terminology.

11.2.1 Voltage Domains and Islands

Traditional designs are greatly simplified by using a single voltage level to represent a logical "1" value. In fact, if it weren't for the increasing demand for energy efficient designs it is doubtful that any designer would opt to so increase the complexity of a design by using multiple voltages. Most of the complexity of using multiple voltages shows up on the "boundaries". The question is, on the boundaries of what? In this point, we need to be clear in our terminology. The boundaries that we will concentrate our attention on are "domains" and "islands".

Islands are defined as a set of cells, or a group of HDL (hardware description language) or ESL statements with common rail connections. These rail connections consist of supply voltages V_{DD} {1..n}; ground voltage V_{SS}; auxiliary supply rail voltage for sequential elements V_{RET}; voltage to footer sleep transistors SLP_N; voltage to header sleep transistors SLP_P; NMOS body bias voltage V_{BBN}; and PMOS body bias voltage V_{BBP}. This is illustrated in Figure 11.1.

Domains are defined by the driving voltage, V_{DD}, that defines a logical "1" level. Note it is possible to have multiple "islands" within a "domain". Many of the checks described later in this chapter rely on the classification of islands and domains.

Wires that connect between two different islands are referred to as *cross-overs* because the signal "crosses over" an island boundary. It is important that all wires are taken into consideration in this regard. It is not only signal wires communicating between blocks that need attention but also clocks, scan chains and other wires that may only be used in special modes (e.g. reset). The differences between the two islands will determine the necessary action that needs to be taken to ensure correct behavior.

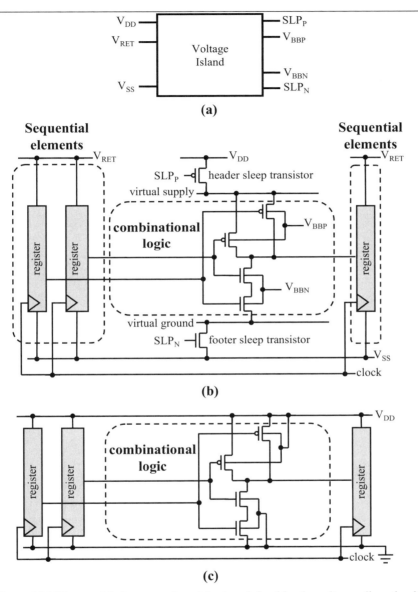

Figure 11.1 Diagram (a) shows a voltage island as defined by the voltage rails and voltage signals that go to it; (b) illustrates this in more detail using a simple example of a NAND2 gate with input and output registers. In contrast, diagram (c) shows the voltage rails that are used in typical static CMOS in ASICs – NMOS wells are tied to ground (i.e. $V_{BBN} = V_{SS} = 0V$); PMOS wells are tied to V_{DD} ($V_{BBP} = V_{DD}$); there are no sleep transistors for power gating; and there is no need for a separate retention voltage for the sequential elements as V_{DD} is not power gated.

11.2.2   Level Shifting

A signal may originate in one domain and then crossover into another domain with a different V_{DD} level. There are two possibilities to consider: 1) the source V_{DD} is greater than the destination V_{DD}; or 2) the destination V_{DD} is greater than the source V_{DD}.

In the first case, the signal may "overdrive" the input at the receiving end. For example on an inverter at the receiving end, this would typically cause the output fall times to decrease and the output rise times to increase (a common mistake is the assumption that both would decrease). This may be acceptable and the designer may choose not to insert any level shifting on this signal. If the design libraries have been characterized for this type of operating condition (i.e. characterized for $V_{DD,in} > V_{DD,gate}$, as well as $V_{DD,in} = V_{DD,gate}$), it is possible to handle these timing conditions during synthesis and the rest of the design implementation.

In the second case, more attention is necessary. If V_{DD} from the source (V_{DDL}) is significantly lower than V_{DD} in the destination domain (V_{DDH}), a V_{DDL} "1" signal will forward bias the PMOS transistors it connects to in the V_{DDH} domain, increasing subthreshold leakage current ($V_{DDL} - V_{DDH} > V_{th,p}$, the PMOS transistor threshold voltage which is negative) or possibly leaving the transistor on ($V_{DDL} - V_{DDH} \leq V_{th,p}$) and causing even more substantial short circuit current. Secondly, the noise margin on the V_{DDL} signal is reduced, and it may not be able to drive the input strong enough to produce a valid output signal (V_{DDL} less than the transition threshold).

An example of a signal from a lower voltage domain driving a gate with a higher supply voltage is shown in Figure 11.2. Domain 1 uses a V_{DD} of 0.65V, while Domain 2 uses a V_{DD} of 1.3V. In this case it is very likely that the output of Domain 1 isn't strong enough to sufficiently drive gates in Domain 2. As the figure shows, this can also lead to internal short circuit paths being created in the gates in Domain 2.

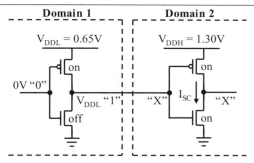

Figure 11.2 This circuit shows a low supply voltage V_{DDL} signal driving a high supply voltage V_{DDH} gate. The PMOS transistor of the V_{DDH} gate is forward biased sufficiently that it is on, resulting in short circuit current I_{SC}. The V_{DDL} "1" signal in the low supply voltage domain results in unknown "X" values in the high supply voltage domain.

In cases where the source domain won't create output signals with sufficient strength to properly drive the destination domain, or to reduce leakage of gates driven by lower voltage, it is necessary to insert voltage level shifters to properly drive the signal from the source to the destination. Voltage level shifters are essentially buffers that pass the same logical signal from input to output but scale the output to the necessary voltage for the logical signal in the destination domain. Designers can define a level shifter insertion threshold based on a percentage of the destination V_{DD} to indicate when it is desirable to insert a level shifter into the destination. For example, if the source V_{DD} is less than 0.8 of the destination V_{DD}, a level shifter would be inserted. The threshold actually chosen would depend on desired circuit performance and the operating conditions characterized for the libraries.

A possible third case also exists when the source and destination islands have independently varying V_{DD} levels. In this case, it may at times be necessary to up level shift the voltage when the source V_{DD} is lower than the destination V_{DD} and down level shift when the source V_{DD} is higher than the destination V_{DD}. In these cases it is necessary to insert an up/down level shifter on each crossover signal to ensure the desired operating characteristics.

11.2.3 Isolation (Shutdown/Sleep)

Implementing domains with different voltages is a powerful technique for reducing energy consumption in active portions of the design. Reducing the voltage also reduces the leakage currents, but to a lesser extent than the savings to dynamic energy consumption. A significant penalty with leakage is that it occurs whether the circuitry is actively switching to perform useful work or just sitting idle. It can be viewed as wasted energy overhead that doesn't significantly contribute to the work performed for useful computation.

A way to significantly reduce leakage is to remove the circuitry from its power source. While this isn't very helpful while the circuit is doing useful computation, it can be extremely beneficial when a block of logic isn't being actively used. However, removing a block from its power source raises some serious issues about how the design will continue to operate properly. The first concern is whether the state of the block needs to be retained and if so, how to accomplish it. When a block is powered down there are three options for handling the state values: 1) scan the state out and store it in a memory external to the block that will remained powered; 2) store the state locally with special circuitry that will remained powered; or 3) if the state isn't needed, throw it away and reinitialize if necessary when the block is powered back up again. Each option has its own set of advantages and disadvantages.

In the case of scanning out the state, it may be possible to reuse the test scan chains for accessing the state and restoring its value later. This may imply a small overhead in the control circuitry to implement this solution. A disadvantage of this approach is the extra time needed to scan the state out and then back in again. Extra energy is also spent to scan the state out and back. During this period, the block needs to retain power. In the case of scanning the state back in, it may delay the start of useful activity. The total cost of the solution may also need to include additional memory to store the state if the memory is otherwise not available.

The second option entails the use of special registers to locally retain the state of the block. This has the advantage of keeping the state closely associated with its circuitry thus making it quicker to save and restore the state. Blocks using this type of implementation may have more power down opportunities than blocks using a scan approach. The downside of this approach is that the retention registers are significantly larger and higher power than non-retention based registers, and it may also be necessary to route a separate power source to the retention portion of the registers.

The third option is the simplest and cheapest to implement. For some designs, every time a block is powered up it is initialized to the same state, making it unnecessary to store the previous state of the block.

Another consideration when shutting down or sleeping blocks is the values that will be placed on the signals that are sourced from the block being shutdown. If the values are allowed to merely float as the turned off circuitry reaches some equilibrium point, there could be serious energy consequences at the destination blocks. As was shown in Figure 11.2 for the case of inserting level shifters between different voltage domains, it is possible to create short circuit paths at the receiving end if appropriate logic levels are not maintained on crossover signals. For blocks that are shutdown, placing isolation cells at the outputs ensures that the crossover signals that are sourced from the shutdown block will maintain valid logic values and not create energy robbing short circuits paths at the destinations. The isolation cells are of 3 basic types: 1) isolate to "0", 2) isolate to "1", or 3) retain the last value before shutdown. The signal to provide isolation can be either active low or active high depending on the application. Some ASIC libraries may contain cells to perform these functions while others need to have them created and incorporated.

Isolation cells can be placed at the destination side with the appropriate control logic. If the isolation is to occur at the output of the source block then it is necessary to route the appropriate power to ensure that the isolation cell itself doesn't get powered off with the rest of the source block.

Shutting down the power to the cells of a block can occur internally or externally to the block. If the power gating switches (sleep transistors in Figure 11.1(b)) are placed in the block, then shutting off the power may be referred to as "sleeping" the block. In this case, there is a signal explicitly sent to the block that will control whether power is available. Note that sleep signals are part of the island definition. For instance, two blocks that operate at the same V_{DD} in the same domain, but have different controlling sleep signals, will form two separate islands both within the same domain. The sleep signal will control power gating switches that interrupt the flow of current either at the V_{DD} rail, V_{SS} rail or both. Figure 11.3(a) show an example of the power gating being controlled from the V_{DD} rail. In this case, the power gating switch is referred to as a "header". Likewise, Figure 11.3(b) shows a case where the power gating switch is on the V_{SS} rail and is referred to as a "footer" switch. Generally, either a header or a footer sleep transistor is required to cut off the leakage current path from V_{DD} to V_{SS}.

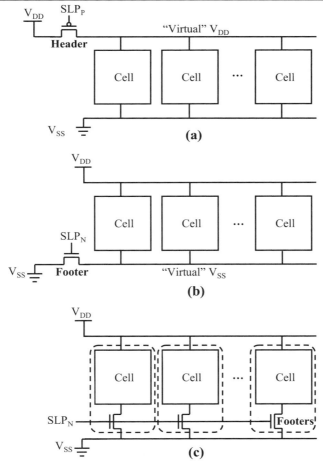

Figure 11.3 Different power-gating configurations are shown here. In (a) there is a shared header sleep transistor; in (b) there is a shared footer transistor; and in (c) each standard cell has a separate footer within it.

There is some voltage drop across the sleep transistor when it is on, so there is a trade-off between the area penalty and power up penalty for larger sleep transistors versus a larger voltage drop across a smaller sleep transistor. In some designs, a header and footer may both be used simultaneously with opposite polarity applied to the controlling sleep signals for the headers and footers respectively. In many layouts, the power gating switches are placed so that the rails that they control take the place of the usual V_{DD} or V_{SS} rails. In this case, the controlled rails are referred to as "virtual" rails.

It is also possible to include the header or footer transistor directly into each cell as shown in Figure 11.3(c). An advantage to such an approach is that it simplifies the analysis and implementation for the design. The effects of the added transistor can now be incorporated directly into each cell's characterization. Any additional delays incurred due to the header or footer transistors show up explicitly in the cell library's timing tables. It also simplifies the question of where and how often to place the power gating switches since they are now distributed directly into the cells. One downside to this approach is the increased area penalty that is incurred by the cells in the library to accommodate the extra transistor and cell input.

A variation on this approach to gain back some performance is to add an additional voltage to the design [7]. Each cell incorporates a footer transistor that is a high performance low threshold voltage transistor. In order to reduce leakage, the cells are designed to have a voltage lower than V_{SS} on the footer gate. This allows better performance and lower leakage with the additional cost of design complexity to provide a voltage lower than V_{SS} to turn off the footers in each cell.

On the other hand, using power gating switches (header or footer transistors) external to the cells allows the use of existing libraries. The overhead for the power gating transistors is typically less than that for the in-cell approach and the usual V_{DD} and V_{SS} connections to the cells are used. The downside is that analysis must be performed to ensure that the IR-drop on the virtual rails is within expected tolerances, otherwise the library has to be re-characterized for the reduced voltage swing. It may be necessary to place more power gating cells to guarantee that the IR-drop is adequately limited.

11.3   DESIGN EXAMPLES

This section describes increasingly complex design examples that demonstrate proper checking for isolation and level shifting as well as dynamic simulations that show the impacts of varying the voltage under different implementations for the following designs.

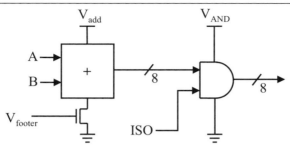

Figure 11.4 Two different domains are used for the adder and isolation. The isolation is contained in the receiving domain (V_{AND}).

To demonstrate these concepts the first simple example, shown in Figure 11.4, consists of an adder and a set of AND gates. The adder and AND gates are in separate voltage domains with the adder in V_{add} and the AND gates in V_{AND}. (Similar to the two inverters shown in Figure 11.2.) This is representative of a block (the adder) that can be slept or shutdown and the isolation circuitry (AND gates) residing in the receiving domain. The adder is power-gated by an nMOS footer transistor controlled by the V_{footer} signal. When the Voltage on V_{footer} drops below the threshold voltage of the footer transistor, the adder module is said to be in sleep mode.

11.3.1 Sleep

The example design in Figure 11.4 helps to demonstrate the importance of checking the dynamic sequencing of voltage controlling signals for behavioral accuracy. The waveforms in Figure 11.5 show a typical RTL simulation that is unaware of any changes in the operation of power-gating transistors controlling a group of gates. The logic definition of the cells isn't changed since the power-gating transistor, in this case a footer, isn't part of the cell. As the voltage V_{footer} drops to 0V, the logical outputs of the AND gate are computed as if the power-gating transistor didn't exist. To accurately simulate the effects of the footer transistor, it is necessary to catch events on the V_{footer} signal and properly assert correct values at the outputs of the effected cells.

Today's logic based simulators do not take into account the effects of changes on voltage rails or the impact of sleeping parts of the design. Figure 11.5 shows a typical RTL simulation where the voltage on V_{footer} drops from 1.2V to 0.0V. The output waveforms show no impact from the adder going into sleep mode. This illustrates a risk in only performing a static check for isolation and not having a dynamic verification technique for checking the FSM controller sequences.

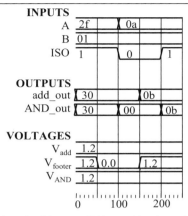

Figure 11.5 The effect of changing V_{footer} to 0V is notably absent in this waveform diagram.

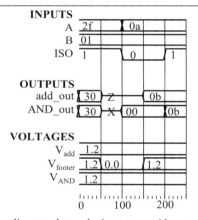

Figure 11.6 This waveform diagram shows the impact on add_out as V_{footer} goes to 0V.

The waveform diagram in Figure 11.6 shows the effects of applying 0V on V_{footer} and sleeping the adder. Note that the when V_{footer} changes to 0V that the output of the adder goes to a high impedance state "Z". An error in the control sequencing now becomes readily apparent.

It is important to note that a simple check for the existence of isolation circuitry is not sufficient to find this type of control sequencing error. A static check would "pass" indicating that the necessary isolation circuitry (AND gates) is present but would fail to detect the dynamic run time error.

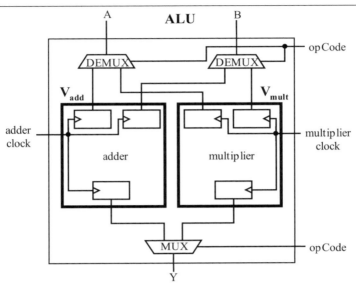

Figure 11.7 This arithmetic logic unit (ALU) consists of an adder and multiplier that may be in different domains with different voltages.

11.3.2 Shutdown

Typically, shutdown involves forcing the driving voltage rail to V_{SS}. This can be accomplished using an on-chip or off-chip voltage regulator. Shutting down a block eliminates any leakage current in that block but typically takes longer to bring back into operating mode. Other considerations include "in-rush" currents that may be excessive and have problematic peak power characteristics as well as reliability issues. Designers will often stage smaller portions of the design to ramp back up in sequence to avoid the issues around creating a large instantaneous current draw.

There are two important types of verification to be performed. The first is to check that the connectivity has been properly handled in the design, and the second is to check that the signals are sequenced properly to ensure correct behavior. For example, once it has been determined that the necessary level shifting and isolation have been inserted, the controlling signals have to be properly asserted in time in order for the circuit to function correctly. A dynamic simulation of the sequencing of the control signals and voltage variations can demonstrate whether the circuit exhibits correct time based behaviors.

To illustrate the points in this section, a simple example with multiple islands is used. A block diagram of the example is shown in Figure 11.7. The design is a simple ALU (arithmetic logic unit), consisting of an adder and a multiplier. The adder and multiplier are their own respective islands, and based on their voltages may be in their own respective domains as well. There are two inputs to the ALU, A and B and one output Y. Note only one of the adder or multiplier is performing useful computation at any given time. Effectively, their inputs are transparently latched in the 1:2 demultiplexers (DEMUX) controlled by the "opcode". The opcode also steers the appropriate module output via the 2:1 multiplexer (MUX) to the ALU output at Y. The following two waveform diagrams show the differences between using a typical RTL simulator and one that is multi-voltage aware.

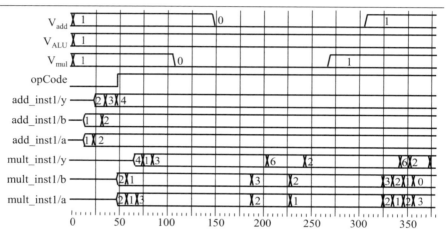

Figure 11.8 Typical RTL simulation ran for the ALU that neglects any changes due to shutdown of the adder or multiplier in the circuitry in Figure 11.7.

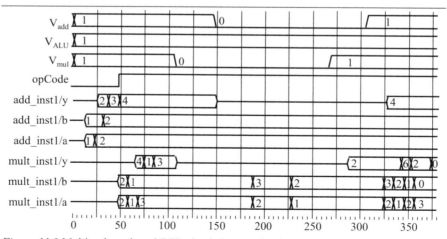

Figure 11.9 Multi-voltage based RTL simulation ran for the ALU reflecting changes due to shutdown of the adder or multiplier in the circuitry in Figure 11.7.

A pure logic simulation without voltage information is shown in Figure 11.8. Note that because of the register inserted at the inputs and outputs of the adder and multiplier that there is an additional delay from when the inputs appear at the adder and multiplier and when the output values change.

11.3.3   System Level, Adaptive Voltage Scaling

For this section, the example used in Figure 11.4 is again used to illustrate the main concepts. A necessary feature for AVS designs is the ability to control the voltage rails in an analog-like fashion.

Figure 11.10 shows a diagram for a simple VRM. The output voltage (V) is controlled by the input powerState. A binary voltage indicator is used to signal when the requested voltage is stable at the voltage output. This model enables a continuous change in the voltage in simulation time. A new voltage level is requested by powerState and then the VRM will set the indicator to "0" and move the output voltage to the requested level. Once the new voltage level is reached, the indicator is set to "1". This signal can be used by other power management circuitry to control the behavior of the design. One possible use is to have the indicator signal assist in any control in clock frequencies. If there is a request to raise the voltage and frequency for higher performance, typically the voltage will first be raised to the new value and then the clock frequency will be raised to its new value. The voltage indicator in this case could be used to signal that it is safe to increase the clock frequency once the requested voltage has been achieved.

The modification made to the design in Figure 11.4 is to use a VRM to control the driving voltage (V_{add}) for the adder. Instead of using a footer to sleep the adder as in Section 11.3.1, the driving voltage is reduced to demonstrate the effects of dynamically varying the power rail.

This example brings into play more advanced concepts of multi-voltage design. One important aspect is the determination of good logic "1" and "0" values. If the driving voltage of a sending block drops below a certain level, the receiving block will not be able to recognize a valid "1" on the corresponding input. For this example, a valid "1" is considered to be 70% of the receiving block's Vdd. In this case, since all driving voltage rails except V_{add} are at 1.2V, the output of the adder must be greater than or equal to 0.84V. If the V_{add} drops below 0.84V then the other blocks will consider a "1" output from the adder module as an "X" at their input.

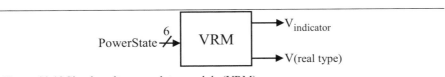

Figure 11.10 Simple voltage regulator module (VRM).

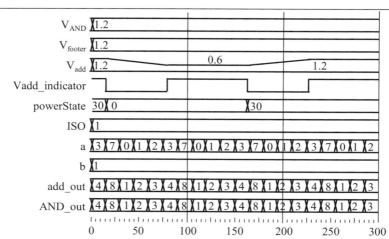

Figure 11.11 Typical RTL simulation ran for the adder and isolation circuitry shown in Figure 11.4.

Another concept covered here is the voltage at which a block still retains its state but no longer functions properly at speed, referred to as the "standby" voltage. In this case for the adder module, that voltage is set to 0.6V. In other words, at 0.6V the adder module retains its state, so that if the block is also clock-gated at that time and the voltage is then later raised and the clock re-enabled, the block will come back in a good operating state. If the block is clocked while the driving voltage is at 0.6V or below, the values could be corrupted.

Figure 11.11 shows an RTL simulation where the driving voltage for the adder is varied between 1.2V and 0.6V. The signal Vadd_indicator is used to indicate when the VRM has reached the appropriate output voltage level. The powerState variable has a range of 0 to 32 and indicates 20mV increments above the baseline voltage of 0.6V. In this example, V_{add} goes to standby voltage but no lower. Setting powerState = 30 (0x1E) corresponds to a requested voltage of 0.6V + (30 × 0.02) = 1.2V which is where the simulation starts. powerState is then set to 0 and V_{add} starts to drop to 0.6V. After a period of time, powerState is again set to 30 and V_{add} returns to 1.2V with the voltage indicator in both cases initially going to "0" and then returning to "1". Since the RTL simulator isn't multi-voltage aware, the outputs of the adder and AND gates are unaffected by the changes in the voltage level of V_{add}.

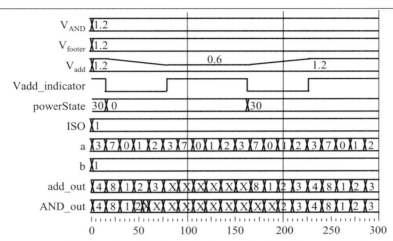

Figure 11.12 Multi-voltage based RTL simulation ran for the adder and isolation circuitry shown in Figure 11.4.

Figure 11.12 shows the same example this time simulated with a multi-voltage based simulator. It is interesting to note that the output values of the AND gates go to "X" before the output of the adder module starts to produce any "X" outputs. In this case, the AND gates are sensitive to the voltage level of any incoming "1" signals. Since V_{add} drops below 70% of 1.2V shortly after 50 time intervals in Figure 11.12, the output values of the AND gates start to go to "X". Once V_{add} drops to 0.6V and the inputs continue to change, the outputs of the adder also go to a value of "X". As the voltage V_{add} begins to increase, the adder once again starts to produce good logic outputs but it is not until V_{add} reaches 0.84V that the AND gates again produce good logic output values.

11.4   SUMMARY

The push to more energy efficient designs is becoming more prevalent. Given a thermal envelope constraint defined by the package that an IC will reside in, more often it is becoming the case that the performance and operating frequency are defined by the power characteristics of the IC. In these cases, more effort is being placed on computational efficiency with respect to power. Witness the current de-emphasis in clock speeds and the move to multi-core designs for processors ranging from laptops and notebooks to high performance servers.

One of the most promising variables that designers have at their disposal is the voltage. As designs with variable voltage, power gating and shutdown become more popular, there will be an increasing need for tools to help design and verify them. Dynamic simulation that accurately models the functional impact of varying the voltage in a design was described in Section 11.3. Tools at this level need to handle voltage as a real variable in the design and be able to incorporate the effects of its dynamic variations. Inability to do so will leave holes in any chip's verification strategy and lead to chips that will have costly errors that will only be found late in the design cycle and likely after tape out.

The work described here can be extended to the gate level as well. The EDA industry will need libraries with power and timing information for a wider range of voltages versus the typical +/– 10% fast/slow corners usually available. The island definitions are valid throughout the flow but the libraries will need to include more data to accurately reflect the realities at lower levels of abstraction.

11.5   REFERENCES

[1] Lackey, D., et. al, "Managing Power and Performance for SOC Designs Using Voltage Islands," *Proceedings of the International Conference on Computer-Aided Design*, 2002, pp. 195-202.
[2] Usami, K. and Igarishi, M., "Low-Power Design Methodology and Applications Utilizing Dual Supply Voltages," *ASP Design Automation Conf.*, 2000, pp. 123-128.
[3] Martin, S.M., Flautner, K., Mudge, T. and Blaauw, D., "Combined Dynamic Voltage Scaling and Adaptive Body Biasing for Lower Power Microprocessors under Dynamic Workloads," *Proceedings of the International Conference on Computer-Aided Design*, 2002, pp. 721-725.
[4] Shigematsu, S., Mutoh, S., Matsuya, Y., Tanabe, Y. and Yamada, J., "A 1-V High-Speed MTCMOS Circuit Scheme for Power-Down Application Circuits," *IEEE Journal of Solid-State Circuits*, vol. 32, no. 6, June 1997, pp. 861-869.
[5] Shigematsu, S., Mutoh, S., Matsuya, Y. and Yamada, J., "A 1-V High-Speed MTCMOS Circuit Scheme for Power-Down Applications," *IEEE Symposium on VLSI Circuits Digest of Technical Papers*, June 8-10, 1995, pp. 125-126.
[6] Mehra, R. and Pangrle, B., "Synopsys Low-Power Design Flow," Chapter 40, CRC Low-Power Electronics Design, CRC Press, 2004, pp. 40-1 to 40-20.
[7] Hillman, D. and Wei, J., "Implementing Power Management IP for Dynamic and Static Power Reduction in Configurable Microprocessors using the Galaxy Design Platform at 130nm," Boston Synopsys Users' Group (SNUG), 2004.
[8] Nowka, K.J., Carpenter, G.D. and Brock, B.C., "The Design and Application of the PowerPC 405LP Energy-Efficient System-On-A-Chip," *IBM Journal of Research and Development*, vol. 47, no. 5/6, 2003.

Chapter 12

WINNING THE POWER STRUGGLE IN AN UNCERTAIN ERA

Murari Mani, Michael Orshansky
Department of Electrical and Computer Engineering
University of Texas at Austin
Austin, TX, 78712, USA

12.1   INTRODUCTION

The growth of process variability in scaled CMOS requires that it is explicitly addressed in the design of high performance and low power ASICs. This growth can be attributed to multiple factors, including the difficulty of manufacturing control, the emergence of new systematic variation-generating mechanisms, and the increase in fundamental atomic-scale randomness – for example, the random placement of dopant atoms in the transistor channel. Scaling also leads to the growth of standby, or leakage power [7]. Importantly, leakage depends exponentially on threshold voltage and gate length of the device. The result is a large spread in leakage power in the presence of process variations.

Recently, considerable research efforts have focused on developing statistical approaches to timing analysis, including the models and algorithms accounting for the impact of delay variability on circuit performance. These techniques concern themselves with eliminating the conservatism introduced by employing traditional worst-case timing models in predicting the timing yield of the circuit. In view of the importance of variability, new methods are needed to evaluate the power-limited parametric yield of integrated circuits and guide the design towards statistically feasible and preferable solutions. This can be achieved through the migration to statistical optimization techniques that account for both power and delay variability.

In this chapter we examine the impact of variability on power, along with the strategies to counter its detrimental effect and improve performance and parametric yield. In Section 12.2 we provide an overview of process variability trends and discuss their impact on power and parametric yield. Section 12.3 deals with analytical techniques for evaluating circuit parametric yield considering leakage and timing variability. Section 12.4 presents an overview of optimization strategies for yield improvement. In Section 12.5 we discuss in detail, an efficient algorithm that targets power minimization under probabilistically specified timing and power constraints.

12.2 PROCESS VARIABILITY AND ITS IMPACT ON POWER

Several factors contribute to the growth in process variability [2][3][24] [34]. While the continued need for more performance necessitates rapid technology scaling, there are severe limitations to our capacity to improve manufacturing tolerances [22]. This is manifested in the rise of such effects as channel length variation due to the optical proximity effect [13][17]; systematic spatial gate length variation due to the aberrations in the stepper lens [38]; and variation in interconnect properties caused by non-uniform rate of chemical-mechanical polishing (CMP) in layout regions of different pattern density [10][39]. Scaling also brings about parameter uncertainty of a fundamental atomic-level nature. This is best exemplified by variability in transistor threshold voltage due to random dopant fluctuations (RDF). As transistors scale, the transistor channel contains fewer dopant atoms whose precise number and location cannot be controlled, while even small fluctuations can impact threshold voltage significantly [8][16][42].

The patterns of variability are also changing: the intra-chip component of variation grows as a percentage of total variability in key process parameters such as channel length and threshold voltage [4][26]. It is this change that is largely responsible for the need to develop new approaches to timing analysis and optimization, as the traditional methods fail in the presence of uncorre-lated intra-chip variability.

The increase in leakage power with scaling, and the strong dependence of leakage on highly varying process parameters, raises the importance of statistical leakage and parametric yield modeling. There are several reasons for increased leakage power consumption. Supply voltage scaling requires the reduction in threshold voltage (V_{th}) in order to maintain gate over-drive strength. Threshold voltage reduction causes an exponential increase in subthreshold channel leakage current. To make matters worse, aggressive scaling of gate oxide thickness leads to significant gate oxide tunneling current [41].

For transistors in the weak inversion region, the subthreshold current can be expressed as:

$$I_{sub} \propto e^{(V_{gs}-V_{th})/\eta V_{thermal}} (1 - e^{-V_{ds}/V_{thermal}}) \qquad (12.1)$$

where V_{gs} and V_{ds} are gate- and drain-to-source voltages, $V_{thermal}$ is the thermal voltage, and η is the subthreshold slope coefficient [41]. For the purpose of statistical analysis, the exponential dependence of subthreshold current on process parameters is better captured by an empirical model in terms of the variation in effective channel length (ΔL) and the variation in threshold voltage (ΔV), taken to be stochastically independent of channel length [32]:

$$I_{sub} \propto e^{-(\Delta L + a_2 \Delta L^2 + a_3 \Delta V)/a_1} \qquad (12.2)$$

where a_1, a_2, and a_3 are process-dependent parameters. The gate tunneling current strongly depends on the oxide thickness (T_{ox}) and can be described as [18]:

$$I_{ox} \propto e^{(c_1 V_{gs} - c_2 T_{ox}^{-2.5})} + e^{(c_1 V_{gd} - c_2 T_{ox}^{-2.5})} \qquad (12.3)$$

where c_1, c_2, are the process-dependent fitting parameters, and V_{gs} and V_{ds} are the gate-to-source and gate-to-drain voltages respectively. A simple empirical model captures the dependence of I_{ox} on the variation in the oxide thickness (ΔT) [32]:

$$I_{ox} \propto e^{-\Delta T / b} \qquad (12.4)$$

where b is the process-dependent parameter.

The models indicate that both subthreshold and gate leakage currents are exponential functions of highly-variable process parameters, specifically effective channel length, threshold voltage, and oxide thickness. This strong dependence causes a large spread in leakage current in the presence of process variations (Figure 12.1), with subthreshold leakage depending primarily on L_{eff} and V_{th}, and gate leakage depending on T_{ox}. Historically, T_{ox} has been a well-controlled parameter, and as a result, it had smaller impact on leakage variability. However, this is rapidly changing as technology approaches the limits of thin film scaling. While leakage power exhibits exponential dependencies on process variables, chip frequency has a near-linear dependency on most parameters [32]. This difference in magnitude of variation is easily observed in measurements. Figure 12.1 shows that a 1.3× variation in delay between fast and slow die could potentially lead to a 20× variation in leakage current [3].

Leakage power is inversely correlated with chip frequency. Slow die have low leakage, while fast die have high leakage (Figure 12.1). The same parameters that reduce gate delay – shorter channel length, lower threshold voltage, thinner gate oxide – also increase the leakage. Moreover, the spread in leakage grows as the chip becomes faster. In characterizing chips according to their operating frequency, it has been observed that a substantial portion of chips in the fast bins have unacceptably high leakage.

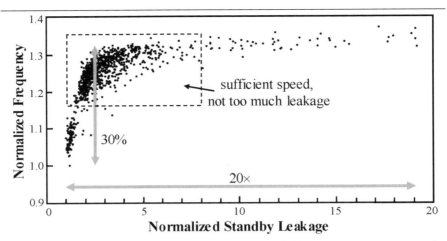

Figure 12.1 Exponential dependence of leakage current on 0.18um process parameters results in a large spread for relatively small variations around their nominal value. Figure courtesy of the authors of [3]. (© 2003 ACM, Inc. Included here by permission.)

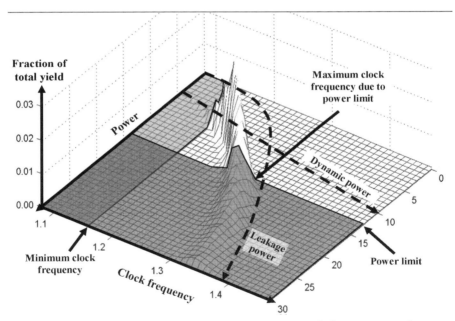

Figure 12.2 Inverse correlation between leakage power and frequency contributes to parametric yield loss. The maximum frequency of usable chips is reduced because chips in what would be the "fast" bin exceed power limits. This data was generated with a normal distribution for the clock frequency, and thus channel length [32], which exponentially affects leakage, and then a log normal distribution for leakage about these points. A scatter plot of leakage vs. frequency shows a similar distribution to Figure 12.1. In contrast, dynamic power increases linearly with clock frequency (i.e. switching activity).

In the absence of substantial leakage power, parametric yield is determined by the maximum possible clock frequency. Switching power is relatively insensitive to process variation. When the leakage power typical of current CMOS technologies is added, the total power starts approaching the power limit determined by the cooling and packaging considerations. Crucially, the exponential dependence of leakage on process spread means that the total power may cross the cooling (power) limit well below the maximum possible chip frequency, since chips operating at higher frequencies have exponentially higher leakage power consumption. Thus, due to the inverse correlation between speed and leakage, yield is limited both by slower chips and chips that are too fast, because they are too leaky.

This is further illustrated in Figure 12.2. The leakage-delay correlation and the resulting dual squeeze on parametric yield is one of the reasons why new methods that can simultaneously estimate timing-limited and power-limited yield need to be utilized.

12.3   PARAMETRIC YIELD ESTIMATION

It is possible to get a fairly reliable estimate of the chip's parametric yield early in the design flow, at the design exploration phase, based on a very small number of chip parameters: the total chip area, the number of devices, the nominal and statistical technology parameters, and the supply and threshold voltages. The estimate can then be used to optimize the technology and design parameters before the design is fully specified. Both subthreshold and gate oxide leakage components can be accounted for [32].

In the estimation of parametric yield, we can safely assume that chip frequency is most strongly influenced by global channel length (L_g) variation. This assumption is validated by both simulation and by industrial practice where microprocessor speed binning is strictly correlated with the gate length variability [32]. Relying on Equations (12.2) and (12.4), the process parameters that impact leakage components are decomposed into their local (ΔL_l, ΔV_l, ΔT_l) and global (ΔL_g, ΔV_g, ΔT_g) contributions.

Because the variation of path delay is primarily defined by the global ΔL_g variation, when estimating yield, it is convenient to express I_{sub} as an explicit function of ΔL_g. The impact of local and global variability on the leakage distribution is evaluated separately. The chip leakage variation due to local variability is a sum of current contributions from all devices on the chip. Because of that, the impact of local variability of all parameters on leakage can be captured by their impact on the *mean* leakage (at fixed values of global parameters). Specifically, the impact of local variability on leakage at a given value of L_g is to shift the mean of the distribution (due to ΔV_g, ΔT_g) by the amount that depends on the *variance* of local variability due to all components. For example, the increase of mean leakage caused by local variability ΔL_l is:

$$S_L(\sigma_{L_i}^2) = (1 + \tfrac{2\lambda_2}{\lambda_1}\sigma_{L_i}^2)^{-0.5} \exp[\frac{\sigma_{L_i}^2}{(2\lambda_1^2 + 4\sigma_{L_i}^2\lambda_1\lambda_2)}] \qquad (12.5)$$

where λ_1, λ_2, and λ_3 are process-dependent variables, and σ_L is the standard deviations of intra-die components L_l. It is easy to see that $S_L \geq 1$, and $S_L = 1$, when local variation is absent. Similar expressions can be derived for scaling factors S_V and S_T that capture impact of local variation of ΔV_l and ΔT_l on the mean of the leakage. The total chip leakage, as a function of global variation terms, is obtained by weighting the leakage contribution of individual gates by their widths, W [32]:

$$I_{total} = \sum W\left(S_L S_V I_{sub}^0 e^{(-(\Delta L_g + a_2 \Delta L_g^2 + a_3 \Delta V_g)/a_1)} + S_T I_{gate}^0 e^{(-\Delta T_g/b)}\right) \qquad (12.6)$$

where S_L, S_V, and S_T are the scaling factors to capture the effect of local variability in L_{eff}, V_{th} and T_{ox} respectively; I_{sub}^0 and I_{gate}^0 are the nominal values of subthreshold and gate leakage respectively; and a_1, a_2, a_3 and b are fitting coefficients. Here ΔL_g, ΔV_g, and ΔT_g are treated as independent normal random variables.

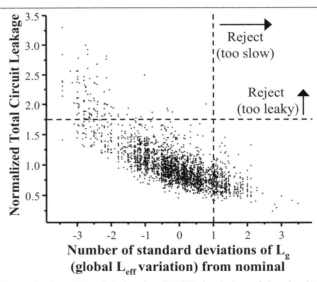

Figure 12.3 Monte Carlo scatter plot showing SPICE simulation of the circuit leakage for a 64-bit adder with 100nm Berkeley predictive technology model [9]. Variability in V_g and T_g are responsible for "local" spread in leakage causing ~27% yield loss in the highest performance bin. Figure courtesy of the authors of [32]. (© 2004 ACM, Inc. Included here by permission.)

Consider the power and delay variability of an adder shown in Figure 12.3. For each value L_g, which corresponds to a specific frequency bin, the spread of leakage is caused by the variation in V_g and T_g. A consequence of this spread is that even though the frequency of a chip confirms to specifications, it may still contribute to parametric yield loss due to its unacceptably high leakage power consumption. The analytical framework developed above enables the estimation of the leakage yield corresponding to a specific leakage constraint, or the leakage current corresponding to any yield quantile. Because power yield can be computed for every specific value of ΔL_g, the estimate of joint timing-limited and power-limited yield can be thus easily found.

12.4 OPTIMIZATION TECHNIQUES FOR YIELD: AN OVERVIEW

The previous section considered analysis methods to evaluate chip-level and circuit-level parametric yield. We now discuss the optimization strategies that can be employed to improve parametric yield. Traditional circuit optimization techniques are insufficient for the purpose of parametric yield improvement in nanometer scale integrated circuits. In the past, case-files have been used effectively with the traditional deterministic algorithms while guaranteeing a specific yield point. Typically, these case files would be worst case, nominal, and best case process corners combined with the worst case, nominal, and best case operating (voltage, temperature) corners. The effect of variability was captured in these case files by modifying the device SPICE model parameters to correspond to a specific percentile of the parameter distribution. Analyzing and optimizing the circuit with these parameters guaranteed that it would meet the performance constraints at a specific percentile of probability [25]. However, this approach works only when variability is predominantly inter-chip, causing differences in the chip-to-chip properties, with parameter variation in devices within a chip being neglected. In nanometer scale technologies, intra-chip variation is significant. Also, deterministic optimization makes the tacit assumption that circuit performances of different gates have identical sensitivities to the variation of process parameters. The highly non-linear and non-additive responses of performance variability make this premise untenable [28]. This results in the breakdown of the case-file based approach to handling variability in optimization as it becomes impossible to come up with a case file that will guarantee a specific yield point.

Circuit-level variability is directly dependent on the decision variables: for instance, the standard deviation of threshold voltage depends inversely on the square root of transistor area [14]. Statistical algorithms that explicitly account for the variance of objective and constraint functions during optimization are expected to perform much better. In contrast, deterministic algorithms lack the notion of parameter variance and parametric yield, preventing design for yield as an active design strategy. An algorithm that does not comprehend the dynamic changes in performance variability arising from threshold voltage dependency on sizing is unlikely to be successful in parametric yield optimization. Instead, if a worst case process corner is assumed to ensure sufficient yield the circuit gets over-designed resulting in worse power consumption and lower performance. Thus, the introduction of rigorous statistical power-optimization has a potentially significant impact on circuit performance and parametric yield.

Optimizing the parametric yield metric directly seems computationally very difficult because of its numerical properties (yield is an integral of the probability distribution function). For that reason, most yield-improvement strategies map yield into other metrics that are more convenient computationally. For the sake of discussion, the known optimization approaches for yield improvement can be classified into two categories: those that model the impact of variability on timing yield only, and those that consider timing and power limited parametric yield simultaneously.

A variety of strategies has been proposed for considering the impact of variability on timing yield. One effect of variability on the behavior of high-performance well-tuned circuits is the spreading among the timing paths from the "wall" of critical paths generated by circuit tuning [1]. The more the paths pushed against the wall, the bigger is the detrimental impact of variability in pushing out the performance. We could improve timing yield by reducing the height of the path delay "wall", since it is simply an artifact of mathematical optimization, which is hard to justify considering the practical design limitations. A penalty function can be introduced in the circuit tuner to prevent such path build-up [1]. This is an indirect strategy for yield improvement however, since the true path delay variance is not used to guide optimization.

It is possible to formulate a general statistical gate sizing problem that can be described by analytical but non-linear functions and solve it directly using a general non-linear solver [15]. The objective and constraints are expressed as explicit functions of the mean and variance of gate delays. However, the techniques relying on non-linear optimization tend to be excessively slow which would greatly limit the capacity for large-scale circuit optimization. In [33] an extension to the Lagrangian relaxation based approach [11] is proposed. Here, the gate sizing problem is solved iteratively while updating the required arrival time constraint using information from a statistical timing analyzer. The notion of timing yield is incorporated by making the delay target be defined at a quantile value. More efficient formulations based on geometric programming are also possible. In [29], the fact that sizing problems have fairly flat maxima is exploited by utilizing heuristic techniques to compute the "soft-max" of arrival times. Statistical static timing analysis is then used to guide the optimization in the right direction. The algorithm based on geometric programming presented in [35] models parameter variations using an uncertainty ellipsoid, and proceeds to construct a robust geometric program, which is solved by convex optimization tools. Efficient algorithms based on the special structure of convex problems, such as conic programming, have also been used for statistical gate sizing [19].

However, as we have argued earlier in the chapter, in the nanometer regime, parametric timing yield alone is not a sufficient metric as it ignores variability in leakage power. This necessitates the development of computationally efficient statistical optimization techniques to minimize parametric yield loss resulting from power and delay variability [24][20] [37]. The early work [37] extends to a statistical setting the well-known iterative coordinate-descent algorithm, best exemplified in the electronic design automation area by TILOS [12]. Specifically, it performs leakage power minimization using the power-reduction potential provided by a dual-V_{th} technology and by gate sizing. In the deterministic approach, the initial configuration is one which meets timing constraints and has all gates set to low-V_{th}. Gates are subsequently swapped from low-V_{th} to high V_{th} based on the following sensitivity measure s:

$$s = \left| \frac{\Delta p}{\Delta d} \right| \delta \tag{12.7}$$

Here Δp and Δd are the changes in power dissipation and delay of the gate if it is swapped to high-V_{th} and δ is the slack of the gate (see Equation (6.4) in Section 6.2 for more details). If the timing constraints are violated after a swap is made, gates are upsized depending on their efficiency to convert the additional power accrued due to resizing to reduction in delay.

In the statistical counterpart of this optimization strategy, statistical timing analysis is used to determine if timing constraints are met. Additionally, the first and second moments of sensitivities are used instead of nominal sensitivity values. However, one major limitation of using a greedy sensitivity based approach described above, is that it may make sub-optimal decisions as it views one gate at a time. This is illustrated in a deterministic setting in Chapter 6, where larger power savings can be achieved by adopting a framework that has a global view of the circuit.

The power savings enabled by this statistical algorithm, as compared to its deterministic counterpart, range from 15% to 35%. The algorithm is computationally expensive, however. While it is based on coordinate-descent algorithms that have proved their practical utility in gate sizing, the extension to the statistical setting causes the run-time to grow considerably. This may become a concern when using the algorithm on large circuits. The optimization approach discussed in the next section is about an order of magnitude faster than the approach in [37], due to the efficient statistical optimization problem formulation as a second order conic program (SOCP).

12.5   EFFICIENT STATISTICAL PARAMETRIC YIELD MAXIMIZATION

The primary limitation of existing statistical CAD techniques is their high computational cost. This makes the application of such algorithms to industrial-size circuits a difficult task. In this section, we focus thoroughly on a statistical yield enhancement technique that achieves high computational efficiency, while treating both timing and power metrics probabilistically [20].

12.5.1 Power Minimization by Delay Budgeting

In order to enable an efficient computational formulation, the problem of parametric yield maximization in this algorithm is converted into that of statistical leakage minimization under probabilistic timing constraints. It uses a two phase approach based on optimal delay budgeting and slack utilization, akin to [27]. The delay budgeting phase is formulated as a robust version of the power-weighted linear program that assigns slacks based on power-delay sensitivities of gates. The notion of variability in delay and power due to process variations is explicitly incorporated into the optimization, by setting up an uncertain robust linear program. The statistical (robust) linear program is cast into a second order conic program that can be solved efficiently. The slack assignment is inter-leaved with the configuration selection which optimally redistributes slack to the gates in the circuit to minimize total power savings.

Post-synthesis circuit optimization heuristics for sizing and dual-V_{th} allocation are effective in reducing leakage, and have been widely explored in a deterministic setting [27][36][43]. While relying on different implementation strategies, all these techniques essentially trade the slack of non-critical paths for power reduction by either downsizing the transistors or gates or setting them to a higher V_{th}.

Since the joint sizing and dual-V_{th} assignment optimization problem is computationally hard, it is convenient to move into the power delay configuration space as described below. The deterministic algorithm for power minimization is a two-phase iterative relaxation scheme. The input to the first phase is a circuit sized for maximum slack using a transistor (gate) sizing algorithm, such as TILOS [12], with all the devices set to low V_{th}. This circuit has the highest possible power consumption of any circuit realization. The available slack is then optimally distributed to the gates based on the power-delay sensitivities: that is, the slack is allocated in a way that maximizes the power reduction. The second phase consists of a local search among gate configurations in the library, such that slack assigned to gates in previous phase is utilized for power reduction.

The idea of using power-delay sensitivity of a circuit as an optimization criterion is itself well known [21]. A linear measure of a gate's power-delay sensitivity is power reduction per unit of added delay:

$$s = -\frac{\partial P}{\partial D} \qquad (12.8)$$

The power reduction for gate i with an added delay $d(i) \geq 0$ is then linearly approximated by $s(i)d(i)$. A unit of added slack to a node with a higher sensitivity will lead to the greater power reduction. This concept is extended to efficient optimization based on large-scale linear programming by converting a power minimization problem into a power-weighted slack redistribution. This is similar to the metric in [37], Equation (12.7), but instead of looking at each gate individually and greedily picking the gate with the best trade-off, a linear program is used to assign the added delays with a global view of the power savings that may be achieved.

Let a gate configuration be any valid assignment of sizes and threshold voltages to transistors in a gate in the library. For any fixed load, a set of Pareto points in the power-delay space can be identified among all the possible configurations (Figure 12.4). A power optimal solution will contain only the Pareto-optimal gate configurations. The trade-offs between delay and both leakage and dynamic power can be captured in tables, parameterized by the capacitive load. For each of the Pareto-optimal gate configurations, the decrease in power consumption (ΔP) and the change in delay (ΔD) are calculated. For example, one may compute the sensitivity of changing the gate from all transistors having low V_{th} to the configuration where all transistors have high V_{th}.

Using this framework, a linear program can be formulated to distribute slack to gates with the objective of maximizing total power reduction while satisfying the delay constraints on the circuit. This can be expressed as [27]:

$$
\begin{aligned}
\text{maximize} \quad & \sum_i s_i d_i \\
\text{subject to} \quad & t_i \geq t_j + d_i^0 + d_i, \text{ for all } j \in \text{fanin}(i) \\
& t_k \leq T_{\max}, \text{ for all } k \in \text{primary outputs} \\
& 0 \leq d_i \leq \delta d
\end{aligned} \qquad (12.9)
$$

Here t_i is the arrival time at node i, $T_{\max}$ is the required arrival time at the primary output, d_i^0 is the delay of the gate i in the circuit configuration obtained by sizing for maximum slack, d_i is the additional slack assigned, and δd is the maximum allowed slack increment.

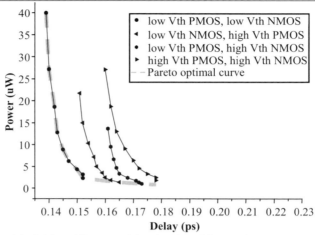

(a) With a 5fF capacitive load and input slew 0.2ns.

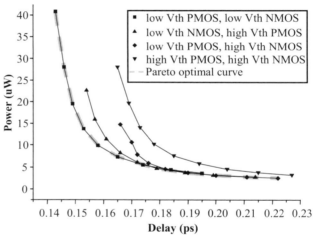

(b) With a 15fF capacitive load and input slew 0.2ns.

Figure 12.4 The power-delay space for a NAND2 gate driving two different capacitive loads. The Pareto frontier is depicted by the dashed gray lines. A power-optimal circuit will consist exclusively of Pareto-optimal gate configurations. The points on a curve correspond to the nine different gate sizes in the library. SPICE simulations were used for analysis with a 70nm process using the Berkeley Predictive Technology Model [9].

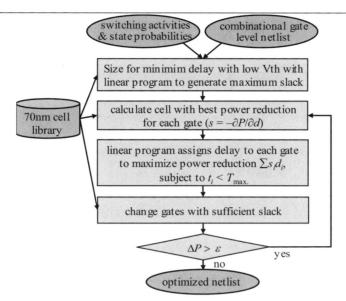

Figure 12.5 Flowchart illustrating the iterative relaxation algorithm for power minimization.

The algorithm is constructed as an iterative relaxation method. Its core is an interleaved sequence of (i) optimal slack-redistribution using linear programming, and (ii) the local search over the gate configuration space to identify a configuration that will absorb the assigned slack (Figure 12.5). It has been shown that when the configuration space is continuous, and delay is a monotonic and separable function, such a procedure is optimal for small increments of slack assignments δd [40]. As the sensitivity vector (s_i) is a first order linear approximation, it is only accurate within a narrow delay range, which also requires moving towards the solution under small slack increments.

The library consists of two discrete threshold voltages (0.1V and 0.2V), and a continuous range of gate sizes, with piecewise linear interpolation of delay and power versus load capacitance from SPICE characterization. Assuming a continuous range of gate sizes is reasonable, given that good low power standard cell libraries should have finely grained gate sizes or use a "liquid cell" sizing methodology. Even though the configuration space generated by V_{th} assignments is discrete, the ability to size transistors in a continuous manner permits treating the delay range for a cell as continuous. This ensures that a configuration maximally utilizing the slack allotted in the slack assignment phase can be found. The value for δd is chosen heuristically – as long as δd is small enough (relative to T_{max} of the circuit), the approach produced good results. A typical value chosen for δd was 2% of the clock period T_{max}.

Table 12.1 Low V_{th} devices exhibit a higher leakage spread, while high V_{th} devices exhibit a higher delay spread.

	Delay		Leakage	
	Nominal	99th percentile	Nominal	99th percentile
Low V_{th} (0.1V)	1.00	1.15	1.00	2.15
High V_{th} (0.2V)	1.20	1.50	0.12	0.20

12.5.2 Statistical Delay Budgeting using Robust Linear Programming

We now describe how the statistical equivalent for the power minimization problem under variability is reformulated as a robust linear program. This will permit using interior-point methods that are highly efficient for solving convex optimization problems. In order to make the presentation specific, we assume that there are two primary sources of variability: effective channel length (L_{eff}) and gate-length independent variation of threshold voltage (V_{th}). These parameters have significant impact on timing (L_{eff}) and leakage power (V_{th}). In general, more sources of variation can be used. An additive statistical model that decomposes the variability, of both L_{eff} and V_{th}, into the global and local variability components is used. For gate length the model is

$$L_{eff} = L_0 + \Delta L_g + \Delta L_l \quad (12.10)$$

A similar model is used for V_{th}. Consistent with empirical data, both L_{eff} and V_{th} are assumed to be Gaussian random variables. Under the leakage models described earlier in the chapter, the leakage power (Equation (12.6)) is a log-normal random variable. In contrast, assuming a fixed clock frequency, it was observed that the dynamic power was only a weak function of process variability in L_{eff}. It can be shown that and the sensitivity coefficient also follows a log-normal distribution. The modeling framework gives the ability to account for the different values of parameter variability in low-V_{th} and high-V_{th} gates: low-V_{th} gates exhibit higher variation in leakage, while high-V_{th} gates exhibit higher delay variability. This is illustrated in Table 12.1 for a 70nm process.

Robust optimization is concerned with ensuring the feasibility and optimality of the solution under all permissible realizations of the coefficients of the objective and constraint functions [5]. The novelty of the described algorithm is that it sets up a rigorous statistical equivalent of the slack assignment using the notion of robust linear programming and explicitly incorporates uncertainty in a formulation that is amenable to highly efficient computation. When formulating a statistical power minimization problem, an equivalent formulation of Equation (12.9), which places the power weighted slack vector into the constraint set, is more convenient. Suppose that P_{max} is the initial maximum power, $\hat{P}$ is the optimal power achieved by Equation (12.9) at a specific T_{max}, and $\hat{d}_1$ the vector of optimal allocated slacks. The following optimization problem is equivalent to Equation (12.9):

$$\text{minimize} \quad \sum d_i$$
$$\text{subject to} \quad \sum s_i d_i \geq P_{max} - \hat{P}$$
$$t_i \geq t_j + d_i^0 + d_i, \text{ for all } j \in \text{fanin}(i) \quad (12.11)$$
$$t_k \leq T_{max}, \text{ for all } k \in \text{primary outputs}$$
$$0 \leq d_i \leq \delta d$$

That is, if $\hat{d}_2$ denotes allocated slacks for Equation (12.11), it can be shown that $\hat{d}_1 = \hat{d}_2$, and $P(\hat{d}_1) = P(\hat{d}_2)$ is a minimum power solution at the specified T_{max}. Equation (12.11) forces the linear program to place a premium on the total slack and assign more slack to gates with higher sensitivity in order to meet the power constraint. The statistical equivalent of Equation (12.11) is now formulated by probabilistically treating the uncertainty of the sensitivity vector and of timing constraints:

$$\text{minimize} \quad \sum d_i$$
$$\text{subject to} \quad \Pr\left(\sum s_i d_i \geq P_{max} - \hat{P}\right) \geq \eta$$
$$t_i \geq t_j + d_i^0 + d_i, \text{ for all } j \in \text{fanin}(i) \quad (12.12)$$
$$\Pr\left(t_k \leq T_{max}\right) \geq \zeta, \text{ for all } k \in \text{primary outputs}$$
$$0 \leq d_i \leq \delta d$$

Here, the deterministic constraints have been transformed into probabilistic constraints, where Pr( ) denotes the probability of the expression inside the brackets. These probabilistic constraints set respectively the power-limited parametric yield η, and the timing-limited parametric yield ζ. Based on the formulation of the model of uncertainty, they capture the uncertainty due to process parameters via the uncertainty of power and delay metrics.

The above probabilistic inequalities have to be reformulated such that they can be efficiently handled by available optimization methods. The challenge is to handle these inequalities analytically, in closed form. The probabilistic timing constraints in Equation (12.12) are transformed such that the resulting expression still guarantees achieving the specified parametric yield level using the quantile (percent point) function:

$$D_i + \phi^{-1}(\zeta)\sigma_{D_i} \leq T_{max} \quad (12.13)$$

where σ_{D_i} is the standard deviation of the ith path with delay D_i at primary output i. In order to reduce the number of constraints and increase the sparsity of the constraint matrices, the path-based constraints are further transformed into node-based constraints. A heuristic method of modeling the node delays with $d_i^0 + \phi^{-1}(\zeta)\sigma_{d_i^0}$, where $\sigma_{d_i^0}$ is the standard deviation of the gate delay, worked well in practice, but more sophisticated mappings can be introduced. This permits the formulation of the probabilistic timing constraint as:

$$t_k \leq T_{max}, \text{ for all } k \in \text{primary outputs}$$
$$t_i \geq t_j + d_i^0 + \phi^{-1}(\zeta)\sigma_{d_i^0} + d_i, \text{ for all } j \in \text{fanin}(i) \qquad (12.14)$$

Using the fact that sensitivity is a lognormal random variable, the power constraint can be transformed into one which is linear in the mean and variance of $s_i d_i$:

$$\text{minimize} \quad \sum d_i$$
$$\text{subject to} \quad \overline{s}^T d + \kappa(\eta)(d^T \Sigma_s d)^{1/2} \leq \ln(\Delta P)/\lambda(\eta)$$
$$t_i \geq t_j + d_i^0 + d_i, \text{ for all } j \in \text{fanin}(i) \qquad (12.15)$$
$$t_i \geq t_j + d_i^0 + \phi^{-1}(\zeta)\sigma_{d_i^0} + d_i, \text{ for all } j \in \text{fanin}(i)$$
$$0 \leq d_i \leq \delta d$$

Here, η and ζ are the power and timing-limited parametric yields; $s \sim LN(\overline{s}, \Sigma_s)$ is the log-normal sensitivity vector with mean $\overline{s}$ and co-variance matrix Σ_s; and $\lambda(\eta)$ and $\kappa(\eta)$ are the fitting functions dependent on η. The mean, variance and covariance of leakage and delay are characterized via a Monte-Carlo simulation for all the cells in the library. The statistical properties of the power-delay sensitivity of the cell can then be computed analytically.

The above problem has a special structure that can be exploited to per-form very fast optimization. The reason is that the constraints are second-order conic functions that can be efficiently optimized by interior point methods [31]. Because the second-order conic programs are convex [5], they guarantee a globally optimal solution to this slack redistribution formulation that considers variation. The reliance on interior-point methods means that the computational complexity of solving this non-linear program is close to that of linear programming, and this is confirmed by experiments. The second phase of the power minimization algorithm is linear in the number of alternatives in the gate configuration space.

12.5.3 Evaluating the Effectiveness of Statistical Power Optimization

The above algorithm was implemented in C as a pre-processing module to interface with the commercial conic solver in MOSEK [23]. The benchmark circuits were synthesized to a cell library that was characterized for a 70nm process using Berkeley Predictive Technology Model [9] .

The gates present in the library are NOR2, NOR3, NOR4, NAND2, NAND3, NAND4 and inverter. Gates have eight discrete sizes, ranging from $1\times$ to $8\times$ the minimum size, and were characterized for a fixed input slew of 20ps. To permit treating the configuration space as continuous, an interpolating function was used to obtain the delay and leakage of gate sizes between the SPICE characterized sizes. Gate delay (average of worst case rise and fall delay) and internal power were specified by lookup tables for each value of load capacitance. Switching power was calculated as $\alpha f C_L V_{dd}^2$, where α is the activity factor, f is the clock frequency, C_L is the load capacitance, and V_{dd} is the supply voltage. The activity factors and state probabilities were determined by random simulation. Leakage power was computed for each input state and the state probabilities were used to obtain the average leakage. The delay analysis can be extended to include separate timing arcs and slews as in the linear programming formulation in Chapter 6.

It is assumed that granularity of V_{th} allocation is at the NMOS/PMOS stack level. For NMOS (PMOS) transistors, the high threshold voltage is 0.20V (–0.20V) and the low threshold voltage is 0.10V (–0.10V). Different levels of variability in L_{eff} were explored ranging from 3% to 8% of σ/μ. Pelgrom's model [30] is used to describe σ_{Vth} dependence on transistor size. The assumed magnitude of V_{th} variability is $\sigma/\mu = 7\%$. An equal breakdown of variability into global and local components was used. Spatial correlation of local variability was not considered, but could be incorporated into the algorithm if needed.

The fundamental reason for the reduction in power enabled by statistical optimization is the ability of the statistical algorithm to explicitly account for the variance of constraint and objective functions. Because of this statistical optimization allots slack more efficiently in that it penalizes allocation of slack to gates with high power variance. As a result, the spread of the leakage distribution is reduced and the mean is shifted towards lower values. Figure 12.6 shows the probability distribution function of the static power obtained by Monte Carlo simulation of the circuit configurations produced by the statistical and deterministic optimizations. The figure indicates that the static power savings increase at higher percentiles. Another manifestation of the greater effectiveness of statistical optimization is the fact that it can assign more transistors to a high V_{th}. For example, for the c432 ISCAS'85 benchmark [6] optimized for a target delay of 0.55ns for 99.9% timing and power yields, the number of transistors set to high V_{th} by the statistical algorithm is 20% more than the corresponding number for the deterministic algorithm.

The comparison of statistical optimization and deterministic optimization results is further illustrated in Figure 12.7. Under the same power and timing yield constraints ($\zeta = \eta = 99.9\%$), statistical optimization produces uniformly better power-delay curves. The improvement strongly depends on the underlying structure of physical process variation. As the amount of uncorrelated variability increases, i.e. the local component grows in comparison with the global component, the power savings enabled by statistical optimization increase. The power savings at the 95th percentile are 23%, and those at 99th percentile are 27% respectively. The ability to directly control the level of parametric power- and timing-limited yield permits choosing a "sweet spot" in the power-delay space.

Figure 12.8 and Figure 12.9 show a set of power-delay curves for one of the benchmarks, c432. Figure 12.8 plots the total power vs. delay at the output obtained by running the statistical optimization for various timing yield levels (ζ), with the power yield set at 99.9%. It can be observed that at tight timing constraints the difference in power optimized for different yield levels is significant. Figure 12.9 confirms that optimizing the circuit for a lower power yield will lead to higher total power consumption and longer delay. For the same yield, the trade-off between power and arrival time is much more marked at tighter timing constraints.

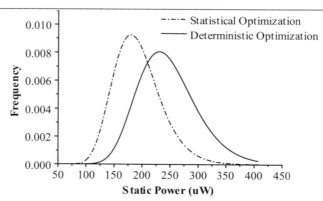

Figure 12.6 The probability distribution functions of static (leakage) power produced by a Monte Carlo simulation of the benchmark circuit (c432) optimized by the deterministic and statistical algorithms.

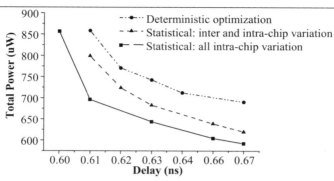

Figure 12.7 Power-delay curves for 99.9% timing and power yield. Statistical optimization does uniformly better. For the case of mixed inter- and intra-chip variability, an equal breakdown is assumed.

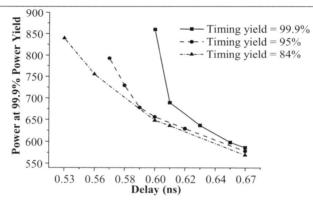

Figure 12.8 Power-delay curves at different timing yield levels for the c432 benchmark. At larger delay, the power penalty for higher yield is smaller.

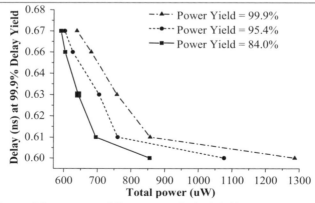

Figure 12.9 Power-delay curves at different power limited yields

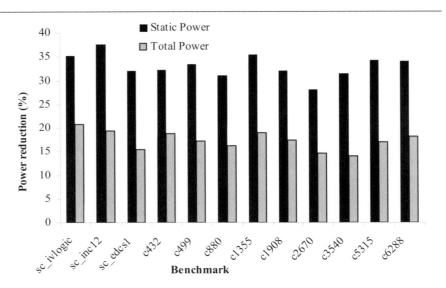

Figure 12.10 Savings in total and leakage power enabled by the statistical algorithm across the benchmark circuits (the ISCAS'85 benchmark circuits [6] and IBM benchmarks courtesy of A. Devgan). Average savings of 33% in leakage power and 17% in total power are obtained.

Figure 12.10 captures the savings in power obtained by employing the statistical optimization algorithm outlined in this section. The average leakage power savings are 33%, which can be achieved without the loss of timing or power yield by statistical optimization, as opposed to the deterministic approach.

Figure 12.11 shows the run-time behavior of the algorithm. The characteristics of the circuits on which the algorithm was tested are shown in Table 12.2. The algorithm was run on a dual core 1.5GHz AMD Athlon workstation with 2GB of RAM. The optimization problems were solved using the interior point optimization package MOSEK [23]. A single SOCP optimization run of c6288 for slack assignment takes about 11 seconds. It can be seen that the run-time is roughly linear in circuit size making the algorithm scalable to large industrial blocks. Note that quadratic runtime growth has been observed by other authors for linear programming (LP) in some cases (see Section 6.6).

The formulation of dual Vth assignment and gate sizing based on SOCP is more than an order of magnitude faster than a coordinate descent algorithm based on [37]. This speedup is obtained due to the special structure of the SOCP program, which is not available to general nonlinear problem solvers, enabling the optimization problem to be solved extremely efficiently. We observed that the constraint matrix of the SOCP formulation is quite sparse – this makes the solution of the SOCP problem quite efficient.

Table 12.2 Circuit characteristics and run time.

Circuit	Number of gates	Number of Inputs	Number of Outputs	Logic Depth	Run Time (s)
sc_ivlogic	40	8	6	9	9
sc_inc12	78	16	9	8	10
sc_edcs1	258	28	12	8	30
c432	261	36	7	23	31
c499	641	41	32	23	52
c880	615	60	26	22	47
c1355	685	41	32	18	56
c1908	1,238	33	25	29	122
c2670	2,041	233	140	25	153
c3540	2,582	50	22	44	171
c5315	3,753	178	123	27	241
c6288	2,704	32	32	88	273

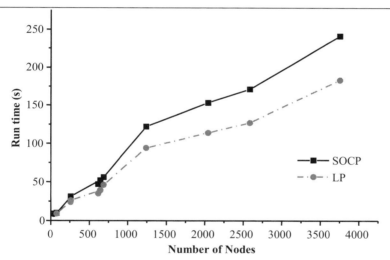

Figure 12.11 Run time behavior of the statistical total power optimization algorithm (SOCP). This is compared to the runtime for solving the deterministic linear programming (LP) problem. Runtime grows linearly with circuit size.

12.6   SUMMARY

In this chapter, we have analyzed the impact of variability on power and its impact on circuit performance and yield. In the recent past it was sufficient to model the impact of variability on timing. With high-end designs experiencing a double-sided squeeze on parametric yield due to the power-dissipation limits, power variability needs to be explicitly taken into account. This requires the adoption of new analysis and optimization methodologies that incorporate the notion of power-limited parametric yield loss. While there are currently no commercially available CAD tools for parametric yield optimization, the area of parametric yield analysis and optimization is rapidly developing, and it can be expected that such tools will soon appear on the market. Continued progress in this tool development area will help ASIC designers deal with variability in a far more effective fashion.

12.7   REFERENCES

[1]   Bai, X., et al., "Uncertainty aware circuit optimization," in *Proc. of Design Automation Conference*, 2002, pp. 58-63.

[2]   Boning, D., and Nassif, S., "Models of Process Variations in Device and Interconnect," *Design of High-Performance Microprocessor Circuits*, A. Chandrakasan (ed.), 2000.

[3]   Borkar, S., et al., "Parameter variation and impact on Circuits and Microarchitecture," in *Proc. of Design Automation Conference,* 2003, pp. 338-342.

[4]   Bowman, K., and Meindl, J., "Impact of within-die parameter fluctuations on the future maximum clock frequency distribution," in *Proc. of IEEE Custom Integrated Circuits Conference*, 2001, pp. 229-232.

[5]   Boyd, S., and Vandenberghe, L., *Convex Optimization*, New York, NY, Cambridge University Press, 2004.

[6]   Brglez, F., and Fujiwara, H., "A neutral netlist of 10 combinational benchmark circuits and a target translator in Fortran," in *Proc. International Symposium on Circuits and Systems*, 1985, pp. 695-698.

[7]   Brodersen, R., et al., "Methods for True Power Minimization," in *Proc. of International Conference on Computer Aided Design*, 2002, pp. 35-40.

[8]   Burnett, D., et al., "Implications of Fundamental Threshold Voltage Variations for High - Density SRAM and Logic circuits," in *Proc. Of Symposium on VLSI Technology*, 1994, pp. 15-16.

[9]   Cao, Y., et al., "New paradigm of predictive MOSFET and interconnect modeling for early circuit design," in *Proc. of IEEE Custom Integrated Circuits Conference*, 2000, pp. 201-204.

[10] Chang, E., et al., "Using a Statistical Metrology Framework to Identify Systematic and Random Sources of Die- and Wafer-level ILD Thickness Variation in CMP Processes," in *Proc. of International Electron Devices Meeting*, 1995, pp. 499-502.

[11] Chen, C., Chu, C., and Wong, D., "Fast and exact simultaneous gate and wire sizing by Lagrangian relaxation," in *Proc. of International Conference on Computer Aided Design*, 1998, pp. 617-624.

[12] Fishburn, J., and Dunlop, A., "TILOS: A Posynomial Programming Approach to Transistor Sizing," in *Proc. of International Conference on Computer Aided Design*, 1985, pp. 326-328.

[13] Fitzgerald, D., "Analysis of polysilicon critical dimension variation for submicron CMOS processes," *M.S. thesis, Dept. Elect. Eng. Comp. Sci., Mass. Inst. Technol., Cambridge*, June 1994.

[14] Hakim, N., "Tutorial on Statistical Analysis and Optimization," *Design Automation Conference,* 2005.

[15] Jacobs, E., and Berkelaar, M., "Gate sizing using a statistical delay model," in *Proc. of Design Automation Conference*, 2000, pp. 283-290.

[16] Keshavarzi, A., et al., "Measurements and modeling of intrinsic fluctuations in MOSFET threshold voltage," in *Proc. of International Symposium on Low Power Electronics and Design*, 2005, pp. 26-29.

[17] Lenevson, M., Viswanathan, N., and Simpson, R., "Improving resolution in photolithography with a phase-shifting mask," *IEEE Transactions On Electron Devices*, vol. 29 (11), pp. 1828-1836, 1982.

[18] Lee, D., Blaauw, D., and Sylvester, D., "Gate Oxide Leakage Current Analysis and Reduction for VLSI Circuits," *IEEE Transactions on Very large Scale Integration (VLSI) Systems*, vol. 12(2), pp.155-166, February 2004.

[19] Mani, M., and Orshansky, M., "A new statistical optimization algorithm for gate sizing," *Proc. of International Conference on Computer Design*, 2004, pp. 272 – 277.

[20] Mani, M., Devgan, A., and Orshansky, M., "An Efficient Algorithm for Statistical Minimization of Total Power under Timing Yield Constraints," in *Proc. of Design Automation Conference*, 2005, pp. 309-314.

[21] Markovic, D., et al., "Methods for true energy-performance optimization," *IEEE Journal of Solid-State Circuits*, vol. 39, no. 8, pp. 1282-1293, Aug. 2004.

[22] Mehrotra, V., et al., "Modeling the effects of manufacturing variation on high-speed microprocessor interconnect performance," in *International Electron Devices Meeting Technical Digest*, 1998, pp. 767-770.

[23] MOSEK ApS, *The MOSEK optimization tools version 3.2 (Revision 8), User's manual and reference.* http://www.mosek.com/documentation.html#manuals

[24] Nassif, S., "Delay Variability: Sources, Impact and Trends," in *Proc. of International Solid-State Circuits Conference*, 2000, pp. 368-369.

[25] Nassif, S., "Statistical worst-case analysis for integrated circuits," *Statistical Approaches to VLSI*, Elsevier Science, 1994.

[26] Nassif, S., "Within-chip variability analysis," in *International Electron Devices Meeting Technical Digest*, 1998, pp. 283-286.

[27] Nguyen, D., et al., "Minimization of dynamic and static power through joint assignment of threshold voltages and sizing optimization," in *Proc. of International Symposium on Low Power Electronics and Design*, 2003, pp. 158-163.

[28] Orshansky, M., Chen, J., and Hu, C., "A Statistical Performance Simulation Methodology for VLSI Circuits," in *Proc. of Design Automation Conference*, 1998, pp. 402-407.

[29] Patil, D., et al., "A New Method for Design of Robust Digital Circuits," in *Proc. of International Symposium on Quality of Electronic Design*, 2005, pp. 676-681.

[30] Pelgrom, M., Duinmaijer, A., and Welbers, A., "Matching properties of MOS transistors," *IEEE Journal of Solid-State Circuits*, Vol. 24, pp. 1433-1439, Oct. 1989.

[31] Prekopa, A., *Stochastic Programming*, Kluwer Academic, 1995.

[32] Rao, R., et al., "Parametric Yield Estimation Considering Leakage Variability," in *Proc. of Design Automation Conference*, 2004, pp. 442-447.

[33] Seung, P., Paul, B., and Roy, K., "Novel sizing algorithm for yield improvement under process variation in nanometer technology, "in *Proc. of Design Automation Conference, 2004*, 2004, pp. 454-459.

[34] Semiconductor Industry Association, International Technology Roadmap for Semiconductors, 2001.

[35] Singh, J., et al., "Robust gate sizing by geometric programming," in *Proc. of Design Automation Conference*, 2005, pp. 315-320.

[36] Sirichotiyakul, S., et al., "Stand-by power minimization through simultaneous threshold voltage selection and circuit sizing," in *Proc. of Design Automation Conference*, 1999, pp. 436-441.

[37] Srivastava, A., Sylvester, D., and Blaauw, D., "Statistical optimization of leakage power considering process variations using dual-V_{th} and sizing," in *Proc. of Design Automation Conference*, 2004, pp. 773-778.

[38] Stine, B., Boning, D., and Chung, J., "Analysis and decomposition of spatial variation in integrated circuit processes and devices," *IEEE Transactions On Semiconductor Manufacturing*, vol. 1, pp. 24-41, Feb. 1997.

[39] Stine, B., et al., "A Closed-Form Analytic Model for ILD Thickness Variation in CMP Processes," in *Proc. of CMP-MIC*, pp. 266-273, 1997.

[40] Sundararajan, V., Sapatnekar, S., and Parhi, K., "Fast and Exact Transistor sizing Based on Iterative Relaxation," *IEEE Transactions on Computer Aided Design*, vol. 21, no. 5, pp. 568-581, May 2002.

[41] Taur, Y., and Ning, T., *Fundamentals of Modern VLSI Devices*, Cambridge Univ. Press, 1998.

[42] Takeuchi, K., Tatsumi, T., and Furukawa, A., "Channel Engineering for the Reduction of Random-Dopant-Placement-Induced Threshold Voltage Fluctuations," in *International Electron Devices Meeting Technical Digest*, 1997, pp. 841-844.

[43] Wang, Q., and Vrudhula, S., "Static power optimization of deep submicron CMOS circuit for dual V_{th} technology," *Proc. of International Conference on Computer Aided Design*, 1998, pp. 490-496.

Chapter 13

PUSHING ASIC PERFORMANCE IN A POWER ENVELOPE

Leon Stok, Ruchir Puri, Subhrajit Bhattacharya, John Cohn
IBM Research, Yorktown Hts, NY
IBM Microelectronics, Essex Jn, VT
leonstok,ruchir,sbhat,johncohn@us.ibm.com

Dennis Sylvester, Ashish Srivastava, Sarvesh Kulkarni
EECS, University of Michigan,Ann Arbor, MI
dmcs,ansrivas,shkulkar@umich.edu

Power dissipation is becoming the most challenging design constraint in nanometer technologies. Among various design implementation schemes, standard cell ASICs offer the best power efficiency for high-performance applications. The flexibility of ASICs allow for the use of multiple voltages and multiple thresholds to match the performance of critical regions to their timing constraints, and minimize the power everywhere else. We explore the trade-off between multiple supply voltages and multiple threshold voltages in the optimization of dynamic and static power.

The use of multiple supply voltages presents some unique physical and electrical challenges. Level shifters need to be introduced between the various voltage regions. Several level shifter implementations are discussed. The physical layout needs to be designed to ensure the efficient delivery of the correct voltage to various voltage regions. More flexibility can be gained by using appropriate level shifters.

To conclude this chapter, we present a semi-custom design methodology which illustrates the benefit of a subset of these low-power optimization techniques using a DSP (digital signal processor) chip for satellite communication. Chips for satellite communications have very stringent requirements on power dissipation but require significant processing capability. These classes of chips are therefore an excellent test for a methodology that brings many of the low power optimizations together.

13.1   INTRODUCTION

Power efficiency is becoming an increasingly important design metric in deep submicron designs. ASICs have a significant power advantage over other implementation methods. A dedicated ASIC will have a significantly better power-performance product than a general purpose processor or regular fabrics such as FPGAs. For designs that push the envelope of power and performance, ASIC technology remains the only choice. However, the cost pressures in nanometer technologies are forcing designers to push the limits of design technology in order to fully exploit increasingly complex and expensive technology capabilities. In this chapter, we discuss technology, circuit, layout and optimization techniques to improve the power delay product. We focus on the issue of pushing ASIC performance in a power envelope by exploiting the use of multiple supply voltages (Vdd) and multiple device thresholds (Vth). In Section 13.2, we discuss the trade-off between multiple Vdd and multiple Vth options to optimize power. In Section 13.3, we present novel design techniques to physically implement fine-grained generic voltage islands for multiple-Vdd implementations. In the context of multi-Vdd implementation, we also present some novel level conversion circuits which can be used to implement very flexible voltage island schemes. Finally, we present a design case study to show the relative impact of some design techniques in a low-power ASIC methodology.

13.2   POWER-PERFORMANCE TRADE-OFF WITH MULTI-VDD AND MULTI-VTH

This section explores the trade-off between multiple supply voltages and multiple threshold voltages in the optimization of dynamic and static power. From a dynamic power perspective, supply voltage reduction is the most effective technique to limit power. However, the delay increase with reducing Vdd degrades the throughput of the circuit. Similarly, to reduce static power an increase in Vth provides exponential improvements, again at the expense of speed. To counter the loss in performance, dual Vdd [5][33] and dual Vth [22][25][34] techniques have been proposed. These approaches assign gates on critical paths to operate at the higher Vdd or lower Vth and non-critical portions of the circuit operate at lower Vdd or higher Vth, reducing the total power consumption without degrading performance (held fixed as a constraint). These techniques have been successfully implemented, but most of the existing work focuses on one of these techniques in isolation as opposed to jointly.

Previous work [11] estimates the optimal Vdd and Vth values to be used in multi-voltage systems to minimize either dynamic or static power respectively. They confirm earlier work [32] claiming that, in a dual Vdd system the optimal lower Vdd is 60-70% of the original Vdd. In general,

[10][32] have found optimized multi-Vdd systems to provide dynamic power reductions of roughly 40-45%. In [29], it is shown that intelligently reducing Vth in multi-Vdd systems can offset the traditional delay penalties at low-Vdd with lessened static power consequences (due to both the reduced Vdd and the off-state leakage current levels). In order to explore the achievable design envelope in a joint multiple Vdd and Vth environment, we abstract a generic CMOS network as a set of non-intersecting parallel paths. We then formulate a linear programming problem to minimize power by assigning capacitance (representing gates) on these paths to a combination of supply and threshold voltages (assuming a known initial path delay distribution) [26].

We perform a path-based analysis of a generic logic network to estimate the power improvement obtained by applying multiple supply voltages and multiple threshold voltages. To simplify the problem, we assume node and edge disjoint paths, as stated above. We also assume that it is possible to apply a combination of supplies and thresholds to any fraction of the total path capacitance. This is equivalent to stating that extended clustered voltage scaling (ECVS) is used, which allows for asynchronous level conversion anywhere along a path [32]. While we do not explicitly consider overhead due to level conversion in most of this work, we describe various level converter topologies and the impact of their power and delay penalties.

Consider Vdd_1 and Vth_1 to be the supply and threshold voltages in a single Vdd/Vth system. If C_{total} is the total path capacitance of a path, then the total dynamic power dissipation is simply expressed as

$$P_{dynamic} = fC_{total}Vdd_1^2 \qquad (13.1)$$

where f is the frequency of operation.

Considering the same path implemented in an n-Vdd/m-Vth design, we define $C_{i,j}$ as the capacitances operating at a supply voltage Vdd_i and threshold voltage Vth_j. If we define the capacitance (C_i) to be the capacitance operating at a supply voltage Vdd_i, it can be expressed as

$$C_i = \sum_{j=1}^{m} C_{i,j} \qquad (13.2)$$

The total dynamic power dissipation can now be expressed as

$$P_{dynamic} = f\left(\left[C_{total} - \sum_{i=2}^{n} C_i\right]Vdd_1^2 + \sum_{i=2}^{n} C_i Vdd_i^2\right) \qquad (13.3)$$

The first term in Equation (13.3) corresponds to the capacitance operating at Vdd_1 and is obtained by subtracting the sum of the capacitances operating at voltages other than Vdd_1 from the total path capacitance $C_{1,1}$. Now the ratio of the dynamic power dissipation to the original design, obtained by dividing Equation (13.3) by Equation (13.1), can be expressed as

$$Gain_{dynamic} = 1 - \frac{1}{C_{total}} \sum_{i=2}^{n} \left(C_i \left[1 - \left(\frac{Vdd_i}{Vdd_1} \right)^2 \right] \right) \tag{13.4}$$

The static power can be expressed similarly. If W_{total} is the total device width (both PMOS and NMOS), then the static power dissipation due to subthreshold leakage, with Vdd_1 and Vth_1 only, is of the form

$$P_{static} = kVdd_1 W_{total} 10^{-Vth_1 / S} \tag{13.5}$$

where S is the subthreshold swing (typically given in units of mV/decade), and k is a constant depending on device parameters and temperature. The reduction in static power in low-Vdd devices is due to: DIBL; the lower Vdd itself; and other complex device-related phenomena such as the relationship among doping, Vth, and S [15]. DIBL occurs because the drain bias (V_{ds}) creates a large drain/substrate depletion region, leading to a reduced Vth. The typical model for DIBL is linear with V_{ds}:

$$Vth = Vth_0 - \eta V_{ds} \tag{13.6}$$

In this model η is the DIBL coefficient and is typically in the range of 60 to 110mV/V, and V_{th0} is the nominal long-channel threshold voltage in the absence of DIBL. Since V_{ds} is Vdd in typical leakage scenarios, a reduction in Vdd for a given device leads directly to a rise in Vth and an exponentially smaller subthreshold leakage current. To capture these effects we assume that static power is proportional to the square of the supply voltage rather than the linear relationship expressed in Equation (13.5). If $W_{i,j}$ is the device width (both PMOS and NMOS) at supply voltage Vdd_i and threshold voltage Vth_j then in an n-Vdd/m-Vth design, the static power can be expressed as

$$P_{static} = k \left(\left[W_{total} - \sum_{(i,j) \neq (1,1)} W_{i,j} \right] Vdd_1^2 10^{-Vth_1 / S} + \sum_{(i,j) \neq (1,1)} W_{i,j} Vdd_i^2 10^{-Vth_j / S} \right) \tag{13.7}$$

The gain in static power is given by

$$Gain_{static} = 1 - \frac{1}{W_{total}} \sum_{(i,j) \neq (1,1)} W_{i,j} \left(1 - \left[\frac{Vdd_i}{Vdd_1} \right]^2 10^{-(Vth_j - Vth_1)/S} \right) \tag{13.8}$$

While our results use Equation (13.8) to reflect the relationship between I_{off} and Vdd, experiments using a linear (Vdd_i/Vdd_1) term rather than quadratic to represent static power gains showed only minor changes in the overall power reductions and optimal Vdd/Vth values.

The change in delay D when Vdd or Vth is changed is estimated using the alpha-power law model [23]:

$$D_{i,j} = \left(\frac{Vdd_i}{Vdd_1}\right)\left(\frac{Vdd_1 - Vth_1}{Vdd_i - Vth_j}\right)^\alpha \qquad (13.9)$$

To obtain the minimum power dissipation condition we note that at the minima

$$\frac{\partial(P_d + P_s)}{\partial x} = \frac{\partial P_d}{\partial x} + \frac{\partial P_s}{\partial x} = 0 \qquad (13.10)$$

where P_d is the dynamic power dissipation and P_s is the static power dissipation, and x represents a design variable such as Vdd or Vth. Let P_{d0} represent the dynamic power in the initial design and P_{s0} represent the static power consumption of the initial design. If we minimize a weighted sum of the gains, where the gains are as expressed in equations (13.4) and (13.8), we obtain

$$\frac{\partial(K(P_d / P_{d0}))}{\partial x} + \frac{\partial(P_s / P_{s0})}{\partial x} = 0 \qquad (13.11)$$

which can be expressed as

$$\frac{KP_{s0}}{P_{d0}}\frac{\partial P_d}{\partial x} + \frac{\partial P_s}{\partial x} = 0 \qquad (13.12)$$

Comparing equations (13.12) and (13.10), we infer that if we minimize a weighted sum of the gains in power and define the weighting factor K as the ratio of dynamic and static power at the initial design point (i.e., $K=P_{d0}/P_{s0}$) we minimize the total power dissipation as well.

As shown in [11], the capacitance and transistor width along a path are largely proportional to the path's delay. Hence the ratios of widths in Equation (13.8) can be replaced by ratios of capacitance. At this point the problem of power minimization for given voltages and thresholds can be formulated as a linear programming problem with the ratios of capacitances as the variables.

For an n-Vdd/m-Vth design, there is a corresponding design space over the allowed range of values for these supply and threshold voltages other than the initial supply and threshold voltage. For example, for a 2-Vdd/3-Vth design, points are of the form (Vdd_2, Vth_2, Vth_3), where $Vdd_i \in [0.6V, 1.2V]$ and $Vth_i \in [0.08V, 0.3V]$, and we assume that Vdd_1 is fixed at 1.2V and Vth_1 is fixed at 0.3V. For each of these design space points (step size 0.01V between points), the problem is formulated and the ratios of capacitance corresponding to different path delays are obtained as a solution of the linear program. The ratios of capacitance are then integrated over the path-delay distribution to obtain the total capacitance operating at each combination of Vdd and Vth. Again, we define the weighting factor K as the ratio of the dynamic to static power in the original single Vdd/Vth design (e.g., $K = 10$

implies that 10/11 of the total initial power was dynamic). As described above, total power minimization is achieved by minimizing a weighted sum of the static and dynamic power. Hence the goal of total power reduction can now be expressed as

$$\text{maximize} \quad K \cdot Gain_{dynamic} + Gain_{static}$$
$$\text{subject to} \quad \left(1 + \sum_{i,j} \frac{C_{i,j}}{C_{total}} \left(D_{i,j} - 1\right)\right) t \le 1 \tag{13.13}$$

which can be simplified to

$$\text{maximize} \quad K \cdot Gain_{dynamic} + Gain_{static}$$
$$\text{subject to} \quad t \sum_{i,j} \frac{C_{i,j}}{C_{total}} D_{i,j} \le 1 \tag{13.14}$$

where t is the original path delay normalized by the critical path delay (i.e. $t \le 1$). The constraint in Equation (13.13) is obtained by multiplying the delay contributed by the fraction of capacitance $C_{i,j}$ by the factor $(D_{i,j} - 1)$, which reflects the increase in delay. This increase in delay is added to the original path delay to obtain the final path delay. The constraint forces the delay of each path to be less than the critical delay of the network (which is normalized to 1), thus we maintain the operating frequency of f. Since paths are independent of each other, minimizing the power dissipation on each of the paths will lead to the minimum power of the complete logic network.

To determine the power savings that may be achieved, we weight the occurrence of paths by how often a given delay t occurs, $p(t)$. Any generic $p(t)$ can be used within this framework to estimate the achievable power reduction using multiple supply and threshold voltages. Note that $p(t)$ plays a key role in the optimization procedure through the constraint in Equation (13.14), although it does not actually appear in either $Gain_{dynamic}$ or $Gain_{static}$. These gain terms only serve to compute the power reductions for a given Vdd and Vth assignment; they do not consider the validity of each given assignment with respect to the timing constraint. See [26] for further details.

This general framework is similar to [11], but enables several key enhancements: 1) minimizes total power consumption, defined as the sum of static and dynamic components, 2) simultaneously optimizes both Vdd and Vth to achieve this goal, and 3) considers DIBL (drain-induced barrier lowering), which strongly limits the achievable power reduction in a multi-Vdd, single Vth environment. Our results indicate that the total power reduction achievable in modern and future integrated circuits is on the order of 60-65% using the dual Vdd/Vth technique (Figure 13.1 and Figure 13.2). A key factor when optimizing a multi-Vdd/Vth system is the parameter K which is the ratio of dynamic to static power in the original single Vdd/Vth design, i.e., $K = P_{dynamic} / P_{static}$.

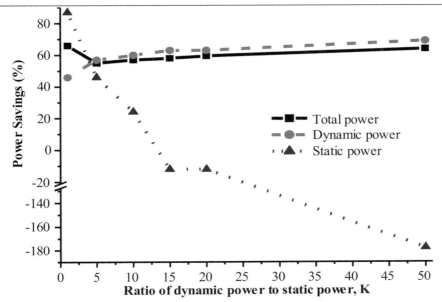

Figure 13.1 Breakdown of total power savings into static and dynamic components with dual Vdd/dual Vth, where $Vdd_1 = 1.2V$ and $Vth_1 = 0.3V$.

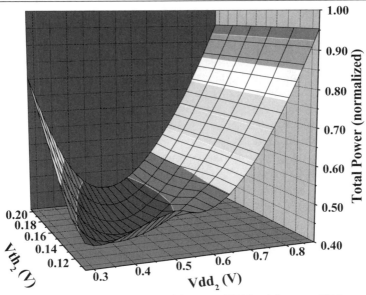

Figure 13.2 Power reduction as a function of the second Vdd and the second Vth.

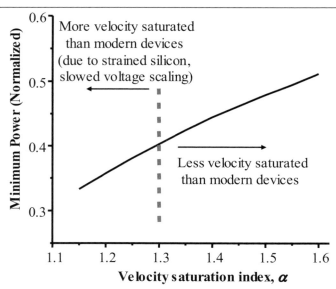

Figure 13.3 Future devices may be more velocity saturated, resulting in lower power consumption.

Larger K values push the optimization towards lower Vdd and lower Vth to address the dominant dynamic power. An important finding is that the optimal second Vdd in multi-Vth systems is about 50% of the higher supply voltage, in contrast with a lower Vdd value of 60%-70% of the higher Vdd for single Vth designs as found previously. An implication of this finding is that level converter structures must be capable of converting over a larger relative range. This seems feasible provided the level converters themselves leverage multiple threshold voltages. However, the delay associated with the level converters themselves limits the amount of achievable power reduction. The inclusion of level conversion delay penalties demonstrates the trade-off between allocating available slack to level conversion and achievable power reductions. Typically, one to two asynchronous level conversions per path are tolerable in designs with larger logic depths (30+ FO4 delays) with <15% power penalty. Also, continued aggressive channel length scaling (without commensurate supply voltage reductions) and new device structures such as strained-Si channels point to increasingly velocity saturated (α closer to 1 in Equation (13.9)) devices that are ideal for voltage scaling (Figure 13.3), since the drive current of a gate, and hence the gate delay, becomes less sensitive to reduction in supply voltage.

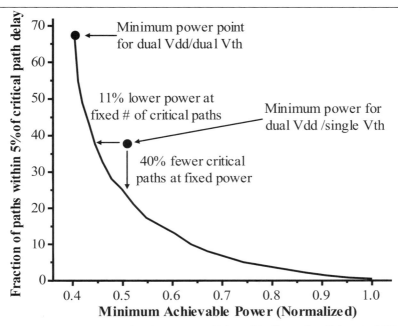

Figure 13.4 Dual-Vdd/Vth provides better power/delay criticality trade-off than dual-Vdd for same power.

Additionally, we note the relationship between power savings and critical path density (which is defined to be the fraction of paths within 5% of the critical path delay); this is important since a rapidly increasing number of critical paths combined with rising process variability increases design times and emphasizes a need for incremental statistical timing analysis tools. Dual Vdd/Vth offers better control of the slack-power trade-off compared to dual Vdd only as shown in Figure 13.4.

In future designs that are both power and variability-constrained, the design space of Figure 13.4 may become crucial. For designs that do not demand ultra low power, designers can avoid the physical design issues associated with the use of multiple supply voltages on a chip by aggressive scaling of a single Vdd combined with multiple device threshold voltages (as illustrated by the case study in Section 13.4). For instance, the use of 1.2V as Vdd for 130nm technologies is commonplace and assumed in the above discussion. However, the use of a single 0.9V supply with a small subset of gates using an ultra-low Vth to maintain speed may yield lower overall power. To investigate this possibility, we use the same design space exploration tool as above to look at the efficacy of single Vdd/multi-Vth design. Again, we normalize power to the single Vdd, single Vth design point.

Table 13.1 This table shows the power consumption that may be achieved using single Vdd and dual Vdd with dual Vdd and dual Vth, compared to an initial design point with single Vdd of 1.2V and single Vth of 0.3V. The columns on the right hand side of the table show the optimal supply and optimal threshold voltages (where the first threshold voltage was fixed at 0.3V) for the single Vdd results.

K	Minimum achievable power (normalized to single Vdd/Vth)			Optimal Supply Voltage (V) for		Optimal Threshold Voltages (V) for	
	Dual Vdd/ Dual-Vth	Single Vdd/ Dual-Vth	Single Vdd/ Triple-Vth	Single Vdd/ Dual-Vth	Single Vdd/ Triple-Vth	Single Vdd/ Dual-Vth	Single Vdd/ Triple-Vth
1	0.34	0.54	0.48	1.20	1.10	0.44	0.25, 0.42
5	0.45	0.67	0.62	0.93	0.87	0.19	0.16, 0.23
10	0.43	0.63	0.56	0.89	0.81	0.17	0.14, 0.21
15	0.42	0.61	0.52	0.89	0.75	0.17	0.12, 0.19
20	0.41	0.58	0.49	0.83	0.75	0.15	0.12, 0.19
50	0.36	0.50	0.41	0.77	0.69	0.13	0.10, 0.17

In Table 13.1 we see that the potential improvements from a single Vdd/multi-Vth system can be quite substantial especially when K is large. For a reasonable K value of 10, a single Vdd system can provide 65-77% of the gains that dual Vdd/Vth shows depending on the number of threshold voltages used (2 or 3). Furthermore, the numbers for dual Vdd/Vth do not include level conversion penalties so can be considered as best-case power reductions. Contrary to the dual Vdd case, the inclusion of a third Vth when a single optimized (flexible) supply voltage is used provides appreciable gains beyond the dual-Vth system. Since each extra mask step for an additional Vth level increases the wafer fabrication cost by 3%, use of multiple supply voltages by itself remains a very attractive choice for power-reduction. In the following section, we discuss the electrical and physical design issues of multiple Vdd implementations.

13.3   DESIGN ISSUES IN MULTI-VDD ASICS

Design of ASICs with multiple supply voltages presents some unique electrical and physical design challenges. In this section, we present some novel solutions to these challenges.

13.3.1   Circuit Design Issues

Electrically, to avoid excessive static power consumption between the low and high voltage regions, voltage level converters need to be inserted. Minimizing the overhead of level converter insertion while meeting interfacing constraints presents a significant challenge. In this section, we describe some novel level converter circuits which not only provide efficient delay and power characteristics but also enable very flexible physical design of multi-Vdd schemes.

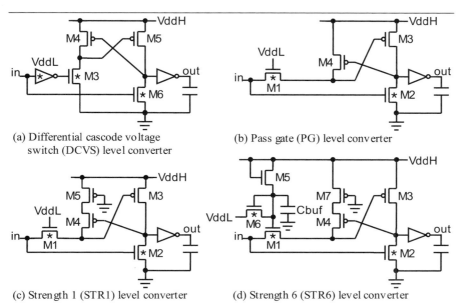

Figure 13.5 Different dual-supply voltage designs for level converters. Transistors and the inverter with ★ indicate low-Vth devices.

We have developed several versions of the low-energy asynchronous pass-gate (PG) based level converter from [10]. Figure 13.5 shows the two existing level converters (DCVS and PG) and the new level converters (STR1 and STR6) [18]. The first, STR1, relies on a known high-performance dynamic logic technique of splitting the keeper into two devices to minimize the capacitive load on the actual gate. STR6, while including the technique used in STR1, employs the threshold drop of M5 to create a higher gate voltage for the pass-transistor and effectively speed it up. Transistor M6 is added to ensure that the gate voltage of M1 does not exceed $VddL + Vth_{M1}$ which would yield reverse leakage current into VddL (where VddL is the lower Vdd). In comparison to the DCVS (Differential Cascode Voltage Swing) level converter, STR6 is up to 25% faster at the optimal delay point or consumes up to 60% less energy at fixed delay. STR1 has a simpler design and enables 30- 40% lower energy than DCVS and 15-30% lower energy than the PG structure. Furthermore, we investigated the use of STR1 for embedded logic functionality and found that it is 4% faster with 55% lower energy than a 2-input NAND DCVS gate when VddL is 0.8V (VddH=1.2V, where VddH is the higher Vdd).

Maintaining robustness is an important concern when circuits are operated at low voltages. Also, the circuits discussed above have a pass transistor at the input. They may thus appear to have more susceptibility to noise because of the lack of input isolation. However, as we explain below, this is not the case with these circuits since the exposed pass transistor is always tied high.

The proposed circuits were found to be closely comparable in robustness to the DCVS circuit and other standard logic gates such as inverters. While typical pass-transistor circuits require input isolation as they may pass erroneous values that are sampled on the output, the PG-based level converters in this work only use their pass transistor to pass the input voltage to an internal node that is connected to the gate of another MOSFET. Since the pass transistor is always ON, there is no chance of a noisy signal being sampled (i.e., disconnected from the input) and stored on the internal node. Thus, from a noise perspective the circuit becomes similar to the case where the input is tied directly to the gate of the pull-up PMOS (e.g., M3 in PG in Figure 13.5(b)). In particular, the problematic 'Pass 0' noise source [4] where a negative noise pulse on the input can turn ON an NMOS device with 0V at its gate and mistakenly pass a 0 to the output, cannot occur here since the input to the pass transistor is tied high. We studied and compared the robustness of the various level converters by adopting the following methods to represent typical on-chip switching behavior.

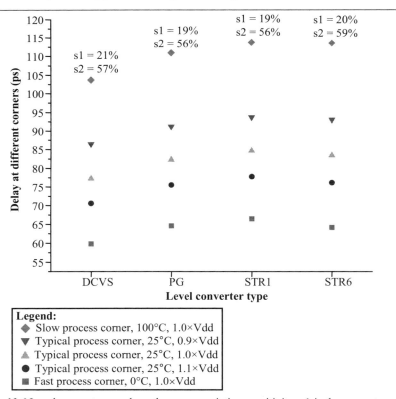

Figure 13.6 Level converter supply and process variation sensitivity. s1 is the percent spread of delay at ±10% Vdd corners measured from the typical corner with 1.0×Vdd. s2 is the percent spread of delay at fast and slow process corners from the typical corner with 1.0×Vdd.

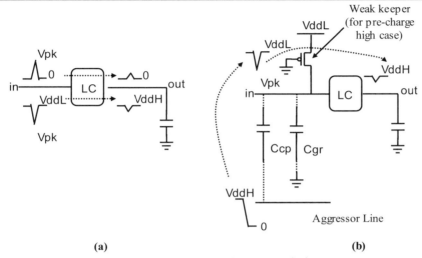

| (a) | (b) |

Figure 13.7 Circuits for voltage level converter robustness analysis.

We first studied the performance of the level converters at different process corners and with varying power supply voltage and temperature. This study gives insight into the sensitivity of each of the circuits to such variations. Results using VddL = 0.8V are shown in Figure 13.6. We studied the delay of all level converters with ±10% DC supply noise on both VddL and VddH, at 25°C and the typical process corner; with nominal Vdd at 0°C and the 130nm fast process corner; and with nominal Vdd at 100°C and the 130nm slow process corner. The delay variation is nearly the same for all level converters and shows acceptable spread. For comparison, the FO4 inverter delay in this technology varies by 18% and 51% for ±10% Vdd variation and fast/slow process respectively with these numbers rising to 20% and 56% at reduced voltages.

In addition, triangular noise pulses with base width of 80ps (2 FO4 inverter delays) and peak magnitude of 0.3V (25% of VddH and 37.5% of VddL in this case) were applied as inputs to each of the level converters when they were sized for optimal speed. In all cases, there was no output glitching whatsoever, implying that these asynchronous level converters are tolerant of substantial input noise. The static voltage transfer characteristics all show large gain in their transition regions which are within 50mV of VddL/2 in all cases.

Since circuit robustness is expected to be worst for the lowest supply voltages (VddL = 0.6V), we further investigated the robustness at such low voltages. We applied more pessimistic triangular noise pulses of width equaling 120ps (twice the FO4 delay at VddL = 0.6V) and varied the amplitude (Vpk) until the circuit failed (i.e., the output reaches 0.5 × nominal_output_high_voltage; the nominal_output_high_voltage for the level converters in our studies is 1.2V, while for the inverter being studied for comparison here, it is 0.6V). Figure 13.7(a) shows this setup. We compared the DCVS and STR6 level converters to an inverter (with similar input and output capacitance) and observed that the circuit robustness of these circuits compares closely to standard logic gates such as inverters. Table 13.2(a) reports our results for this study. Here we have only reported numbers for STR6, since STR6 is expected to be more susceptible to noise because of the raised pass transistor voltage.

We also studied a scenario where the level converter is a part of a larger dynamic circuit (Figure 13.7(b)). The input of the circuit under test acts as the victim line (a dynamic node with a weak keeper) and a capacitively coupled aggressor (operating at VddH) is considered as the coupling noise source. For a fixed ground capacitance of the victim line (10fF), the coupling capacitance was increased until the circuit failed. Table 13.2(b) summarizes our results for this study. The capacitance reported in the table is the coupling capacitance at which the circuit failed. A higher capacitance thus implies superior robustness. Under this scenario too, we found the level converters to be at least as robust as the inverter (i.e., required a larger amount of coupling capacitance and hence coupled noise).

The scenario described by Figure 13.7(a) was also examined in the presence of +10% DC supply noise on both VddH and VddL to test the circuits under even more aggravated noise conditions. Table 13.2(c) reports results for this study. Again, we observe that the level converters are comparable in robustness to the inverter.

Table 13.2 Level converter robustness analysis with VddL of 0.6V, compared to an inverter.

Glitch Type	Inverter	DCVS	STR6
Positive-going (higher value means more robust)	0.48V	0.53V	0.53V
Negative-going (lower value means more robust)	0.06V	0.14V	0.16V

(a) The failure voltage of the circuits is tabulated below for both polarities of noise glitches at the input (positive glitch starting and settling at 0V, e.g. 0V → 0.48V → 0V, and negative glitch starting and settling at VddL, e.g. 0.6V → 0.06V → 0.6V).

Aggressor Swing Direction	Inverter	DCVS	STR6
VddH to 0V (higher value means more robust)	6.5fF	5.8fF	7.6fF
0V to VddH (higher value means more robust)	9.4fF	11.2fF	10.2fF

(b) The failure coupling capacitance is tabulated below for both swing directions (VddH to 0V, and 0V to VddH) of the aggressor.

Glitch Type	Inverter	DCVS	STR6
Positive-going (higher value means more robust)	0.51V	0.56V	0.56V
Negative-going (lower value means more robust)	0.09V	0.18V	0.19V

(c) The analysis in (a) above is repeated in the presence of +10% VddL and VddH variation.

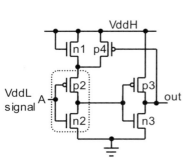

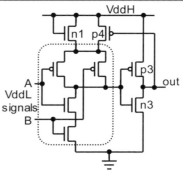

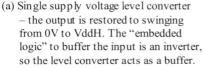

(a) Single supply voltage level converter – the output is restored to swinging from 0V to VddH. The "embedded logic" to buffer the input is an inverter, so the level converter acts as a buffer.

(b) The same level converter design approach with embedded NAND2 logic. The output is level restored and inverted, i.e. AND2.

Figure 13.8 Voltage level converters that require only a single VddH supply rail.

Level converters presented above require both a high and low power supply for level conversion. This limits the physical placement of such level converters to the boundary of high and low voltage designs which restricts the physical design flexibility. To address this, we developed a novel asynchronous level-converter, which requires only one supply (VddH) to convert the incoming low voltage signal to the higher voltage making its placement much more flexible [21] in the entire high voltage regions. In addition to the single supply advantage, this converter exhibits a significantly improved power dissipation compared to the traditional DCVS converter.

Figure 13.8 shows the new voltage level converter that requires only a single supply rail. We utilize the threshold drop across the n-channel MOSFET n1 to provide a virtual low-supply voltage to the input inverter (p2,n2).

Section 13.2 discussed the optimal low-supply voltage in a dual-supply design, which was generally found to be 40% below the high supply voltage. However, scaling of Vdd is limited to by how low Vth can be scaled – otherwise drive current is degraded (e.g. consider reducing Vdd in Equation (13.9) without reducing Vth – the delay gets worse). To maintain good CMOS performance characteristics, it is desirable to have the ratio of Vth/Vdd below 0.3 [31]. Scaling of Vth is limited due to exponentially increasing subthreshold leakage as Vth is reduced (see Equation (13.5)). To prevent excessive power consumption due to subthreshold leakage, the threshold voltage is limited to about 0.2V at 100°C [31]. Thus in sub-100nm technologies, the supply voltage cannot be scaled much below 1V. Typically, the low supply in sub-100nm designs will be limited to 25-30% below the high-supply voltage.

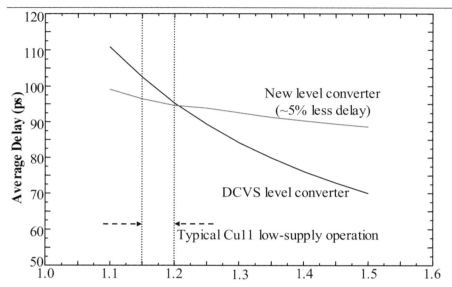

Figure 13.9 Comparison of the delay of the DCVS level converter with the single-supply level converter versus the voltage for the low supply.

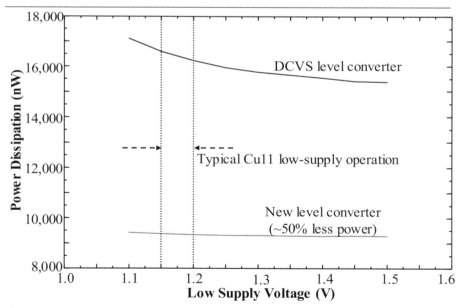

Figure 13.10 Comparison of the total power of the DCVS level converter with the single-supply level converter versus the low supply voltage, with a switching activity of 0.1 (where a switching transition is 0-1-0 or 1-0-1).

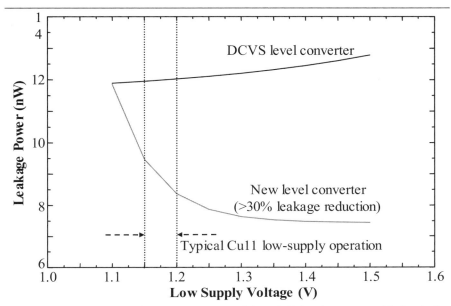

Figure 13.11 Comparison of the leakage power of the DCVS level converter with the single-supply rail voltage level converter versus the low supply voltage value.

Figure 13.9, Figure 13.10 and Figure 13.11 show that when compared to the traditional DCVS level converter (in 130nm Cu-11 technology with nominal Vdd=1.5V), the new converter achieves up to 5% less delay, and consumes approximately 50% less total power and 30% less leakage power, in the nominal operating range of the low-voltage supply. The biggest advantage of this level converter is its flexible placement which enables efficient physical design of fine-grained voltage islands as discussed in the following section.

13.3.2   Physical Design Issues

Most of the previous work [35] in multi-Vdd designs has mainly focused on Clustered Voltage Scaling by Usami et al. [33]. Unfortunately, this methodology enforces a rigid circuit row based layout of high and low voltage cells. This can be overly restrictive as it may require significant perturbation in location of timing critical cells thereby degrading performance. In this section, we present some physical implementation schemes based on voltage islands which have more flexibility in their layout.

13.3.2.1   Macro based Voltage Islands

Recently, a new voltage island methodology to enable multiple supply voltages in systems on chip (SoC) was introduced [19] which allows various functional units of the ASIC/SoC to operate at different voltages. This voltage island methodology can be used in variety of designs. For example, in an SoC that integrates a processor core with memory and control logic, the performance critical processor core requires highest voltage to maximize its performance. However, the on chip memory and control logic may not require the highest voltage operation and can be operated at a reduced voltage to save significant active power without compromising system performance. In addition, voltage flexibility at unit level allows pre-designed standard components from other applications to be reused in a new SoC application. Voltage islands can also facilitate power savings in battery powered applications which are more sensitive to standby power. Traditionally, designers use power gating [16] to limit leakage current in quiescent states. The use of voltage islands at functional unit level in a SoC provides an effective physical design approach to gate the power supply of the entire macro in order to completely power it off.

13.3.2.2   Fine-Grained Generic Voltage Islands

The macro-based voltage island methodology is targeted towards an entire macro or functional unit being assigned to a different voltage. For designs that are highly performance critical as well as severely power constrained, it is useful to have a finer grained control over the supply voltages for ASICs or even within a macro/core in an SoC. We propose a flexible physical design approach that allows generic voltage islands and enables a fine grained implementation of the dual-supply voltage assignment in a placement driven synthesis framework [6]. A generic voltage island structure with power grid is shown in Figure 13.12, where we can assign different voltages at both macro and cell levels. It has more freedom in terms of layout style by allowing multiple voltage islands within the same row. A generic design flow is built on top of IBM's placement driven synthesis (PDS) design closure tool [8]. PDS integrates logic synthesis, placement, buffering, gate sizing, and multiple threshold voltage optimization [20].

The overall flow with generic voltage islands is as follows. First, PDS timing closure is run with the entire circuit timed at VddH. For deep sub-micron circuits, interconnect delay dominates the gate delay. Thus we need rough placement information to identify critical versus non-critical cells. Once PDS reaches a later stage of optimization, e.g., global placement is determined and timing is more or less closed, we can perform the generic voltage island generation, by assigning non-critical cells to a lower supply voltage.

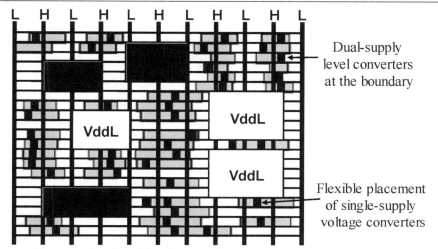

L H L H L H L H L H L

Dual-supply
level converters
at the boundary

VddL

VddL

VddL

Flexible placement
of single-supply
voltage converters

Figure 13.12 Generic voltage island layout style.

To minimize the physical design overhead, we consider two kinds of adjacencies during VddL macro/cell selection. One is the logic adjacency, i.e., the low voltage cells are as contiguous as possible in signal paths to minimize the number of level shifters. The other is the physical adjacency, i.e., low voltage cells are physically close to each other, so that it is easy to form voltage islands.

Since the generic voltage islands are implemented within the framework of PDS, we can employ various optimization engines during voltage assignment, e.g., to trade-off gate sizing with voltage assignment. After voltage assignment, low and high voltage cells are clustered to form the fine grained generic voltage islands. The clustering step requires the knowledge of power grid topology which is co-designed with this placement in order to enable a flexible placement of fine grained voltage islands. We first define the power grid patterns to facilitate the placement movement. They are computed based on the cell locations that are assigned to high and low voltage cells. Then we will move cells locally (while trying to maintain the original cell order) to form VddL and VddH islands.

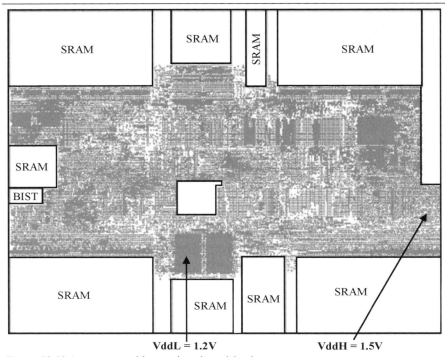

Figure 13.13 A processor with generic voltage islands.

Traditionally, a dual-supply DCVS level converter is used to interface signals across VddL and VddH voltage islands Since DCVS level converters require both VddL and VddH supplies, their placement is limited to the boundary of low and high voltage islands where both the supplies are easily available. To remove this placement restriction on level converter, we utilize the single supply voltage level-converter (Figure 13.8). Since this converter requires only VddH supply, it can be placed anywhere in the VddH voltage islands, thereby enabling much more flexible placement. This results in significantly smaller physical design overhead for level converter insertion as the converters can be inserted in uncongested regions. We have applied this generic voltage island approach to an IBM processor core in 130nm Cu-11 technology with approximately 50,000 cell instances with VddH = 1.5V and VddL = 1.2V. Figure 13.13 shows the layout of this processor designed using generic voltage islands which resulted in 8% total power savings without any delay or area penalty.

13.3.3   Issues in using multiple threshold voltages

Using cells with multiple threshold voltages has power-performance benefits, as discussed in Section 13.2. Even though using multiple threshold voltages in a design is relatively easy, it is not free by any means, and does require changes to library creation and the design flow. We discuss some of the issues in using multiple threshold voltages in this section.

As has been mentioned in Section 13.2, each additional Vth level increases fabrication cost by 3%. Two to three Vth levels are common in today's technologies. Cost sensitive ASICs often use two threshold levels, or even a single threshold level. Using three threshold levels is more common in high-performance processor designs. A low threshold device in 90nm technology can have leakage more than 30 times the leakage of a regular threshold device with the same area but has only 15% better performance. Hence it should be clear that only a small percentage of the total devices in the design can be low threshold devices.

From a design point of view, using multiple threshold voltages has only a small effect on the design flow, which is a big advantage, unlike voltage islands which require major changes such as modifying the power grid and the introduction of level converters. Mixing devices with different thresholds does introduce extra placement constraints between the devices. But the constraints are usually enforced during library design in the layout of the library cells. Hence no additional constraints need to be enforced in the rest of the design flow.

Having multiple threshold devices increases the library size by 2× or 3×. Since a larger library increases runtimes of synthesis tools, a typical design flow will use only regular-Vth cells during the synthesis and placement phase even if multiple threshold voltage cells are available. A second reason for not using multiple thresholds during the initial synthesis phase is that most tools are not leakage aware. Allowing leakage-insensitive tools to use low Vth cells will lead to a design with high static power dissipation. Hence low threshold cells are usually used in a post-processing step by path-delay optimization tools which are leakage aware.

It should be pointed out that an insensitive partitioning of the design flow into a first phase using only regular Vth cells and post-processing steps with low Vth cells may not be always wise. If the cycle time is aggressive, and only regular Vth cells are available, tools can increase the power by upsizing the regular Vth cells needlessly to try to achieve the cycle time in the first phase. A smarter modified methodology could be to use a less aggressive cycle time during the initial phase which uses only regular Vth cells, and pushing for the aggressive cycle time during the post-processing phase with low Vth cells.

In the next section, we discuss a power and performance critical design for satellite applications that was designed using a semi-automatic design flow using two supply voltages and two threshold voltages.

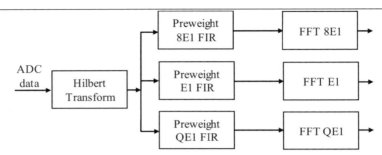

Figure 13.14 The seven macros of the DEMOD ASIC chosen for the case study and their data flows. The ADC is the analog to digital converter; the FIR filters are finite input response filters; and FFT is the fast Fourier transform.

13.4   CASE STUDY

Modern communications satellites, as well as many military applications, require significant on-board digital signal processing (DSP) capabilities, enabled by application-specific integrated circuits (ASICs). Such systems are also driven by severe size, weight, and power constraints. For satellites, power is most critical due to limitations on generation and heat removal, as well as need for high reliability. Such systems are equally hard driven by cost and schedule.

In this section, we focus on the application of semi-custom design techniques for high-performance, yet power efficient DSP ASICs. We evaluate the feasibility of significant improvements over today's state-of-the-art near custom chip performance with an ASIC-like cost and schedule. Specifically, we will discuss a synthesis and physical design methodology to reduce the performance (delay) × power metric for DSP ASICs.

The classes of DSPs we are looking at are the fixed function, real-time, distributed (FRD) DSPs. In a FRD DSP, processing elements and memory are allocated exactly where needed to execute a fixed data flow algorithm. At the other extreme of DSP architectures is the software programmable DSP, with centralized compute and memory resources. Our experience, as well as studies by University of California Berkeley (UCB) [7], finds that the FRDs are about 50× more power efficient than the processors. While the ASIC designs investigated in this case study are "mission specific", the design techniques are suitable for a wide range of applications.

We identified a representative FRD DSP class circuit from existing Boeing communications satellite ASICs for intensive benchmarking of the semi-custom design methodology. Specifically, we chose a subset of the DEMOD ASIC, a critical component of the SPACEWAY™ communication satellite DSP unit [28]. The original chip is about 2.3 million gates in complexity and implemented in IBM's 0.18um SA-27 ASIC technology. For this work we decided to use IBM's Cu-11 0.13um technology, and we focused on the critical seven-macro subset consisting of Hilbert Transform, FIR filter (3), and FFT (3) macros shown in Figure 13.14. The subset requires about 240,000 logic gates and 42 KB of register array —about 20% of the full DEMOD design. To ensure that the design optimization work was sufficiently challenging, we scaled the target clock rate from 83 MHz in SA-27 to 175 MHz in Cu-11. After re-mapping the design to Cu-11, we ran the baseline flow, followed by application of a subset of the semi-custom techniques to optimize for low power.

13.4.1 The Relative Power Performance Metric

To evaluate the contribution of the various design steps on the quality of the design as well as evaluating the final design, we used the *Relative Power Performance (RPP)* metric. The *Power Performance (PP)* metric is defined as the product of the delay or performance of the design, and the power of the design, i.e., PP = power × performance. The initial design point has a RPP of 1.0, and any other design point has a RPP which is given by $PP_{initial\_design\_point}/PP_{new\_design\_point}$. A higher RPP implies a faster or less power-hungry implementation, i.e. a more efficient implementation.

To measure the delay, the netlist was placed and Steiner routing was performed. Thus realistic wire delay and load models were used for timing closure and for measuring the path delays. Accurate load calculation is important for selecting the sizes of the gates, and this in turn affects area, path delays and power consumption of the circuit.

A power estimation methodology is fundamental to exploring power-performance tradeoff. We consider both active and leakage power components while estimating the power consumption. Active power dissipation depends on the total capacitance being switched, the switching factor, the clock frequency f and the operational voltage V_{dd}. The following equation gives the details of the power calculation:

$$P_{total} = fV_{dd}^2 \left(\sigma_{logic\_nets} \sum_{\forall\ logic\_nets} C_{net} + \sigma_{clk\_nets} \sum_{\forall\ clk\_nets} C_{net} \right)$$
$$+ P_{macros} + P_{leakage} \qquad (13.15)$$

where

$$C_{net} = \sum_{\forall\ pins\_on\_net} C_{pin} + \sum_{\forall\ wires\_on\_net} C_{wire} + C_{internal(source\_gate)} \qquad (13.16)$$

C_{net} is the total capacitance of each net comprising of the pin capacitance C_{pin}, the wire capacitance C_{wire}, and the internal gate capacitance of the driver $C_{internal(source\_gate)}$. σ_{logic_nets} is the average switching factor of the logic nets, and σ_{clk_nets} is the average switching factor of clock nets. P_{macros} is the average power of hard macros (e.g., arrays), and $P_{leakage}$ is the leakage power.

In our ASIC design experience, Steiner tree length correlates relatively well to post-routing net length, especially when the same Steiner algorithm is used through various routing stages. Therefore we extracted detailed parasitics on the Steiner routing estimates to calculate total wire capacitance C_{wire} loading the gates.

We assumed a value of 0.1 for the average switching factor of the logic nets σ_{logic_nets}, and 1.0 for the average switching factor of the clock nets σ_{clk_nets} (in this chapter by switching activity we refer to a complete switching transition, i.e., 0-1-0 or 1-0-1). Since the focus is on exploring power-performance tradeoffs, relative power comparisons among various design points are the primary focus rather than the absolute accuracy of total power. Thus using approximate values for the switching factors is justified.

Since leakage power, $P_{leakage}$, increases exponentially with decreasing threshold voltages, this component plays a crucial role in deciding the amount of lower threshold voltage cells we could accept in the design in order to gain performance. The power numbers are computed at the worst case process corner for power which is not necessarily the worst case process corner for delay. We consider leakage power in our calculations by averaging over the input state space, i.e., for a two input gate, the average power over input values "00", "01", "10" and "11" is used.

13.4.2   Baseline Flow

We used a traditional flow to establish a baseline against which we could compare a semi-custom flow targeting low power implementations. For the baseline design flow we deployed IBM's BooleDozer [27] logic synthesis system, Cplace [13] placement program and Xrouter [12] for routing. In this "traditional" baseline flow, synthesis and physical design are separate steps, interconnect estimation is based on wireload models during the synthesis stage, and there is no automated post-placement timing correction. The design was partitioned into seven regions for floorplanning based on the top level macros mentioned in Section 13.4. The floorplan is shown in Figure 13.15. The design from the baseline flow can be run at a frequency of 94MHz and dissipates 106mW (column 2 of Table 13.4).

13.4.3   Semi-custom flow for low power designs

The proposed semi-custom flow is illustrated in Figure 13.16. In the pre-synthesis stage, bitstack components are inferred. Gain-based synthesis is used to take advantage of finer grained libraries, and to avoid the use of wire-load models during the synthesis stage. After logic optimizations in synthesis, the final netlist is placed and routed (using Steiner tree routes) by our Placement Driven Synthesis tool (PDS) [8]. PDS is IBM's optimization tool that combines placement, synthesis and global wire optimization to do timing-driven placement and synthesis. In the PDS stage advanced custom logic techniques such as low-Vth and voltage scaling were applied to minimize the area and power terms in the DAP metric. The various steps of the semi-custom flow are explained in more detail in the following sub-sections.

Figure 13.15 The placed design for the baseline flow. The seven macros are shaded gray or white alternately. Compare this figure with Figure 13.4.

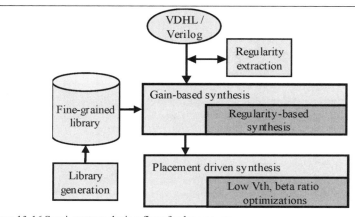

Figure 13.16 Semi-custom design flow for low power.

13.4.4   Arithmetic Optimizations

A detailed study of the critical paths revealed many adders in series, often coming out of multiplier structures. The arithmetic expression optimizations using the IBM behavioral synthesis tool Hiasynth [3], including tree-height balancing and carry-save adder (CSA) implementations, resulted in significantly improved area and delay. The clock frequency increases from 94 to 145MHz, as can be seen in column 3 of Table 13.4 (see Section 13.4.10), due to the critical path reduction in the arithmetic trees. The area for the unoptimized and optimized arithmetic circuits is almost equal at the beginning of logic synthesis. But since the former has much longer paths, logic synthesis tries to meet the timing constraint for the former by using larger cells and larger buffers. A direct implication of using smaller sized cells and a smaller number of buffers for the arithmetic optimized circuit is that power consumption is significantly reduced, from 106mW to 86mW at 94MHz, but increases to 134mW at the best frequency of 145MHz (comparing columns 2 and 3 of Table 13.4).

13.4.5   Semi-Custom Bitstacks

Since the FRD DSPs have many adders and multipliers, we investigated fast implementations of such circuits including carry-lookahead adders and Wallace tree multipliers [14].

However, bitstacked implementations (Figure 13.17) go one step further by paying attention not only to the number of levels of logic required to implement the operations, but also creating an implementation which can have a compact placement with very short wires. We mapped the adders and multipliers in the design to the bitstack implementations for IBM's Cu-11 technology. The bitstack generator also takes in an argument that controls the drive strength of the unit that is generated. The size chosen ensured that the output cells of the bitstacks have sufficient strength to drive the loads at the outputs of these bitstacks. By not applying synthesis on the bit-stacks, we guarantee that the bitstacks can be placed exactly in their row/column scheme. Bitstacking regular arithmetic units can have a significant impact on delay, area and power. Near zero slack was reached after inserting the bitstacks while power consumption at the higher frequency of 177 MHz was only 107mW compared to 134mW for the arithmetic optimized only design (columns 3 and 4 of Table 13.4).

However we noticed several things about the DEMOD design which prompted us to optimize the bitstacks using synthesis though this involved sacrificing the built-in regularity of the bitstacks. Several bitstack components had constant signals as inputs. Constant propagation would allow many gates to be optimized away. In addition, redundancy removal could use this information to optimize other portions of the design.

Secondly, the outputs of the bitstacks had significantly different loads. Selecting a bitstack component that would not cause any violation at any of its outputs leads to significant overdesign. To overcome the above sources of sub-optimality, we selected a small size implementation for all the bitstack components and did not protect them from synthesis. This allowed many gates to be optimized away due to constant propagation. It also allowed resizing to close on timing by automatically choosing the most optimal gate sizes. Even though the regularity was lost, PDS used the connectivity and timing constraints effectively to place the bitstacks in close regions as can be seen from the placement view shown in Figure 13.18. This improved the Relative Power Performance metric to 1.86 in column 4 of Table 13.4.

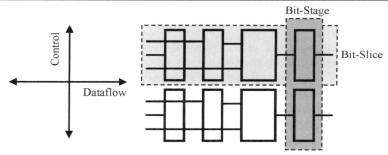

Figure 13.17 An illustration of a bitstack layout. Logic in a bit is placed in a column. Several columns are stacked side-by-side from left-to-right. The control signals are routed vertically.

Figure 13.18 Placement view with the darker regions showing compact placement of three of the bitstack components.

13.4.6 Fine-Grained Libraries

In a full custom methodology, a designer has the option to size each transistor exactly to match the load that it is driving. This allows for delay and especially power and area optimization. In ASIC design using conventional standard cell libraries, limited choices are available in cells sizes for each function. For example, an inverter is only available in four or five standard sizes. An oversized cell is typically chosen to drive a particular load in order to meet the delay and slew (rise time) constraints on the cell outputs since a smaller cell would violate these constraints. However, a large cell reflects a large load back to its inputs, requiring its input to be upsized as well. This works its way all the way back to the inputs of the synthesized partition and all gates are sized larger than necessary. One way to prevent this is to use a fine-grained library with many more sizes for each type of cell. Unfortunately, adding many cell sizes to a library slows down conventional synthesis considerably since most synthesis algorithms resize by looping through all cell sizes and will be penalized with at least a linear slow down. Gain-based synthesis [2] avoids this problem, by using a single delay equation for all sizes of a particular cell type. Only in the final part of synthesis or after placement the actual size of a cell is calculated.

Gain-based synthesis also addresses the wireload problem of traditional synthesis algorithms. In designs like the DEMOD, dominated by low fan-in and low fan-out arithmetic logic gates, the wireload models shipped with a technology which are design independent are on average too pessimistic. When designs are synthesized with overly pessimistic wireload models, large cells are chosen to satisfy the timing and slew constraints for the given wireload. Placement places these larger cells further apart, resulting in longer wires that need to be driven. A better approach therefore is to start with a design that is minimally sized, and leave the final sizing up to a design closure tool like PDS which is able to place and optimize the netlist simultaneously. Gain-based synthesis does not require wireload models for its delay calculations. Since the load has been removed from the delay equation, good delays can be predicted without the use of wireload models. This allows us to create a realistic sized design in synthesis before placement, and to obtain timing closure more quickly. Applying gain-based synthesis without wireload models to our DEMOD macros result in a final power of 93mW and a RPP of 2.15.

13.4.7 Maximizing Frequency

When positive slack is present in the design it can be traded off for power reduction through voltage scaling. To find out how much extra slack existed in the design we re-ran PDS to target the fastest possible design at 1.2V. The fastest design we could obtain was at 193MHz (at a cost of 17%

more power) because of a register file to register file cycle limiting path as shown in column 6 (1.2V fast) in Table 13.4.

13.4.8   Voltage Scaling

Since the maximum frequency is significantly higher (193MHz compared to the required 177MHz), we had the opportunity to trade performance for power. To do this, we reduced the supply voltage from 1.2V to 1.1V. This produced a large number of negative slack paths. We then used PDS to re-close timing at the original cycle time but at 1.1V. The 1.1V design, in column 7 of Table 13.4 uses approximately 16% less power at the target performance of 177 MHz compared to the 1.2V design in column 5. To study the effect of voltage scaling further, we reduced the voltage to 1.0V and 0.9V (the minimum voltage allowed in the Cu-11 technology). However, to keep the performance at 177MHz we had to apply Low Vth transformations as will be discussed in the next section.

13.4.9   Low-Vth Logic and Voltage Islands

The use of lower threshold devices increases device performance along with increasing its sub-threshold current, i.e., leakage power. So, use of low Vth devices is restricted to timing critical paths in order to avoid excessive increase in leakage current, especially in mobile low power applications. In addition, the increase in quiescent current also interferes with IDDQ fault testing. IBM's Cu-11 ASIC library includes a low Vth version of each cell. These elements have the same layout footprint but higher performance than their nominal Vth counterparts. We made use of these low Vth cells to recover some performance lost through voltage scaling with very little power increase by utilizing the low Vth optimization capability in PDS. PDS low Vth optimization substitutes cells on critical paths with their equivalent low Vth versions and dynamically updates the critical paths information. This substitution of low Vth cells is guided by the dynamic analysis of leakage power and can be constrained by a maximum limit on the leakage power increase.

By applying the multi-Vth operations in PDS, we were able to keep the performance at 177MHz and reduce the power to 64mW at a 1.0V operating point compared to 78mW at 1.1V. The total power savings more than offsets the increase in leakage power (leakage increased from 0.24mW to 2.3mW). It should be pointed out that at smaller technology nodes, leakage will be a much higher percentage of total power, and leakage increase will require more attention.

We also used PDS to further lower the supply voltage after selective low Vth substitution. This experiment had some interesting outcomes. PDS low Vth substitution allowed us to lower the supply voltage all the way to 0.9V before any combinational path became critical. This allowed power to be reduced to a very low value of 46mW. However the register arrays demonstrated large performance sensitivity to voltage reduction. The critical path at 0.9V was a register array to register array path, which could not be tuned by PDS and which prevented us lowering the overall voltage to 0.9V while maintaining the desired frequency. In the future we will experiment with placing these sensitive arrays in separate voltage islands [19] which would allow the chip logic supply voltage to be scaled separately from the array supply voltage.

13.4.10 Results

To evaluate the relative contributions of the optimizations, all results are reported after place and route. We turn on the optimizations one-by-one and run the baseline flow for the remainder to get to a final design.

The power consumption for the clock tree, flip-flop data and clock, random logic, and register array are listed in Table 13.3, with the leakage for the whole design. Most of the base line's power (68%) is due to the random logic, but the power for the random logic is more than halved after the optimizations, and contributes all the power savings prior to voltage scaling.

Table 13.4 summarizes a general improvement of the RPP metric that tracks the increasing sophistication of the semi custom flow. The final result of the voltage scaled gain-based PDS flow has a combined metric improvement factor of 3.13. As can be seen from the table, for the case of a single voltage for the whole design, aggressive voltage scaling to 1.0V with multi-Vth optimizations provides the best power-performance tradeoff with an RPP metric of 3.13. Allowing two voltage islands with the majority of the design at 0.9V and the register arrays at 1.0V improves the metric further to 3.47.

The progression of the optimization steps can be summarized from a second viewpoint. To maximize performance and minimize power, the design was divided into two voltage islands. One voltage island had the register arrays operating at 1.0V. The second voltage island had the rest of the logic operating at 0.9V. To drive down the operating voltage of the voltage island with the logic from a 1.2V to 0.9V, aggressive synthesis and selective insertion of low Vth cells were used. The above approach towards design optimization represents a general optimization scenario for industrial designs. The design is split into a small number of voltage islands, usually two, and then each voltage island is optimized aggressively using a combination of low threshold devices and voltage scaling.

Table 13.3 Distribution of power consumption (mW) in the design.

	Base Line	Bit Stack	Finer Grained Library	Voltage Scaling (1.1V)
Clock	11.9	18.4	15.6	13.1
Flip-flop Data	9.3	17.5	14.6	12.2
Flip-Flop Clock	4.0	7.5	7.5	6.2
Logic	72.2	48.1	39.8	33.4
Register Array	7.9	15.0	15.0	12.5
Leakage	0.3	0.3	0.3	0.3
Total	105.7	106.8	92.7	77.7

Table 13.4 Tracking the relative power performance (RPP) metric with the design optimizations.

	Base Line	Arithmetic Optimizations	Bit Stack	Finer Grained Library	fast 1.2V	1.1V	1.0V	0.9V
Power (mW)	105.7	133.9	106.8	92.7	108.1	77.7	63.6	45.7
Performance (MHz)	94	145	177	177	193	177	177	141
Power Savings		-26.6%	-1.0%	12.4%	-2.2%	26.5%	39.9%	56.8%
Relative Power Performance	1.00	1.22	1.86	2.15	2.01	2.56	3.13	3.47

13.5 SUMMARY

In this chapter, we explored the trade-off between multiple supply voltages and multiple threshold voltages in the optimization of dynamic and static power which can result in 60% power savings. Novel solutions to the unique physical and electrical challenges presented by multiple voltage schemes were proposed. We described a new single supply level converter that does not restrict the physical design. A power performance improvement of ×3.13 was obtained by applying some of these optimization techniques to a hardwired DSP test case. In this, electrical optimizations such as voltage scaling, multi-threshold optimization and the use of finer grained libraries enabled 1.7× improvement and the remaining 1.9× was enabled by high level arithmetic optimizations and bitstacking.

13.6 ACKNOWLEDGMENTS

The work on generic voltage islands was done in collaboration with Tony Correale, Doug Lamb, David Pan and Dave Wallach. We thank Lakshmi Reddy for help with PDS experiments.

Figure 13.19 The placed design for the semi-custom flow.

13.7   REFERENCES

[1] Albrecht, C., Korte, B., Schietke, J., and Vygen, J., "Cycle Time and Slack Optimization for VLSI-Chips," proceedings of the International Conference on Computer-Aided *Design*,1999, pp. 232-238.
[2] Beeftink, F., Kudva, P., Kung, D., Puri, R., and Stok, L., "Combinatorial cell design for CMOS libraries," *Integration: the VLSI Journal*, vol. 29, 2000, pp. 67-93.
[3] Bergamaschi, R.A., et al., "High-level Synthesis in an Industrial Environment," *IBM Journal of Research and Development*, vol. 39, 1995.
[4] Bernstein, K., et al., *High Speed CMOS Design Styles*, Kluwer Academic Publishers, Boston, 1998.
[5] Chen, C., Srivastava, A., and Sarrafzadeh, M., "On gate level power optimization using dual-supply voltages," *IEEE Transactions on VLSI Systems*, vol. 9, Oct. 2001, pp.616-629.
[6] Correale, A., Pan, D., Lamb, D., Wallach, D., Kung, D., and Puri, R., "Generic Voltage Island: CAD Flow and Design Experience," *Austin Conference on Energy Efficient Design*, March 2003 (IBM Research Report).
[7] Davis, W.R., et al., "A Design Environment for High Throughput, Low Power, Dedicated Signal Processing Systems," *proceedings of the Custom Integrated Circuits Conference*, 2001, pp. 545.
[8] Trevillyan, L., Kung, D., Puri, R., Kazda, M., and Reddy, L.,"An integrated environment for technology closure of deep-submicron IC Designs," *IEEE Design & Test of Computers*, vol. 21, no. 1, February 2004, pp. 14-22.

[9]   Fishburn, J.P., "Clock Skew Optimization," *IEEE Transactions on Computers*, vol. C-39, 1990, pp. 945-951.

[10] Hamada, M., et al., "A top-down low power design technique using clustered voltage scaling with variable supply-voltage scheme," *proceedings of the Custom Integrated Circuits Conference*, 1998, pp. 495-498.

[11] Hamada, M., Ootaguro, Y., and Kuroda, T., "Utilizing surplus timing for power reduction," *proceedings of the Custom Integrated Circuits Conference*, 2001, pp. 89-92.

[12] Hetzel, A., "A sequential detailed router for huge grid graphs," *proceedings of Design Automation and Test in Europe*, 1998 , pp. 332.

[13] Hojat, S., et al., "An integrated placement and synthesis approach for timing closure of PowerPC™ microprocessors," *proceedings of the International Conference on Computer Design*, 1997, pp. 206.

[14] Hwang, K., *Computer Arithmetic: Principles, Architecture, and Design*, Wiley Publishers, 1979.

[15] Krishnamoorthy, R., et al., "Dual supply voltage clocking for 5GHz 130nm integer execution core," *proceedings of the Custom Integrated Circuits Conference*, 2002, pp. 128-129.

[16] Kosonocky, S., et al., "Low Power Circuits and Technology for wireless digital systems," *IBM Journal of Research and Development*, vol. 47, no. 2/3, 2003.

[17] Kudva, P., Kung, D., Puri, R., and Stok, L., "Gain based Synthesis," International Conference on Computer-Aided Design tutorial, 2000.

[18] Kulkarni, S.H., and Sylvester, D., "High performance level conversion for dual VDD design," *IEEE Transactions on VLSI Systems*, vol. 12, 2004, pp. 926-936.

[19] Lackey, D., et al., "Managing Power and Performance for SOC Designs using voltage islands," *proceedings of the International Conference on Computer-Aided Design*, 2002.

[20] Puri, R., D'souza, E., Reddy, L., Scarpero, W., and Wilson, B., "Optimizing Power-Performance with Multi-Threshold Cu11-Cu08 ASIC Libraries," *Austin Conference on Energy Efficient Design*, March 2003 (IBM Research Report).

[21] Puri, R., Pan, D., and Kung, D., "A Flexible Design Approach for the Use of Dual Supply Voltages and Level Conversion for Low-Power ASIC Design," *Austin Conference on Energy Efficient Design*, March 2003 (IBM Research Report).

[22] Rohrer, N., et al., "A 480MHz RISC microprocessor in a 0.12μm Leff CMOS technology with copper interconnects," *proceedings of the International Solid State Circuits Conference*, 1998, pp. 240-241.

[23] Sakurai, T., and Newton, R., "Alpha-power law MOSFET model and its application to CMOS inverter delay and other formulas," *IEEE Journal of Solid-State Circuits*, vol. 25, no. 2, April 1990, pp. 584-593.

[24] Sakurai, T., and Newton, R., "Delay Analysis of Series-Connected MOSFET Circuits," *IEEE Journal of Solid-State Circuits*, vol. 26, no. 2, February 1991, pp. 122-131.

[25] Sirichotiyakul, S., et al., "Standby power minimization through simultaneous threshold voltage selection and circuit sizing," *proceedings of the Design Automation Conference*, 1999, pp. 436-441.

[26] Srivastava, A., and Sylvester, D., "Minimizing total power by simultaneous Vdd/Vth assignment," *IEEE Transactions on CAD*, vol. 23, 2004, pp. 665-677.

[27] Stok, et al., L., "BooleDozer Logic Synthesis for ASICs," *IBM Journal of Research and Development*, vol. 40, no. 3/4, 1996.

[28] Sunderland, D.A., et al, "Second Generation Megagate ASICs for the SPACEWAY™ Satellite Communications Payload," *NASA Symposium on VLSI Design*, May 2003.

[29] Sylvester, D., and Kaul, H., "Future performance challenges in nanometer design," *proceedings of the Design Automation Conference*, 2001, pp. 3-8.

[30] Szymanski, T. G., "Computing Optimal Clock Schedules," *proceedings of the Design Automation Conference*, 1992, pp. 399-404.

[31] Taur, Y., "CMOS Design near the limit of scaling," *IBM Journal of Research and Development*, vol. 46, no. 2/3, 2002.

[32] Usami, K., et al., "Automated Low-Power Technique Exploiting Multiple Supply Voltages Applied to a Media Processor," *IEEE Journal of Solid-State Circuits*, vol. 33, no. 3, 1998.

[33] Usami, K., and Horowitz, M., "Clustered voltage scaling techniques for low-power Design," *proceedings of the International Symposium on Low Power Electronics and Design*, 1995.

[34] Wang, Q., and Vrudhula, S., "Algorithms for minimizing standby power in deep submicron, dual-Vt CMOS circuits," *IEEE Transactions on CAD*, vol.21, 2002, pp. 306-318.

[35] Yeh, C., et al., "Layout Techniques supporting the use of Dual Supply Voltages for Cell-based Designs," *proceedings of the Design Automation Conference*, 1999.

Chapter 14

LOW POWER ARM 1136JF-S™ DESIGN

George Kuo, Anand Iyer
Cadence Design Systems, Inc.
San Jose, CA 95134, USA

14.1   INTRODUCTION

Methodologies for ASIC design have been seen as lagging behind custom design methodologies for a long time. With process migration to 90nm and below, ASIC design methodologies are fast catching up with custom design methodologies.

An economic driver for ASICs is the increasing demand for mobile and consumer devices. These devices have smaller form factors. They are becoming part of everyday life with high usage, and need to be robust. For example, a device that combines cellular telephony with a PDA is used many times during the day without recharging. This is forcing designers to look at power as an important metric when they design chips for these devices. Such low power designs are becoming more and more common-place.

Low power has always been the forte of custom design methodologies. Whether in system design, process and logic selection or implementation, low power design was handled by specialist designers using custom tools. Increased demand for mobile and consumer applications with high device integration have pressured circuit designers to adopt faster, automated design approaches. Synthesizable application specific designs can meet these time-to-market needs with moderate power consumption, but to achieve lower power custom design approaches must be adopted and automated.

Many EDA companies are addressing this challenge to translate custom low power methodologies into a more generalized methodology. These metho-dologies are validated through designing prototype chips. One such project was completed recently to validate a low power methodology including voltage scaling for controlling power. This chapter talks about this project and outlines the designer choices and the decisions as the project progressed. Emphasis was given to getting maximum power reduction without changing the under-lying architecture nor using a specialized process technology.

14.2   PROJECT OBJECTIVE

A variety of power management techniques have been developed and applied to date, but most of these would require extensive design expertise or manual implementation process. The team developed a design methodology to reduce power dissipation of a typical microprocessor, resulting in an easily adoptable power management solution that neither requires complex archi-tecture nor expensive low power process technology. The team wanted to evaluate the power reduction from the design flow choices independent of any superior architecture or fancy process technology. Design description

14.2.1.1 ARM1136JF-S Based System-on-Chip

The integrated circuit that the team developed, which included an ARM1136JF-S microprocessor and related circuitry, was designed to func-tion in an ARM system development board, ARM RealView® [1]. The major components of this chip include the microprocessor core; the ETB11 and ETM11 trace bus and memory functions; and a multi-level advanced high performance bus (AHB) at the chip level to connect the AHB Lite ports of the core for accessibility from the external pins of the device. The bus structure also allows access to the 128 KB on-chip RAM to enable data transfers from any four ports concurrently [4][6]. The test chip is shown in Figure 14.1.

Blocks shown on the test chip on the left of the ARM1136JF-S in Figure 14.1 are the ARM MBIST (memory built-in self test) logic for testing memories; JTAG (Joint Test Action Group) IEEE standard test access port (TAP) and boundary scan logic; external and internal clock muxing logic, and logic generating the reset signal; and coprocessors for validating the chip.

In the ARM1136JF-S, the ARM11 core is an implementation of the ARMv6 running 32-bit ARM, 16-bit Thumb and 8-bit Jazelle instructions. The vector floating point coprocessor supports scalar and vector arithmetic on vectors with up to four double precision elements [4]. The TLB is the translation look aside buffer that caches which physical memory addresses corresponds to which virtual addresses. The DMA (direct memory access) logic supports peripherals transferring information directly with the memory.

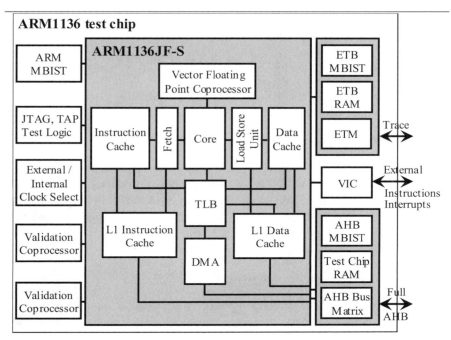

Figure 14.1 ARM1136JF-S test chip block diagram.

On the right of the ARM1136JF-S in Figure 14.1, are the embedded trace macrocell (ETM) for debug and trace and the embedded trace buffer (ETB) for capturing the ETM output and saving it to an on-chip buffer for later access; the vector interrupt controller (VIC) for handling interrupts; and the AHB bus logic.

Two additional co-processors were included to exercise the ARM1136 co-processor interface. These components along with the usual support logic required for manufacturing test and debug in a typical ARM system-on-chip (SoC) device were included at the test chip level. These test components with those detailed above form the main structure of the test chip.

The entire design with the specific memory configuration was verified using the Cadence NC simulation environment with the binary validation testbench kit provided by ARM [2], which totals more than 700 test sets and required several days of run time. The simulation results were also captured in both VCD (Voltage Change Dump) and TCF (Toggle Count Format) for subsequent power optimization and detail power analysis in the flow. In particular, peak power and average power patterns were used as benchmark references for the simulated results and the final silicon measurement. The TCF was used by the RTL Compiler synthesis to better estimate the actual switching activities, to help produce better balanced dynamic power optimi- zation and leakage state probability. The VCD file, being much larger in size (about 4GB for one of the peak power test cases) to capture all relevant logic events, was used in the detail power analysis with Voltage Storm Power Meter™.

As this was a test chip, a 388 pin BGA (ball grid array) package was used to ensure high availability of functional signals. The package was also defined by the requirements of the ARM1136 evaluation platform provided for application development.

14.2.1.2 Technology and libraries

To validate the broad applicability of our approach, a typical process technology was used, the TSMC 90nm G silicon process, and the ARM Artisan® general-purpose physical IP, including SAGE-X™ standard cell libraries and memory generators [5]. As described below, the standard cell libraries were augmented with extended voltage range characterization and cells aimed at enabling power reduction design techniques.

14.2.1.3 EDA Tools and Methodology

Version 4.1 of the Cadence Encounter digital IC design platform was used for implementation of this low power methodology, including RTL Compiler™ synthesis, CeltIC Nanometer Delay Calculator™, and Voltage-Storm for power analysis. The tool flow used in this project is shown in Figure 14.2. The motivation behind the Encounter low power methodology is described in [9].

14.2.2 Project strategy overview

This design tackled three areas of power management challenges: leakage power; dynamic (or active) power; and power integrity of the design as a whole. This design also looked at developing and validating an integrated and effective flow as one of the desired goals.

Leakage power optimization was mainly based on the usage of multiple threshold voltage (V_{th}) cells to balance between timing and power performance. In addition, voltage scaling contributed to the overall reduction in leakage power. The key challenges with the traditional method of multiple-pass synthesis or using post-processing scripts were the complexity added to the design flow, and uncertainty in whether or not the optimal balance of timing and leakage power had been achieved. Both challenges were solved in this design by using single pass-global optimization from Encounter RTL Compiler™ in the synthesis step, followed by SOC Encounter™'s post route leakage optimization to fine-tune the results with more detailed parasitic information. This single pass methodology can be used in a hierarchical design flow if the design capacity or implementation strategy requires it.

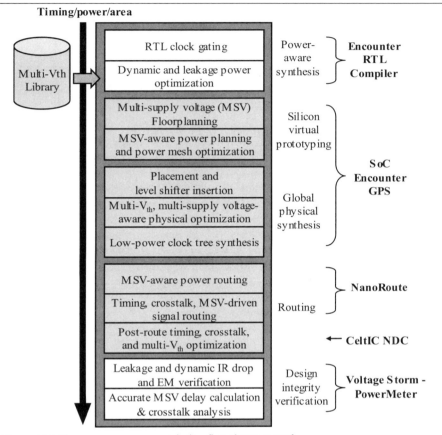

Figure 14.2 The power management design flow that was used.

Minimizing the dynamic power focused on voltage scaling the design into different power domains based on their performance requirements, which the team refer to as a Multiple Supply Voltage implementation (MSV). This is essentially the voltage island approach proposed in [10].

An automated tool flow was critical to meet the four month netlist to tape-out schedule, with the peak engineering resource of four people, or an equivalent of about 12 person-months total. In this MSV flow, two power domains were used, one with 1.0V supply and the other with 0.8V supply. Voltage level shifters were automatically inserted, placed and routed (power and signal) for interface from the 0.8V domain to the higher voltage domain. Besides managing the complex power routing, this design also used SOC Encounter to ensure optimization, routing and analysis across the different operating voltages was performed with sufficient accuracy.

Further reductions in dynamic power were achieved with aggressive implementation of clock gating where possible to reduce the switching activities. The clock gating flow started with allowing Encounter RTL Compiler to automatically identify all opportunities for clock gating in the netlist. The clock gating was then moved automatically to the highest hierarchical starting point of the clock tree, and then the desired clock gating cells were inserted. This flow finished with power aware clock tree synthesis and clock skew balancing. Finally, the impact of traditional timing performance and design integrity challenges, such as signal integrity and IR drop (supply voltage drop due to current across resistance) were amplified due to the voltage scaling. These challenges were solved with the advancement in timing model and improved design practices.

14.3   KEY DECISIONS AND IMPLEMENATIONS

14.3.1   Dynamic power

14.3.1.1   Voltage scaling decisions

In this project, the design team first addressed dynamic power consumption, which can be represented by the equation:

$$P_d = kfCV_{dd}^{2} \qquad (14.1)$$

where k is the toggle rate (the fraction of time that transistors are switching); C is circuit capacitance, including interconnect and transistor capacitance; V_{dd} is the supply voltage to cells; and f is the operating frequency.

As Equation (14.1) indicates, power is proportional to the square of the supply voltage. Consequently, designers can save a significant amount of dynamic power simply by reducing the voltage – an approach called voltage scaling. On the other hand, lowering the supply voltage slows transistor switching speeds (as detailed in Section 4.2). Because this design needed to perform to 350MHz to meet the requirements of ARM's development partners, the team had to be selective in determining which parts of the design could use the voltage scaling technique.

To get a rough baseline for the design performance that can be achieved with the targeted technology, the team did a first pass synthesis check, using the Artisan 90nm SAGE-X library and zero wireload setting (i.e. assuming no wire loads). This information was then fed from Encounter RTL Compiler into SOC Encounter to perform floorplanning and partition exploration.

In this case, the team created a multi-supply voltage (MSV) design, partitioning the design into separate "voltage islands" or "voltage domains", where each domain operates at a different supply voltage depending on its timing requirement.

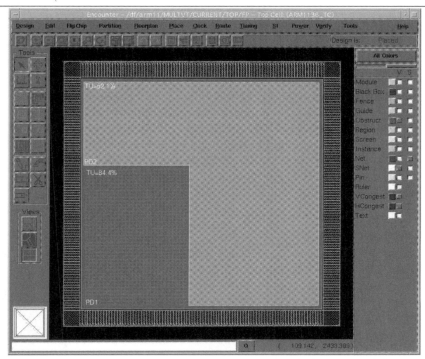

Figure 14.3 Test chip voltage domain layout, where the rectangle in the lower left corner is the low supply voltage domain and the remainder is the high voltage domain.

Timing-critical blocks were put in the high Vdd domain, operating at the standard 90nm supply voltage of 1.0V. Blocks with less critical timing paths were aggregated into a second domain, anticipating that the supply voltage of these blocks would be scaled down to reduced the power. A floorplan of the chip is shown in Figure 14.3. At this step, determining the proper voltage required an analysis of the relationship of cell performance versus Vdd.

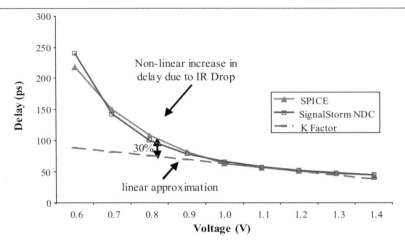

Figure 14.4 Supply voltage versus delay for a buffer in 90nm technology.

For this particular technology, the safe operating range for the supply voltage of the cells was from about 0.7V to 1.2V, with the nominal voltage of 1.0V. However as shown in Figure 14.4, the delay impact of reducing the supply voltage by 30% could be on the order of 2.5×.

In the initial timing exploration of this design, a 400MHz timing target was used to identify timing critical regions of the logic. This was performed without both wire loads and detailed floorplan for a quick analysis. Various memory accesses had surfaced as potentially timing critical, while the ARM1136JF-S core itself was able to meeting the timing requirements with some slack. This suggested a natural grouping of the design into two separate partitions. Though the design was not stressed to determine the maximum slack at 1.0V, the relative fast synthesis execution (about 1.5 hours) and minimum negative slack paths observed seem to agree with continuation of the partitioning exercise. The low supply voltage of 0.8V was selected based on this analysis and with the design performance profile. With the low voltage partition scaled from 1.0V down to 0.8V, a 36% reduction in dynamic power for that portion of the design was expected (as per Equation (14.1)).

Now that the voltages were selected, the next task was to characterize the standard cell libraries for these voltages. The characterization relied on the single physical footprint having two timing views: one for operation at 1.0V operation, and one for operation at 0.8V. The delay of a NAND2 gate at these two different supply voltages is shown in Figure 14.5. The characterization did not take a long time; approximately two days were used to recharacterize the entire standard cell set of about 450 cells, as only two voltages were used inside the chip. This step could be slow and cumbersome if there were many voltage values for the power domains.

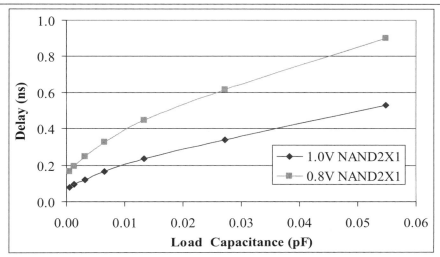

Figure 14.5 NAND2X1 cell delay with 1.0V and 0.8V supply voltage.

A library characterized at many supply voltages is not only huge for analysis, but also can suffer in accuracy from the non-linear nature of the delay effects on power. A new characterization format called Effective Current Source Modeling (ECSM) avoids these problems by recognizing the current waveform of a device to compute the delay through the device. This format helps in analyzing not only timing in multiple voltage scenarios but also in multiple driver situations. ECSM is discussed in detail in Section 14.3.1.4.

14.3.1.2 Voltage level shifters

Once the libraries are characterized for multiple supply voltages, the design team had to create voltage level shifters for signals crossing between these two blocks. A voltage level shifter translates the interface signal from one voltage level to another, e.g. from a low voltage swing of 0V to 0.8V to a full (high) voltage swing of 0V to 1.0V.

Voltage level shifters were used for several reasons. Firstly, the crosstalk due to high Vdd (VddH) signals on low Vdd (VddL) signals can be significant, so it is best not to route VddH signals within the VddL domain. Secondly, a high Vdd gate driven by a low Vdd input has forward biased PMOS transistors which cause substantial leakage current. Thirdly, standard cell libraries are typically characterized assuming the same voltage swing on the inputs as the supply voltage for the cells, rather than say characterizing VddH gates with VddL inputs.

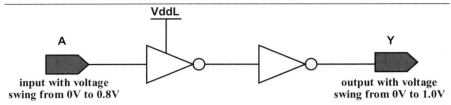

Figure 14.6 The basic design of the low Vdd to high Vdd voltage level shifter that was used. The supply rail for the level shifter is high Vdd, and the first inverter in the buffer used a low Vdd supply connection as well.

In principal, a level shifter might be used for both the VddH $\rightarrow$ VddL direction and the VddL $\rightarrow$ VddH direction, where VddH is the high supply voltage and VddL is the low supply voltage. However, in practice, it is typically acceptable in CMOS design to "over-drive" the cells, with slight timing inaccuracy at the lower voltage sink node (i.e. using the low Vdd input to low Vdd cell characterizations for high Vdd input to low Vdd cells). For example, the switching threshold of a 1.0V signal to a gate operated at 0.8V may cause the switching to be slightly faster in one direction. This induced inaccuracy has very small impact on the overall timing, especially if the boundary logic consists of synchronous registers as a general good design guideline in this project; level shifters were inserted only in the upshift direction, VddL $\rightarrow$ VddH.

The level shifter that the design team created was that of a simple buffer function, as shown in Figure 14.6, and was four standard cell rows tall. The default power supply rail was 0.8V. The 1.0V supply connected to a pin of this cell. Both the 0.8V V_{dd} routing and 1.0V V_{dd} routing were done using a power router. Using a prescribed width for routing these different power supplies, the team eliminated any electro-migration failures in the design. The level shifter placement and power routing is shown in Figure 14.7.

It is worth mentioning that to guard against higher risk of latch-up due to power ramp-up/ramp-down between the two voltage domains, the layout of the level shifters cells had additional n-well spacing – beyond what was minimally required by the specific 90nm technology from the foundry.

This project used a prototype version of the voltage level-shifter cell. The level shifter cell was later optimized, as a result of this project, to a much more compact and efficient design in the Artisan Design Component Library. As it was, level shifters account for less than 5% of the overall chip area, and each level shifter consumed about the equivalent power of drive strength X8 buffer (Bufx8). As the level shifter was used to go between domain boundaries, sufficient drive strength of the same buffer equivalent (BufX8) was also maintained.

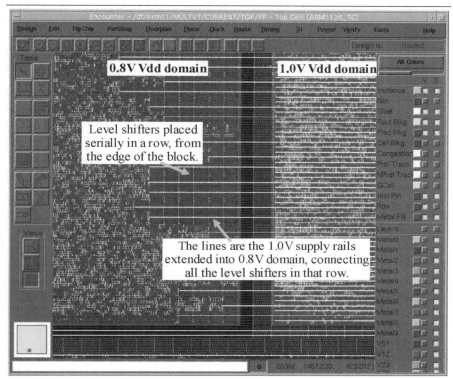

Figure 14.7 Level shifter placement and power supply rail routing.

14.3.1.3 Level shifter insertion and placement

As described in Section 14.3.1.1, this chip was partitioned into two power domains based on performance. Once the power domains were defined, the level shifters were automatically inserted for the signals going across the boundary from the low voltage domain to the high voltage domain. The placement of these level shifters proved to be a challenge. Placement of level shifters needed to take several things in consideration:

1. Level shifters needed to be placed along the natural signal path – any deviation can impact timing.

2. Level shifters need access to both low Vdd and high Vdd – a long route of the alternate supply can result in electromigration violations.

3. Level shifter placement should be isolated from normal standard cell placement – any interaction between low voltage and high voltage signals can increase signal integrity violations.

Typically, the design implementation of level shifters has been performed manually, and in an iterative fashion: first, the routing topology for the intra-block signal's optimal crossing is determined; then the level shifter cells are manually inserted; and then the proper power connection is manually designed. All these steps are labor intensive and error-prone. Especially considering the potential iterations of different floorplan explorations, this manual approach can have a significant impact on the design and its design time.

Automating level shifter insertion and placement was one of the key aspects of this project. The resulting placement is shown in Figure 14.7.

There were 3,300 signals crossing from the 0.8V domain to the 1.0V domain. Automating level shifter insertion became the key to meeting the project schedule, because these specialized cells had added implementation complexity with insertion, placement and power routing,. The level shifters needed to be placed along the boundary of the low voltage block where signals were output to eliminate any timing closure issues. Signal integrity was another concern as the high voltage signals can couple with a low voltage signal causing more cross coupling noise than normal. In addition, signal noise immunity depends on the voltage level. Signals at the higher voltage level have higher noise tolerance than those at a lower supply voltage. At the time of this project, a conservative noise threshold was used to ensure acceptable noise tolerance. The noise threshold was set to 25% of the lower of the two different supplies (i.e. 0.2V). This ensured tight signal integrity acceptance for the 0.8V signals, and a slight margin for the 1.0V signals. The tool was later updated to handle signal integrity issues between voltage domains automatically. At the interface between two blocks, the signals with 1.0V and 0.8V co-exist, and even a moderate strength 1.0V signal can cause noise issues on a 0.8V signal net. To detect this situation, the signal integrity tool uses two techniques. Firstly, the noise tolerance may be different for different signals. For example, a 1.0V signal may have a 0.1V noise tolerance and a 0.8V signal may have 0.08V noise tolerance. Secondly, a SPICE like simulation of the noise waveform through the path using the transfer function of the cells (including level shifters) ensures accurate propagation of the noise waveform. With these two techniques, the signal integrity issues can automatically be analyzed and corrected.

The concurrent placement of both the 1.0V and 0.8V domains included around 100,000 instances in the 0.8V region and 200,000 instances in the 1.0V region.

14.3.1.4 Timing analysis across power domains

Timing paths crossing the power domains pose a challenge to the existing static timing infrastructure. To get around this issue, designers try either to artificially confine their power domains within synchronous boundaries

[11] or abstract various power domains for static timing analysis. The static timing analyzer built on Encounter tools allowed us to perform timing checks for paths crossing the power domains by loading up the correct libraries for the different power domains and using the timing information of the level shifters. The voltage differences are handled through modeling current as opposed to modeling voltage (which a traditional timing analyzer does). The newly developed Effective Current Source Modeling (ECSM) addressed this specifically and allowed an accurate estimate of the delay across the power domains [7].

ECSM models were derived from looking at the cell characterization problem differently. There were two observations made during the characterization process: (1) Modeling the driver as a voltage controlled *current source* provided better accuracy than traditionally modeling it as a voltage controlled voltage source and (2) The C_{eff} (effective capacitance seen at the output of a driver) which is assumed as constant in the normal modeling actually varies during the transition. ECSM models take care of both these issues by storing an I/V characterization using a time quantized C_{eff}. Figure 14.8 shows the characterization space for a given input slew rate.

Modeling for multiple voltages is an easy task with ECSM. For multiple voltage scenarios, there are two ways that ECSM can help the designer:

1. When used in a design with many power domains (many voltage values), in order to obtain accurate delay information, designers need to characterize the timing views of the cells in the design at these various voltage values. This not only increases the characterization effort but also slows down the EDA tool that needs to read these characterized libraries. ECSM requires the cells to be characterized at a subset of voltage values and accurately interpolating the delays - with a bounded accuracy of 5% to SPICE – between these points. For example, a library that is characterized at 0.8V, 1.0V and 1.2V using ECSM models could be used at any voltage point in between 0.8V and 1.2V.

2. ECSM can also be used to compute additional delay due to IR drop on the nominal voltage supply. This additional delay can be used to perform meaningful delay-power trade-offs.

To validate ECSM's ability to provide continuous accurate delay coverage across the entire potential operating range of Vdd levels, a joint study was conducted with ARM Physical IP (Artisan). Delay analysis with the ECSM models and SPICE simulation was compared with five sampled cells, under different loading and slew rates, and a range of voltages from 0.7V to 1.2V [7]. The ECSM model was built with three characterization points of 0.70V, 0.90V and 1.08V (nominal voltage) for each cell, and measured at six different voltage levels (0.70V, 0.80V, 0.90V, 1.00V, 1.08V, 1.20V) with different input slew (0.04ns, 0.20ns, 1.50ns) and output loads (1.7fF, 20fF, 170fF), to collect the comparison data.

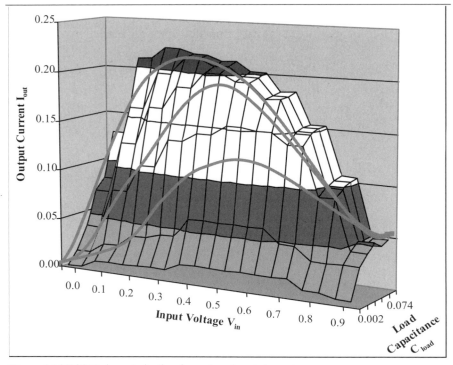

Figure 14.8 ECSM characterization for a given input slew rate.

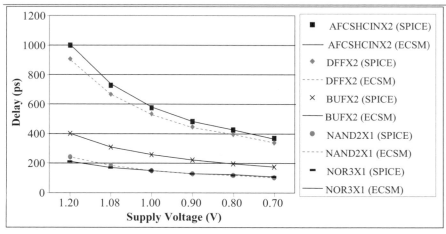

Figure 14.9 This figure shows the gate delay versus supply voltage for the ECSM models and SPICE simulation, which are in good agreement.

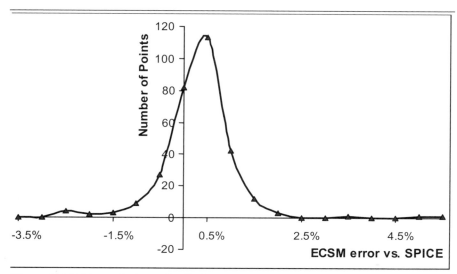

Figure 14.10 This graph shows the distribution of ECSM model error vs. SPICE. The mean error is about 0.5%, and the standard deviation is 0.6%.

Figure 14.9 shows the ECSM and SPICE results at each comparison point with increasing voltage. From this graph, there is little difference between the ECSM prediction and the SPICE results. Looking at the deviation of the approximately 200 comparison points (see Figure 14.10), one sees that the average error of ECSM vs. SPICE is 0.5%, with a standard deviation of 0.6%. This was a remarkable validation of the accuracy of the ECSM models, especially for multiple-supply voltage design.

14.3.1.5 Clock gating

One of the techniques to reduce the dynamic power is to reduce the switching activities (k in Equation (14.1)). As the clock signal transitions twice each clock cycle, one of the major strategies to reduce switching activity in synchronous digital design is to "turn-off" the clock while the logic or the "state" of the synchronous register is not expected to be changing. This is also known as "clock gating".

Traditionally, clock gating has been designed manually, as the designer would be familiar with which portions of the function can be stopped and when. However, not all opportunities for clock gating may be found this way; it may be time consuming, and may not be comprehensive in coverage. For this project, the team decided to implement clock gating wherever possible because of the significant reduction in switching power that can be achieved, and the cost of design implementation can be low. Encounter RTL Compiler was used to automate identification and insertion of clock gating functions, using ARM PIP's integrated clock gating cells (ICG).

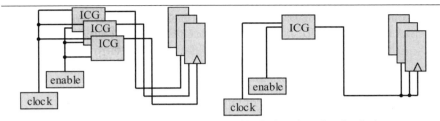

(a) clock gating without de-cloning   (b) clock gating after de-cloning

Figure 14.11 Pruning the number of clock gating cells in (a) by moving clock gating upstream hierarchically (as in (b)) for logic that can share the same clock enable logic.

Automatically identifying clock gating opportunities is done by using the tool to examine the entire netlist, to determine which registers and latches can share the same clock enabling logic. 1,112 clock gating opportunities were identified, in addition to the clock gating that was already coded in the RTL design.

After identifying these new clock gating opportunities, RTL Compiler was used to prune the number of clock gating cells by moving the gating function upstream hierarchically (shown in Figure 14.11). By gating at a higher level in the clock tree, more logic can be turned off. And by having less clock gating cells and branches, the design has a better starting point for clock tree generation. However, this process, also known as de-cloning, needs to be done with caution. Moving clock gating cells up the hierarchy can lead to needing to generate complex gating signals. This can also put undue burden on the setup time at the enable pin of the gating cell. If constrained within a few levels of hierarchy, this can give us an additional 5% to 10% of clock power savings.

14.3.2   Leakage power

Leakage power has become a growing concern with advanced technology nodes of 130nm and below. Why is leakage power a problem?

Power consumption of a CMOS gate has three major components:

$$P_{total} = P_d + P_{sc} + P_{lk} \qquad (14.2)$$

where P_d is the dynamic power, P_{sc} is the short circuit power, and P_{lk} is the leakage power. Before deep submicron processes, P_{lk} was marginal relative to switching power. However, this leakage power grows from less than 5% of the total power budget at 0.25um to 20-25% at 130nm and to 30-40% at 90nm [12]. Below 90nm, the chip could be dissipating almost as much power due to leakage as due to dynamic power.

Table 14.1 Leakage and saturation drain current in TSMC 90nm and 130nm processes, for the low, standard, and high transistor threshold voltage libraries.

TSMC	90nm			130nm	
Transistor threshold voltage	Low	Standard	High	Low	Standard
Leakage current (pA/um)	100	10	1.5	10	0.25
Saturation drain current (uA/um)	755	640	520	590	535

Table 14.2 Threshold voltage value of the 1.0V cells.

	N-Channel	P-Channel
Standard Vth	0.228V	0.165V
High Vth	0.354V	0.333V

To deal with increasing subthreshold leakage currents, semiconductor foundries have added higher threshold voltage transistors that have lower leakage at the cost of greater delay. Standard cell designers can use those transistors to design the same functional gate with different leakage current, but maintaining the same cell footprint. This enables cell swapping to reduce leakage power without impacting the place and route floorplan. The trade-off between leakage and drive strength (saturation drain current) for different Vth values in TSMC's 90nm and 130nm processes is shown in Table 14.1. See Table 14.2 for the standard and high V_{th} values of this design's libraries.

The cost of this leakage optimization is reduction in speed: about a 25% increase in delay for 4× leakage power reduction at 0.8V supply comparing high threshold voltage and standard threshold voltage cells in Figure 14.12 and Figure 14.13. In addition, there is the process expense for the additional implant required if two transistor threshold voltages are used.

In this low power design, the complete RTL was synthesized with the newly developed global optimization synthesis technology, using two V_{th} libraries (standard and high) concurrently optimized for leakage power, timing and area in a single pass strategy. Gates in the high V_{th} library are lower leakage, but slower than gates in the standard V_{th} library. It was important to note that the balance between the different but equally important design targets, such as timing and power, routinely requires trade-offs in cell selection based on dynamic power, delay and leakage. Automation of that optimization in the synthesis tool simplified this design implementation. In the design, standard V_{th} cells were used in timing critical paths, whereas high V_{th} cells were used in other paths to optimize for power.

Having a global view of the design with RTL Compiler helped optimize the entire design by trading timing slack for area/power effect during the initial mapping stage of synthesis. Cells on critical paths are mapped to fast cells that are low V_{th} and are narrow (few inputs) functional cells, which avoids slow pull-up or pull-down series transistor chains in a logic gate. To reduce power, other cells are high V_{th} to reduce leakage, and wide (more inputs) to collapse instances and reduce net count – thereby reducing area and power. It was obvious that the bigger portion of the design the RTL Compiler can see, the better optimization can be realized.

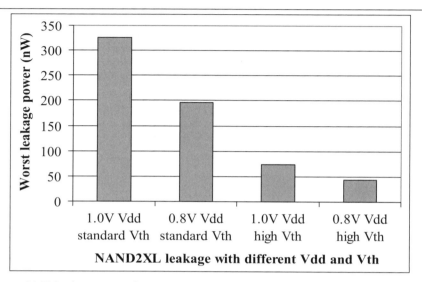

Figure 14.12 Leakage power for a NAND2XL (low power, small NAND2) cell at different supply Vdd and threshold voltages V_{th}.

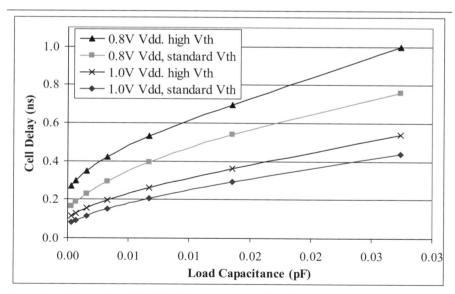

Figure 14.13 Delay for a NAND2XL (low power, small NAND2) cell at different supply and threshold voltages.

Table 14.3 Multi-Vth cell utilization in the low and high supply voltage domains. 8.5% of the cells in the low voltage domain were high V_{th}, and 97% of the cells in the high voltage domain were high V_{th}.

Library	Cell count	Percentage of Total Cells
0.8V supply, high threshold voltage	16,210	5.4%
0.8V supply, standard threshold voltage	174,738	57.8%
1.0V supply, high threshold voltage	108,052	35.8%
1.0V supply, standard threshold voltage	2,710	0.9%

Besides using multiple threshold voltages to reduce leakage, RTL compiler uses other optimization techniques to meet both the dynamic and leakage power design target. Such optimizations include logic restructuring, buffer insertion and removal, pin swapping and gate resizing. Cell sizing and buffer manipulations are aimed at optimizing the switching time, to reduce unnecessary switching activity due to glitching, and therefore reducing dynamic power and maintaining balanced performance. Pin swapping swaps functionally identical input pins, so that the signal with higher switching activity connects to the gate input pin with lower input capacitance. Logic restructuring minimizes the number of logic levels traversed by high switching activity signals to reduce the dynamic power.

The leakage power optimization does not occur only in the synthesis step. As the design went through detailed place and route, more accurate RC (wiring) parasitic information becomes available. This information was used to fine-tune the multi-V_{th} cell selection, balancing the performance goal with the power target. In this project, timing and area target were both fixed, but the power target was set aggressively to understand the potential limit of the power optimization tools. This was performed with SOC Encounter's post route optimization stage, with *optLeakagePower -postRoute –highEffort* (example script command line).

The overall mix of multi-V_{th} cells showed an expected profile between performance and power. 97% of all cells used in the 1.0V domain where high V_{th} cells, after the final optimization (see Table 14.3). Less high V_{th} cells were used in the 0.8V domain to ensure performance targets were met.

14.3.3 Power Integrity Verification

The sign-off power analysis tool based on the Encounter platform recognized the power domains and gave us results per power domain. The analysis results took the level shifters into account, and the power nets were traced through the level shifters into the other domain. Figure 14.14(a) shows the results of power analysis on the shared ground (GND) net; Figure 14.14(b) shows the power analysis on the Vdd net for the 0.8V region; and Figure 14.14(c) shows the power analysis on the Vdd net for the 1.0V region. The design's IR drop was less than 22mV, or about 2% worst case.

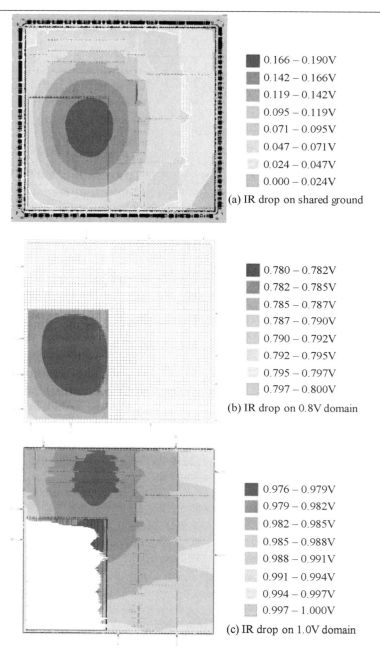

Figure 14.14 This figure shows the IR drop effect (a) on the shared ground for the complete chip, (b) on the 0.8V domain supply voltage, and (c) on the 1.0V domain supply voltage.

The team also performed dynamic power analysis to assess the usage of decoupling capacitances and utilized the what-if capability of the tool to try out various combinations of decoupling capacitance placement. Though multiple supply voltage design does not have any direct impact on the placement of decoupling capacitors, the dynamic power analysis was required to ensure the benefit from the decoupling capacitances would not be neutralized by the additional power penalty induced by these decoupling capacitances.

For this design, an initial switching activity of 30% was assumed to get the early power estimate and budget. As the design implementation progressed, actual gate level functional simulation patterns (derived from the verification vector set described in [2]) for peak-power and typical power were executed to capture the needed VCD and TCF files for more accurate power analysis Since the static timing analyzer uses the absolute voltage value, it was very easy to translate the instance-based IR drop numbers for the placement into delays. Optimization was done to account for this delay and the team could thus trade-off power against timing and area.

14.4   RESULTS

14.4.1   Simulated Results at Tape-Out

To compare and contrast the effectiveness of the power management strategies, this project implemented the same design in two different flows. One is with the traditional timing closure flow; the other is with the power management solution described so far. To make the comparison reasonable, both implementations used the same RTL design, same technology library, the same die size (4mm × 4mm), same floorplan, and most importantly, tapeout at the same target frequency of 355MHz.

Tapeout analyses were done in both worst case corner (slow process, 125°C, 0.9V) and best case corner (fast process, –40°C and 1.1V), with additional leakage power analysis done with fast process, 125°C and 1.1V for the potential worst case for power.

The simulated results at tapeout corresponded well to our expectations. Table 14.4 compares the power savings obtained against the baseline implementation, normalized to the overall power from the baseline.

Recall that the major strategies used to reduce dynamic power were voltage scaling and clock gating. The 1.0V domain has about 12% dynamic power savings due to additional clock gating and other logic optimizations, such as power aware cell selection, pin swapping, gate sizing, buffer insertion and removal, logic restructuring, and reduction of gate counts overall. Note that while power minimization was not performed specifically on the baseline implementation, area minimization was performed, which would have 2nd order effect on some power reduction.

The 0.8V power domain results show much higher dynamic power savings of 50.3%. With supply voltage reduced by 20%, a power saving of 36% (= $1.0^2 - 0.8^2$) was expected. The other 14% power savings was due to clock gating and logic optimization.

The leakage power optimization also correlated well to estimates. As would be expected, the 1.0V domain has much higher performance margin to allow for use of high V_{th}. As noted previously, over 97% of cells used in this region were high V_{th} cells, and there was nearly 70% leakage power savings as expected. (As noted earlier, high V_{th} gives about a ×4 reduction in leakage current, so the team expect reduction in leakage power to 0.25 × 0.97 + 0.03 = 27% of the original power, or 73% leakage power savings.)

In the lower supply voltage domain, there was less opportunity to use high V_{th} as there was less timing slack available due to using low V_{dd}, and less savings were expected. However, leakage power is proportional to the current-voltage product (P_{lk} α $I_{lk}V_{dd}$), so reducing V_{dd} does also reduce the leakage power. The 20% reduction in V_{dd} gives more than 20% reduction in leakage as there are also additional factors affecting subthreshold leakage such as drain-induced barrier lowering (DIBL) which is reduced at lower V_{dd}, and 8.5% of the cells were changed to high V_{th}, giving 33.5% leakage power savings in the 0.8V V_{dd} domain.

The combined overall power saving of 40.3% was a great achievement using mainstream production tools, with a mature and general purpose process and library, and impacting neither the design architecture nor the design implementation flow.

Table 14.4 Power savings of the multi-V_{dd}/multi-V_{th} design versus the baseline design, where both domains had single 1.0V supply and only the standard V_{th}. The total dynamic and leakage power savings were respectively 38.0% and 46.6%.

		Normalized power		Power
		Baseline	Low Power	Reduction
1.0V domain	**Dynamic power**	0.235	0.207	11.9%
(includes RAM)	**Leakage power**	0.097	0.030	69.1%
	Subtotal	0.332	0.237	28.6%
0.8V domain	**Dynamic power**	0.501	0.249	50.3%
(no RAM)	**Leakage power**	0.167	0.111	33.5%
	Subtotal	0.668	0.360	46.1%
Total for both domains		1.000	0.597	40.3%

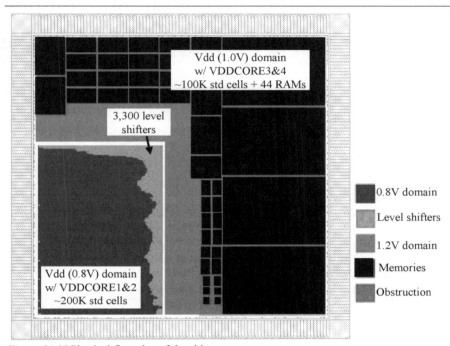

Figure 14.15 Physical floorplan of the chip.

The physical floorplan is shown in Figure 14.15. The die size is 4mm by 4mm, with 360 I/O pads. There are about 300,000 cells in total. The area utilization is about 80%, with memories comprising approximately 60% of the total area. Overall, the area overhead for implementing multiple supply voltages was less than 5%. Note that the same floorplan was used for the baseline. As the cell footprints are the same, there was no significant change in area due to using multiple threshold voltages. There was minimal impact on the design flow and schedule due to our low power design approach comared to the baseline timing closure implementation.

14.4.2   Silicon Validation

To validate the results and correlate to actual silicon behavior, the chip was fabricated (shown in Figure 14.16). The received silicon IC (integrated circuit) parts were packaged in a BGA package, then tested using Inovys Personal Ocelot tester under typical operating conditions (room temperature and nominal supply voltages).

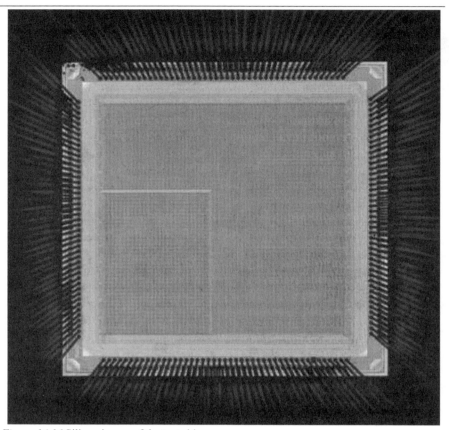

Figure 14.16 Silicon image of the test chip.

The IC achieved functional and electrical design validation with first silicon. Additionally, over 15,000 system-level validation tests have been completed successfully using ARM's RealView® system validation board at speed. The fabricated ARM chip successfully runs the Linux (version 2.4), Windows CE, and SunOS operating systems.

Basic parts screening included normal JTAG, Scan and Memory BIST (built-in self test). Functional patterns were also used to check minimum functionality of the devices. Finally, looping Dhrystone benchmarks were used to measure the power, at different combinations of system clock frequencies from 1MHz to 50MHz on the ATE (Automated Test Equipment) test fixture. The table below summarizes the simulated baseline results, the simulated low power implementation, the measured low power implementation, and a reference power measurement of the same ARM core in 0.13um.

Table 14.5 Active power of the chip for the Dhrystone 2.1 benchmark set, comparing power estimates (at 25°C, 90% of nominal supply voltage, and typical process corner) and measured power for the fabricated chip.

Power Domain	Active Power Dissipation (mW/MHz)			
	Simulated Baseline (90nm)	Simulated Low Power (90nm)	Measured Low Power (90nm)	ARM published 1136JF-S Power in 130nm
Core	0.28	0.14	0.10	0.60
Other	0.36	0.32	0.21	
Total	**0.64**	**0.46**	**0.31**	

The relationship between the simulated baseline and simulated low power implementation at tapeout was elaborated in the last section. The measured silicon results show the correlation between silicon and simulation are consistent, with the simulation being conservative by about 30% to 40%, due to assuming more severe operating conditions and process variances (see Table 14.5). For example, the simulated typical condition assumed 25°C and 10% Vdd variance but the actual silicon measurement would deviate in both the temperature and voltage supply. However, in general the results were better than expected, as indicated by the 0.14mW/MHz estimate against the measured 0.10mW/MHz.

In addition to the implementation comparison, an ARM published power measurement of the ARM1136JF-S core [3] was used to draw reference against the silicon verified 0.1mW/MHz power performance. Though the published number of 0.6mW/MHz was for 130nm technology and under typical operating conditions, it served as a perspective of the power performance achieved with this low power design.

14.5   SUMMARY

This power management project demonstrated the usability in an EDA flow of multiple supply and multiple threshold voltages to reduce power, along with more standard power minimization techniques. This strategy, when applied with EDA tools that can handle these approaches and automate them properly, can realize significant power savings without much impact to the design architecture or process. Voltage scaling and clock gating achieved 38% dynamic power savings while maintaining a high clock frequency. Leakage power was reduced 47% by using multi-Vth cell libraries, again without impacting the timing performance. By creating a comprehensive low power design flow this project has provided mainstream system-on-chip designers the capability to effectively manage power. It is the conclusion of this project that adoption of these techniques should be easy for a main stream ASIC design.

14.6   ACKNOWLEDGMENTS

We would like to thank C. Chu, A. Gupta, L. Jensen, T. Valind, P. Mamtora, C. Hawkins, P. Watson, Huang, J. Gill, D. Wang, I. Ahmed, P. Tran, H. Mak, O. Kim, F. Martin, Y. Fan, D. Ge, J. Kung, V. Shek, for their contribution to the project [8]. Special thanks to D. Le, T. Nguyen, S. Yang, P. Bennet and A. Khan for their contribution to the project and review of this document.

14.7   REFERENCES

[1]   ARM, (Realview) Core Tile for ARM1136JF-S, 2005. http://www.arm.com/products/ DevTools/Versatile/CT1136JF-S.html
[2]   ARM, *ARM1136 ImplementationGuide*, 2002.
[3]   ARM, ARM1136J(F)-S, 2005. http://www.arm.com/products/CPUs/ARM1136JF-S.html
[4]   ARM, ARM 1136JF-S and ARM1136J-S Technical Reference Manual, 2005. http:// www.arm.com/pdfs/DDI0211F_arm1136_r1p0_trm.pdf
[5]   ARM, ARM - Artisan Products: Standard Cells, 2005. http://www.artisan.com/products/ standard_cell.html
[6]   Bennett, P., and Kuo, G., "ARM1136 Low Power Test Chip – case study for 90nm Low Power Implementation," DesignCon, Santa Clara, California, 2005, http://www. designcon.com/conference/2005/3-ta3.html
[7]   Cadence, "Accurate Multi-Voltage Delay Analysis: Artisan Libraries and Cadence Encounter Digital IC Design Platform Enable Low Power Design," technical paper, 2004. http://www.cadence.com/datasheets/ArtisanMSMV_tp.pdf
[8]   Khan, A., et al. "A 90nm Power Optimization Methodology and its Application to the ARM 1136JF-S Microprocessor," *proceedings of the Custom Integrated Circuits Conference*, 2005.
[9]   Kuo, G., and Iyer, A., "Empowering Design for Quality of Silicon: Cadence Encounter Low-Power Design Flow," technical paper, 2004. http://www.cadence.com/datasheets/ lowpower_tp.pdf
[10] Lackey, D., et al.,"Managing Power and Performance for System-On-Chip Designs Using Voltage Islands," *proceedings of the International Conference on Computer-Aided Design*, 2002, pp. 195-202.
[11] Liu, R.H., "How to create designs with dynamic/adaptive voltage scaling," *presentation at the ARM Developers' Conference*, Santa Clara, California, 2004.
[12] Rusu, S., "Trends and Challenges in High-Performance Microprocessor Design," *keynote presentation at Electronic Design Processes*, Monterey, California, 2004. http://www. eda.org/edps/edp04/submissions/presentationRusu.pdf

INDEX

Printed in the United States of America.